J. Willard Gibbs

The Modern Theory of Capillarity

Moderne Kapillaritätstheorie

The Modern Theory of Capillarity

To the Centennial of Gibbs' Theory of Capillarity

Moderne Kapillaritätstheorie

Herausgegeben von

PROF. DR. F. C. GOODRICH †
Potsdam, N.Y. (USA)

und

PROF. DR. A. I. RUSANOV
Leningrad (UdSSR)

unter Mitwirkung von

Prof. Dr. habil. H. Sonntag und
Dr. sc. nat. M. Bülow (Berlin)

Mit 51 Abbildungen und 19 Tabellen

Akademie-Verlag · Berlin
1981

Erschienen im Akademie-Verlag, 1080 Berlin, Leipziger Straße 3—4

Lizenznummer: 202 · 100/471/81
Einbandgestaltung: Rolf Kunze
Gesamtherstellung: VEB Druckerei „Thomas Müntzer“, 5820 Bad Langensalza
Bestellnummer: 7628778 (6599) · LSV 1215
Printed in GDR
DDR 45,— M

Contents

Preface

This book takes its origin in a noteworthy event in the history of science. 1978 marks a century since the publication of the second part of WILLARD GIBBS' classical memoir *"On the Equilibrium of Heterogeneous Substances"* in which he presented his thermodynamic theory of capillarity. The importance of this theory cannot be over-emphasized. Fundamental in nature and developed in detail, GIBBS' theory is the foundation for the subsequent elaboration of the thermodynamics of interfacial phenomena down to the present time, and still plays the central role in modern surface and colloid science.

As the appearance of GIBBS' theory was such an important advance in the development of natural science, it seems appropriate to celebrate its jubilee by the publication of a special international volume of homage. This is the book now presented to the reader. Scientists from East and West have shared in the preparation of this volume, and their contributions reflect various topics of interest in the modern theory of capillarity. The articles by B. V. DERJAGUIN, M. M. DUBININ, S. S. DUKHIN, O. A. PETRY, and A. I. RUSANOV were presented at the conference, dedicated to the centennial of GIBBS' theory of capillarity (Leningrad, February, 1978). All other papers have been written especially for this volume.

Because of the variety of topics treated, it was difficult to adopt any systematic method of arranging the order of the papers, and we have arranged them alphabetically (in the Cyrillic alphabet as in the Russian language edition of the volume) with respect to the authors' names, with the exception of the introductory paper, which appears, of course, at the beginning of the volume.

(During the preparation of the monograph Dr. GOODRICH died, unexpectedly.)

A. I. RUSANOV
F. C. GOODRICH †

The centennial of GIBBS' theory of capillarity

A. I. RUSANOV

(Department of Chemistry, Leningrad State University, Leningrad/USSR)

Abstract

This contribution gives a review of the most important facts relating to the centennial of GIBBS' theory of capillarity. The following topics are discussed: the understanding of, and fresh insight into, some aspects of GIBBS' theory, the development and generalization of GIBBS' theory of capillarity, the rise of new trends in the thermodynamics of interfaces. Furthermore, the introduction of the surface of tension for curved interfaces, the theory of GIBBS' elasticity of films, and the problem of a layer of finite thickness in the thermodynamics of interfaces are under discussion. Special attention is payed to the generalization of the GIBBS adsorption equation and the phase rule. In the category of new trends, the investigation of interfacial layer thickness, the thermodynamics of thin films, and the theory of surface separation processes are considered.

Introduction

A century has passed since GIBBS created the theory of capillarity published in 1878 in the second part of his famous work *"On the Equilibrium of Heterogeneous Substances"* [1]. Distinct from all previous theories and, in particular, from LAPLACE's theory of capillarity, GIBBS' theory was of a thermodynamic character and remains an indispensable part of GIBBS' thermodynamics up to now. GIBBS' theory of capillarity was the first detailed thermodynamic theory of interfacial phenomena.

During the ensuing century, the thermodynamics of interfaces has notably progressed. The following points may be mentioned: a) the understanding, systematization, and new interpretation of some statements of GIBBS' theory of capillarity; b) the development and generalization of GIBBS' theory of capillarity; c) the rise of new trends in the thermodynamics of interfaces, not resulting from GIBBS' theory of capillarity.

The understanding of GIBBS' theory of capillarity

It may be said that GIBBS' theory of capillarity is both very simple and very complicated. It is simple because GIBBS managed to find a method for obtaining the most compact and elegant thermodynamic relationships applicable equally to flat and to curved interfaces. GIBBS wrote: *"One of the principal objects of theoretical research in any department of knowledge is to find the point of view from which the subject appears in its greatest simplicity"* [2]. Such a point of view in GIBBS' theory of capillarity is the concept of dividing surfaces. The use of the clear geometrical image of the dividing surface and the introduction of excess quantities allowed to describe the surface properties in the simplest way and pass over the problem of the structure and thickness of the surface layer, which was not studied in GIBBS' time and up to now remains not completely investigated. GIBBS' excess quantities (adsorption and others) depend on the dividing surface location, and the latter may also be determined on the basis of maximum simplicity and convenience. GIBBS used two main positions of the dividing surface: the position for which the adsorption of one of components is zero (such a position is now called the equimolecular surface) and the position for which there is no explicit dependence of surface energy on surface curvature (such a position was called by GIBBS the surface of tension). GIBBS used the equimolecular surface for considering flat fluid interfaces (and solid surfaces) and the surface of tension for describing curved interfaces. For each choice of dividing surface the number of variables is reduced, and the maximum mathematical simplicity is achieved.

Now about the difficulty in GIBBS' theory. Being very simple mathematically, GIBBS' theory is still intuitively difficult, and this originates from a number of causes. First, it is impossible to understand GIBBS' theory of capillarity separately from all of GIBBS' thermodynamics based on a very general deductive method. Great generality always makes a theory more abstract, which, certainly, lowers its intuitive appeal. Secondly, GIBBS' theory of capillarity itself, based upon broad conventions, requires a unity of perception without neglecting details. It is quite impossible to study GIBBS' works in a dilettante way. Finally, it is undeniable that GIBBS' wrote his works concisely, using a very difficult language. According to RAYLEIGH, this work "*is too condensed and too difficult for most, I might say all readers*" [3]. As GUGGENHEIM wrote, "*it is much less difficult to use* GIBBS' *formulae than to understand them*" [4].

The use of GIBBS' formulae without an adequate understanding led, naturally, to numerous mistakes in the interpretation and application of certain results of GIBBS' theory of capillarity. Many mistakes were associated with the failure to comprehend the necessity of a unique definition of the dividing surface location for obtaining a correct physical result. Mistakes of this kind are found in the analysis of the surface curvature dependence of surface tension, and even BAKKER, one of the founders of the theory of capillarity, did not avoid them. An example of other mistakes is a wrong interpretation of the chemical potentials in considering interfacial phenomena and external fields.

Soon after the publication of GIBBS' theory of capillarity, a need was expressed for a more complete and detailed explanation of GIBBS' work in scientific literature. In his letter to GIBBS cited above RAYLEIGH suggested that GIBBS should take this job upon himself, but the task was not completed until considerably later. A commentary on the whole of GIBBS' theory of capillarity was prepared by J. RICE [5], and some aspects of the theory were reviewed by FRUMKIN, DEFAY, REHBINDER, GUGGENHEIM, TOLMAN, SEMENCHENKO, BUFF and others. Many aspects of GIBBS' theory were clarified and justified by more simple and effective methods.

A typical example is KONDO's paper [6] in which a descriptive and easily understandable method was proposed for introducing the surface of tension by means of imaginary displacements of the dividing surface. If we write the expression for the energy of an equilibrium two-phase system, $\alpha - \beta$ (α denotes the internal phase, β

the external phase), with a spherical interface

$$U = TS - p^{\alpha} V_{\alpha} - p^{\beta} V_{\beta} + \sigma A + \sum_i \mu_i m_i \tag{1}$$

and will in thought shift the dividing surface (i.e. change its radius, r), then, evidently, such physical characteristics as energy, U, temperature, T, entropy, S, pressure, p, the chemical potential, μ_i, and mass, m_i, of the component i, as well as the total volume of the system, $V_{\alpha} + V_{\beta}$, will not change. As for the volume $V_{\alpha} = 4\pi r^3/3$, the area $A = 4\pi r^2$, and the surface tension, these quantities will depend on the dividing surface location, and then, for the above imagined change in r, we obtain from (1)

$$-p^{\alpha}\, dV_{\alpha} + p^{\beta}\, dV_{\beta} + \sigma\, dA + A\, d\sigma = 0 \tag{2}$$

or

$$\left(\frac{d\sigma}{dr}\right)^{*} + \frac{2\sigma}{r} = p^{\alpha} - p^{\beta}\,. \tag{3}$$

Equation (3) defines a nonphysical (this is denoted by asterisk) dependence of the surface tension on the dividing surface location. This dependence is characterized by a unique minimum at a value r which corresponds to the surface of tension. Thus, according to KONDO, the surface of tension is that dividing surface for which the surface tension takes its minimum value.

As is known, GIBBS introduced the surface of tension in another way. He started from the main equation of the theory of capillarity

$$d\bar{U} = T\, d\bar{S} + \sigma\, dA + \sum_i \mu_i\, d\bar{m}_i + C_1\, dc_1 + C_2\, dc_2 \tag{4}$$

(a bar denotes an excess for an arbitrary dividing surface with the principal curvatures c_1 and c_2) and considered a physical (not purely imaginary) process of curving an interface at its given location and fixed external conditions. According to GIBBS, the surface of tension is a dividing surface location for which curving the surface layer at constant external parameters does not influence surface energy and also corresponds to the condition

$$\frac{\partial \sigma}{\partial r} = 0\,. \tag{5}$$

GUGGENHEIM commented on GIBBS' approach above: "*I have found* GIBBS' *treatment difficult, and the more carefully I have studied it the more obscure it appears to me*" [4].

This confession confirms that an understanding of GIBBS' surface of tension is difficult even for specialists in thermodynamics.

As for KONDO's approach, it is understandable at a glance. However, it is necessary to examine whether GIBBS' and KONDO's surfaces of tension are the same. It may be demonstrated, for example, using the hydrostatic definition of the surface tension [7] that

$$\sigma = \frac{1}{r^2}\left[\int_{R^{\alpha}}^{r} (p^{\alpha} - p_T)\, r'^2\, dr' + \int_{r}^{R^{\beta}} (p^{\beta} - p_T)\, r'^2\, dr'\right] \tag{6}$$

where p_T is the local value of the tangential component of the pressure tensor, r' is the radial coordinate, and the radii R^{α} and R^{β} are the surface layer boundaries.

Differentiating (6) with a purely imaginary shifting of the dividing surface at constant physical state (KONDO's approach) leads to Eq. (3), whereas differentiating (6) while curving the surface layer at constant physical state (GIBBS' approach, R^α and R^β are variable in this case) yields

$$\frac{\partial \sigma}{\partial r} = -\frac{2\sigma}{r} + p^\alpha - p^\beta , \tag{7}$$

where the conditions $p_T(R^\alpha) = p^\alpha$ and $p_T(R^\beta) = p^\beta$ are used. It is seen from a comparison of (3) and (7) that the condition (5) is equivalent to the condition $(d\sigma/dr)^* = 0$ and, consequently, the simpler and more descriptive approach of KONDO is equivalent to GIBBS' approach.

As another example, we consider the theory of elasticity of liquid films which was first formulated by GIBBS as part of his theory of capillarity. This kind of elasticity, typically found in surfactant solution films, is now called GIBBS' elasticity (see review [8]).

Two features of GIBBS' formulation may be pointed out: a general description of GIBBS' elasticity and the derivation of a thermodynamic formula for the elasticity modulus of a partly open film. The general description was given by GIBBS in textual form as follows: "*When the film has two or more components of which the potentials are not maintained constant by the contiguous gas-masses, they will not in general exist in the same proportion in the interior of the film as on its surfaces, but those components which diminish the tension will be found in greater proportion on the surfaces. When the film is extended, there will therefore not be enough of these substances to keep up the same volume- and surface-densities as before, and the deficiency will cause a certain increase of tension*". In these words, the origin and the mechanism of the surfactant solution film elasticity is correctly described.

As for GIBBS' formula, it refers to a binary film and has the form [GIBBS' Eq. (655)]

$$E = 4\Gamma_{2(1)}^2 \left(\frac{d\mu_2}{dG_2}\right)_{G_1} \tag{8}$$

where E is the film elasticity modulus, $\Gamma_{2(1)}$ the relative adsorption of the second component, μ_2 its chemical potential, and G_i the amount of the component i per unit film area. As it was pointed out in the literature [6, 7], GIBBS' derivation of this formula is somewhat contradictory. One of the possible causes of it is that GIBBS used the notions of open system thermodynamics elaborated by himself, whereas the system under consideration referred to closed or partly closed systems. In consequence, GIBBS' formula (8) was of no practical use, and there arose the necessity of a reformulation of GIBBS' theory of the elasticity of films [9—14]. The exact formula for GIBBS' elasticity modulus of a film with n components has the form [8, 14]

$$E = 4RT \sum_{i=1}^{n-1} \frac{\Gamma_{i(n)}^2}{c_i} \frac{1 + \dfrac{d \ln f_i}{d \ln c_i}}{h\left(1 - \dfrac{d \ln c_n}{d \ln c_i}\right) + 2\,\dfrac{d\Gamma_{i(n)}}{dc_i}} \tag{9}$$

where c_i and f_i are the concentration and activity coefficients of the component i while h is the film thickness defined as the distance between two dividing surfaces for which the adsorption of the component n (of a solvent) is zero. Formula (9) is convenient for practical use since it explicitly contains the film thickness and all the quantities in (9) may be determined by experiment.

The above-mentioned examples show that the task of understanding GIBBS' theory of capillarity has been a long-standing one and is still topical today. Many papers have been devoted to various methods for the derivation of the GIBBS adsorption equation, to the interpretation of GIBBS' quantities for solids, and to other questions of GIBBS' theory of capillarity.

The development and generalization of GIBBS' theory of capillarity

1. The method of a layer of finite thickness

The method of a layer of finite thickness founded by VAN DER WAALS [15], BAKKER [16], VERSHAFFELT [17], and GUGGENHEIM [4] originally was developed as an independent method of the thermodynamics of interfaces. Later, attention was attracted to the fact that the strict formulation of this method also demands the use of the dividing surface concept, two (but not one) dividing surfaces being used in this case [18]. Still closer connection with GIBBS' method was demonstrated in the formulation of the thermodynamics of curved interfaces by the method of a layer of finite thickness [19, 20] where the notion of the surface of tension was used, as in GIBBS' method.

Let us consider an equilibrium two-phase system $\alpha - \beta$ with a flat interface, the state of which is characterized by the equation

$$dU = T\,dS - p\,dV + \sigma\,dA + \sum_i \mu_i\,dm_i \tag{10}$$

and introduce two dividing surfaces on the sides of the phases α and β, with an arbitrary distance, τ, between them. We imagine the parts of the system separated by the layer of thickness τ to be filled with the bulk phases α and β, so that their state is described by the equations

$$dU^\alpha = T\,dS^\alpha - p\,dV^\alpha + \sum_i \mu_i\,dm_i^\alpha\,, \tag{11}$$

$$dU^\beta = T\,dS^\beta - p\,dV^\beta + \sum_i \mu_i\,dm_i^\beta\,. \tag{12}$$

If we now subtract (11) and (12) from (10), we obtain the equation

$$d\bar{U} = T\,d\bar{S} - p\,dV^\sigma + \sigma\,dA + \sum_i \mu_i\,d\bar{m}_i \tag{13}$$

in which every barred extensive quantity refers to the volume $V^\sigma = A\tau$ and represents the sum of a real quantity for this volume plus excess quantities from the bounding regions of the bulk phases. For instance,

$$\bar{m}_i \equiv m_i^\sigma + (\Gamma_i^\alpha + \Gamma_i^\beta)\,A\,, \tag{14}$$

where m_i^σ is a real amount of the component i in the layer of the thickness τ, Γ_i^α the absolute adsorption from the phase α on the dividing surface adjacent to the phase α, and Γ_i^β the absolute adsorption from the phase β on the dividing surface adjacent to the phase β. Evidently, the form of Eq. (13) does not depend on the location of the dividing surfaces and the value of τ. When $\tau \to 0$, Eq. (13) becomes GIBBS' fundamental equation (4); when $\tau \to \infty$, Eq. (13) is converted into Eq. (10) for the whole two-phase system.

The whole thermodynamic formalism is based on a joint consideration of Eqs. (11)—(13) and other relationships resulting from these equations. This formalism results in GIBBS' theory of capillarity in the limit $\tau \to 0$. In the opposite limit ($\tau \to \infty$), another limiting variant of interfacial thermodynamics is obtained in which the dividing surface concept is absent (this variant was recently analysed by GOODRICH [21]). Thus, we may conclude that the method of a layer of finite thickness is a generalization of GIBBS' method and the most general method of the thermodynamics of interfaces.

2. The GIBBS adsorption equation

The most famous result of GIBBS' theory of capillarity is the GIBBS adsorption equation

$$d\sigma = -\bar{s}\, dT - \sum_i \Gamma_i\, d\mu_i \,. \tag{15}$$

where $\bar{s}$ is the excess entropy per unit area, Γ_i is the absolute adsorption of the component i. This equation was deduced by GIBBS only for fluid interfaces. It refers to the surface of tension and is valid both for flat and curved interfaces.

The GIBBS adsorption equation was generalized many times during the past century, and every new form was a step towards a better understanding of GIBBS' theory. A case in point is a generalized form of the GIBBS adsorption equation for an arbitrary location of the spherical dividing surface [6, 22, 23]

$$d\sigma = -\bar{s}\, dT - \sum_i \Gamma_i\, d\mu_i + (d\sigma/dr)^*\, dr \,. \tag{16}$$

It should be noted that Eq. (16) is only a generalization of the form of the GIBBS adsorption equation and is an exact physical equivalent of Eq. (15). Moreover, being less simple than Eq. (15), Eq. (16) complicates the interpretation of the quantity σ since the analogy with the tension of an elastic membrane is lost. Strictly speaking, the notion "surface tension" is only applicable to the surface of tension.

Another generalized and also physically equivalent form of the GIBBS adsorption equation is the finite thickness layer equation [24]

$$A\, d\sigma = -\bar{S}\, dT + p^\alpha\, dV_\alpha + p^\beta\, dV_\beta - \sum_i \overline{m}_i\, d\mu_i \,, \tag{17}$$

where V_α and V_β are the parts of the interfacial layer volume, V^σ, separated by the surface of tension. For the case of a flat interface Eq. (17) becomes [4, 17, 18]

$$A\, d\sigma = -\bar{S}\, dT + p\, dV^\sigma - \sum_i \overline{m}_i\, d\mu_i \tag{18}$$

and corresponds to Eq. (13).

As it was stated above, Eq. (15) was obtained by GIBBS for a fluid interface. A corresponding equation for a flat solid interface in an isotropic state was derived by ERIKSSON [25]:

$$d\sigma = -\bar{s}\, dT + (\gamma - \sigma)\, d \ln A - \sum_i \Gamma_i\, d\mu_i \,. \tag{19}$$

Here γ is a mechanical analogue to the surface tension of a liquid (the true interfacial tension of a solid) and σ a thermodynamic analog to the surface tension of a liquid (the conventional interfacial tension of a solid). In the more general case of an anisotropic solid interface, the adsorption equation takes the form [26, 27]

$$d\sigma = -\bar{s}\,dT + (\hat{\gamma} - \sigma\hat{1}) : d\hat{e}^\sigma - \sum_i \Gamma_i\,d\mu_i\,, \tag{20}$$

where $\hat{\gamma}$ is the excess surface stress tensor, $\hat{1}$ the unit tensor, and $\hat{e}^\sigma$ the surface strain tensor while the symbol : denotes the scalar product of tensors. The summation in (20) is carried out over all mobile components. As for immobile components forming the lattice of a solid, their chemical potentials are not present in Eq. (20). GIBBS did not employ chemical potentials for immobile components. Such chemical potentials may be defined only conventionally and separately for each lattice plane direction of a solid as the chemical potential in an equilibrium fluid phase in contact with the solid along a given plane. The immobile component chemical potential, μ_j', so defined will depend at each point in the bulk on the direction of the unit normal $\vec{\nu}$ to the imaginary lattice plane. Besides, even for the true equilibrium state, the quantity μ_j' will not be the same for all the points of the lattice plane and, then, when proceeding to interfacial excess quantities, it will be necessary to take the excess of the product of the chemical potential times the mass of an immobile component. For every direction $\vec{\nu}$ on the dividing surface, we may define the quantity

$$\bar{g}_\nu \equiv \sum_j \overline{\mu_j' m_j}/A \tag{21}$$

and therefrom results the relationship [26, 27]

$$\sigma = \gamma_\nu + \hat{g}_\nu\,, \tag{22}$$

where γ_ν is the tension on the surface in the direction $\vec{\nu}$. Substituting (22) in (20) yields the equation [26, 27]

$$d\gamma_\nu = -\bar{s}\,dT + (\hat{\gamma} - \sigma\hat{1}) : d\hat{e}^\sigma - \sum_i \Gamma_i\,d\mu_i - d\bar{g}_\nu \tag{23}$$

which is also a generalization of the GIBBS adsorption equation for the case of a solid interface, but it is formulated in terms of the excess surface stress. For a fluid interface $\hat{\gamma} = \gamma\hat{1} = \sigma\hat{1}$, and Eqs. (20) and (23) are transformed into the GIBBS adsorption equation.

When applied to the surface of a liquid electrode, the GIBBS adsorption equation acquires an additional term connected with the change of an electrical potential. It may be said that, for isothermal-isobaric conditions, this term was obtained by GIBBS himself because GIBBS gave a thermodynamic derivation of the LIPPMANN equation. This problem has been discussed many times in investigations of electrocapillary phenomena (see, e.g. [28—30]). A strict derivation of the GIBBS adsorption equation for the flat surface layer of an electrode was given by PARSONS [31] (see also [20, 32]). We write here the corresponding equation for a curved layer [20, 32]:

$$A\,d\sigma = -\bar{S}\,dT + V_\alpha\,dp^\alpha + V_\beta\,dp^\beta - \sum_i \overline{m}_i\,d\mu_i - A\theta\,d(E + K)\,, \tag{24}$$

where θ is the charge per unit area and E the e.m.f. of the cell formed by combining a given electrode with a reference one, the latter being chosen so as to make the quantity K a function of only temperature and pressure. For a solid electrode the adsorption equation has the following form [26]:

$$d\sigma = -\bar{s}\,dT + (\hat{\gamma} - \sigma\hat{1}) : d\hat{e}^\sigma - \sum_i \Gamma_i\,d\mu_i - \theta\,d(E + K)\,. \tag{25}$$

For ionic systems the quantities μ_i in (24) and (25) should be understood as electrochemical potentials.

A very complicated case of the thermodynamics of interfaces is obtained when considering curved interfaces in external fields. Gibbs himself initiated the analysis of interfacial phenomena in a gravitational field. As for an electric field, the results were obtained considerably later. The origin of the difficulty here depends on whether or not the co-existing phases are conductors or dielectrics. The problem of adjacent conductors is solved easily [33] as compared with the case of dielectrics [34]. Here we cite as an example the generalized adsorption equation for a spherical interface between two dielectric phases in an external electric field (the whole system is supposed to be spherically symmetrical) [34]:

$$d\sigma = -\bar{s}\,dT - \sum_i \Gamma_i\,d\mu_i - \vec{\boldsymbol{P}}\,d\vec{D} + \vec{\boldsymbol{P}}'\,(\partial\vec{D}/\partial r)\,dr\,. \tag{26}$$

In Eq. (26) $\vec{D}$ is the electric induction vector and r the radius of the surface of tension satisfying the condition

$$p_N^\alpha(r) - p_N^\beta(r) = 2\sigma(r)/r\,, \tag{27}$$

where p_N is the normal component of the total pressure tensor in a polarized system [the surface tension is given, as before, by Eq. (6), but with the substitution of $p_T^\alpha(r)$ for p^α and $p_T^\beta(r)$ for p^β]. The quantities $\vec{\boldsymbol{P}}$ and $\vec{\boldsymbol{P}}'$ are defined by the expressions

$$\vec{\boldsymbol{P}} \equiv \int_{R^\alpha}^{r} (\vec{P} - \vec{P}^\alpha)\,dr' + \int_{r}^{R^\beta} (\vec{P} - \vec{P}^\beta)\,dr'\,, \tag{28}$$

$$\vec{\boldsymbol{P}}' \equiv \int_{R^\alpha}^{r} (\vec{P} - \vec{P}^\alpha)\,(1 - r/r')\,dr' + \int_{r}^{R^\beta} (\vec{P} - \vec{P}^\beta)\,(1 - r/r')\,dr'\,, \tag{29}$$

where $\vec{P}$ is the local polarization vector and r' the radial coordinate. For the case of a flat interface, the quantity $\vec{\boldsymbol{P}}$ is the excess dipole moment per unit area. As for the quantity $\vec{\boldsymbol{P}}'$, it is only essential in the case when the surface layer thickness is comparable with the surface curvature radius. The last term in (26) plays the role of a correction which is negligible if the surface curvature radius is much larger than the effective thickness of the surface layer.

An important step in the development of the thermodynamics of interfaces was a generalization of the Gibbs adsorption equation for the case of the absence of adsorption equilibrium. We may first note the works by Defay [35, 36] in which the notion was introduced of the lateral chemical potentials, ε_i, characterizing the dependence of the surface tension on the state of bulk phases, α and β:

$$d\sigma = -\bar{s}\,dT - \sum_i \Gamma_i\,d\mu_i^\sigma + \sum_i \varepsilon_i^\alpha\,dc_i^\alpha + \sum_i \varepsilon_i^\beta\,dc_i^\beta\,. \tag{30}$$

It is assumed in Eq. (30) that thermal and mechanical equilibrium is given while diffusional equilibrium has not yet been established.

The process of establishing adsorption equilibrium includes translational and rotational molecular motions and among them the orientation of non-spherical molecules in the interfacial layer. If orientation proceeds much more slowly than the translational diffusion process, it is possible to imagine a case where the absence of equilibrium in the whole system is due to the molecular (e.g., dipole) orientation process in the interfacial layer. For this case the following generalization of the GIBBS equation was suggested [33]:

$$d\sigma = -\bar{s}\, dT - \sum_i \Gamma_i\, d\mu_i - \frac{1}{A}\left[\sum_{i,x,\alpha} \boldsymbol{A}^{\alpha}_{i,x,0}\, d\langle m^{\alpha}_{i,x}\rangle + \sum_{i,x,\alpha} \boldsymbol{A}^{\alpha}_{i,x,Q}\, d\langle m^{\alpha 2}_{i,x}\rangle\right] \quad (31)$$

where $\langle m^{\alpha}_{i,x}\rangle$ is the average of the x-constituent of the dipole moment of the molecules of component i in phase α, $\langle m^{\alpha 2}_{i,x}\rangle$ is the average of the square of the same quantity, $\boldsymbol{A}^{\alpha}_{i,x,0}$ and $\boldsymbol{A}^{\alpha}_{i,x,Q}$ are the corresponding affinities, and the summation with respect to x and α denotes the summation over all the dipole moment constituents, all the phases, and all the thin elementary (treated as homogeneous) layers inside the interfacial layer. It should be noted that the derivation of Eq. (31) is based on a very conventional hypothesis about the independence of translational and rotational contributions to the adsorption-diffusion process.

3. The phase rule

If the adsorption equation may be considered to be the most famous result of GIBBS' theory of capillarity, the phase rule is indisputably the most widely known result of the whole of GIBBS' paper *"On the Equilibrium of Heterogeneous Substances"*. GIBBS' phase rule is

$$f = n - r + 2\,, \quad (32)$$

where f is the number of degrees of freedom for a system containing n independent components and r phases. Equation (32) refers to open systems. Later it was generalized for systems with various degrees of material isolation [37, 38]. However, here we will be interested only in that aspect of the phase rule which relates to interfacial phenomena.

As it is known, GIBBS established the phase rule outside of his theory of capillarity and did not extend it to interfacial phenomena. This was done later. First, the phase rule for open systems was generalized by DEFAY [35, 36] for the case of a system with flat or curved interfaces. Evidently, the presence of only flat interfaces without two-dimensional phase transitions does not influence the phase rule since a new variable (σ) and a new equation (the GIBBS adsorption equation) appear simultaneously for each interface. However, if there are two-dimensional phase transitions, the number of degrees of freedom diminishes by the difference between the surface phase number, p, and the number of types of interfaces, l:

$$f = n - r + 2 - (p - l)\,. \quad (33)$$

If all the interfaces are curved, equality of pressure in the bulk phases is no longer valid, and the number of variables increases by $r - 1$. In this case

$$f = n + 1\,. \quad (34)$$

If, additionally, two-dimensional phase transitions take place, then

$$f = n + 1 - (p - l)\,. \quad (35)$$

In the general case of the presence of both flat and curved interfaces, the variance is given by the relationship [19, 20]

$$f = n - r_0 - p_0 + z + y + 1\,, \tag{36}$$

where r_0 is the number of bulk phases possessing at least one flat interface, z is the number of multiphase regions, which are uniform in pressure, p_0 and y are two-dimensional analogs of the quantities r_0 and z, respectively. The formulae (33)—(36) were derived for open systems in which all the substances are able to travel freely from the system to the surroundings and return (such substances are called mobile). The further development of the phase rule was connected with the consideration of closed and partly open systems there may be both mobile and immobile components [39]. The most complex case is given when there are, additionally, chemical reactions in a system [40]. Taking chemical reactions into account is trivial for an open system since the number of components, n, in (33)—(36) is defined as the difference between the number of substances, c, and the number of independent chemical reactions, w:

$$n = c - w\,. \tag{37}$$

However, in the case of a partly closed system, taking chemical reactions into account becomes more complicated because of the connection with the material isolation conditions.

Let a system include r bulk and p surface phases and consist of the substances $B_1, B_2, \dots, B_c$, the substances $B_1, B_2, \dots, B_i$ ($i \leqq c$) being immobile. Let it be that in the system there are w reversible, homogeneous and heterogeneous, dependent and independent, chemical reactions:

$$\sum_{j=1}^{c} \nu_{jm} B_j = 0 \quad (m = 1, 2, \dots, w)\,. \tag{38}$$

The set of equations (38) may be characterized by the stoichiometric matrix composed of the coefficients ν_{jm}. Let the rank of this matrix be k. We also introduce the concentration-stoichiometric matrix for immobile components:

$$\begin{pmatrix} x_1^{(1)}, \dots, x_1^{(r+p)}, \nu_{11}, \dots, \nu_{1w} \\ \dots\dots\dots\dots\dots\dots \\ x_i^{(1)}, \dots, x_i^{(r+p)}, \nu_{i1}, \dots, \nu_{iw} \end{pmatrix} \tag{39}$$

where x_j is the mole fraction of the substance B_j and the superscript indicates the phase. Let the rank of this matrix be s^*.

The formula for the variance of the above-mentioned system is [40]

$$f = c - k - i + s^* - \sum_{q=1}^{r} (q - 1)\,\varphi_q - \sum_{t=1}^{p} (t - 1)\,\varphi_t + 1\,, \tag{40}$$

where φ_q is the number of pressure uniform regions consisting of q bulk phases, φ_t is the number of surface tension uniform regions consisting of t surface phases. If every phase belongs to one region, then

$$f = c - k - i + s^* - r - p + z + y + 1\,. \tag{41}$$

All the known formulae for the variance of multicomponent multiphase systems may be obtained from Eq. (41) as particular cases.

In addition to the variance, f, it is interesting to consider the total variance, F, describing not only the state, but also the amounts of coexisting phases. For the above system

$$F = f - s^* + s + r + p\,, \tag{42}$$

where s is the rank of the stoichiometric part of the matrix (39).

The maximum value of the variance is calculated as [40]

$$f = \begin{cases} F, & i > s + r + p\,, \\ F - (s + r + p - i)\,, & i \leqq s + r + p\,. \end{cases} \tag{43}$$

The maximum number of coexisting phases is determined by the expression [40]

$$(r + p)_{\max} = n - k + z + y + 1 \tag{44}$$

and does not depend on the degree of material isolation of a system.

For a closed system, one may also derive a generalization [40] of DUHEM's theorem

$$F = z + y + 1\,. \tag{45}$$

For a closed system under uniform pressure ($z = 1$), without taking interfacial phenomena into account ($y = 0$), Eq. (45) changes into DUHEM's theorem: $F = 2$.

The development of new trends in the thermodynamics of interfaces

1. The estimation of surface layer thickness

In GIBBS' time there was no information about the surface layer thickness, and it might only be supposed, as GIBBS did, that this thickness is very small due to the short range of molecular forces. On this account GIBBS aimed at formulating a thermodynamic method independent of surface layer thickness. Nowadays, however, a number of methods is known for the investigation of surface layer thickness (see review [41]), and very interesting experimental and theoretical results have been obtained in this field. It turns out that thermodynamics itself may give some information about surface layer thickness. This is all the more striking because the method of a layer of finite thickness is equivalent to GIBBS' method, and the main equation of the method of a layer of finite thickness, Eq. (13), is valid for an arbitrary surface layer thickness and, consequently, cannot give any information about this quantity. The situation changes, however, if one takes into account not only the equilibrium conditions given in Eq. 13, but also the stability conditions expressed in thermodynamic inequalities. Indeed, if we consider an interface as a real physical body, we should assume it to satisfy the stability conditions. One of them is the stability condition with respect to the surface layer thickness which describes the ability of the surface layer to reestablish its equilibrium thickness after accidental changes. Strictly speaking, this is the ability of the surface layer to maintain its molecular structure and the equilibrium profiles of all its local properties.

The stability condition with respect to surface layer thickness may be expressed by the inequality [42; 20, p. 49]

$$E' \equiv a\left(\frac{\partial\sigma}{\partial a}\right)_{T,p,x_i} > 0\,, \tag{46}$$

where σ is the surface tension and a the area per unit mass of the surface layer, the derivative is taken at constant temperature, external pressure, and composition of the surface layer. The quantity E' is the conventional elasticity modulus of the surface layer.

When applied to a binary system, the condition (46) makes it possible to prove the inequality [42; 20, p. 151]

$$\left(\frac{dx_1^\sigma}{dx_1^\alpha}\right)_T \geqq 0\,, \tag{47}$$

where x_1^σ and x_1^α are the mole fractions of one of the components in the surface layer and in the bulk phase, respectively. This inequality is the analog to Konowaloff's third law for liquid-vapor equilibrium and is intuitively obvious without any proof. In addition there is the selfevident inequality

$$0 \leqq x_1^\sigma \leqq 1\,. \tag{48}$$

The relative adsorption, $\Gamma_{1(2)}$, determined from experiment may be uniquely related to the quantity x_1^σ if the surface layer thickness, τ, is given. Then, inequalities (47) and (48) impose certain restrictions on the surface layer thickness. For liquid-vapor and liquid-solid systems, these restrictions are expressed by the inequalities (assuming an incompressible liquid phase) [43]

$$\tau \geqq v_{01}\Gamma_{1(2)} - v^\alpha(1 - x_1^\alpha)\,\frac{d\Gamma_{1(2)}}{dx_1^\alpha}\,, \tag{49}$$

$$\tau \geqq v_{01}\Gamma_{1(2)}\,, \tag{50}$$

where v^α is the molar volume of the liquid phase, and v_{01} the partial molar volume of the first component in the liquid phase.

The equality sign in (49) and (50) represents the minimum possible value for the surface layer thickness as a function of solution composition.

Thus, though thermodynamics cannot determine the true effective surface layer thickness, it makes it possible to estimate a lower limit for possible values of the thickness. Application of inequalities (49) and (50) to experimental data on surface tension and adsorption shows that the minimum possible thickness may prove to be very small (smaller than molecular dimensions) and the inequalities are not informative in this case. In other cases, the minimum possible thickness proves to be appreciable or even considerable. In particular, it has been established by this method that the surface layer thickness for the liquid-gas boundary dramatically increases when approaching the critical solution temperature for the liquid phase [44, 41].

2. The thermodynamics of thin films

In his theory of capillarity GIBBS confined himself to the consideration of only thick films for which it is possible to neglect the interaction of the surface layers on the opposite sides of a film. A thin film is in principle different from a thick one in that the surface layers of the thin film may not be considered independently. Actually, it is impossible to distinguish between the surface layers and the bulk phase in a thin film, the latter should be considered as a whole. An important distinct characteristic of a thin film is the disjoining pressure which is manifested in experiments through a change in external pressure during the transition from a thick film to a thin film. The notion of the disjoining pressure was introduced by DERYAGUIN [45] who also carried out the first measurements of this quantity.

There are several equivalent definitions of the disjoining pressure of a flat thin film. First, the disjoining pressure, Π, may be defined as the difference between the external pressure values, p^α, for the thin film and the corresponding thick film:

$$\Pi = p^\alpha(h) - p^\alpha(\infty)\,, \tag{51}$$

where h is the thickness of the thin film. If the thin film has been formed from the phase γ and is kept in equilibrium with this phase (e.g., when a bubble is approaching a solid surface: the phase α is gas and the phase γ is liquid), the disjoining pressure may be defined as

$$\Pi = p^\alpha(h) - p^\gamma\,. \tag{52}$$

Finally, since for a flat film the external pressure is always equal to the normal component of the pressure tensor inside the film, it is possible to give the definition

$$\Pi = p_N - p^\gamma \tag{53}$$

and to formulate it verbally as follows: the disjoining pressure is the difference between the normal pressure inside a film (or the external pressure) and the pressure in the bulk phase of the same nature and at the same values of temperature and the chemical potentials as in the film. The definition (52) was first used in experimental investigations of the disjoining pressure [46—49] while the definition (53) was used for calculations [50].

As a thermodynamic quantity, the disjoining pressure may be related to other thermodynamic parameters, and the corresponding relationships form the thermodynamics of thin films as a special chapter of the theory of capillarity. The thermodynamics of thin films was elaborated in a number of papers (e.g., see [51—61]) and described in detail in the monograph [20, p. 259—310]. The thermodynamics of thin films has found important application in the theories of electrocapillarity [59], adsorption [60—64], and chromatography [65].

Here we demonstrate, as an example, an approach to the thermodynamics of thin films using a pair of dividing surfaces. Let us imagine a film to be formed by thinning a layer of phase γ between phases α and β. Then, having chosen the positions of the two dividing surfaces and taken the excesses from the adjacent regions of phases α

and β, we come for the layer of finite thickness to Eq. (13), which for the case under consideration has the form

$$d\bar{U} = T\,d\bar{S} - p_N\,d(hA) + \gamma\,dA + \sum_i \mu_i\,d\bar{m}_i\,, \tag{54}$$

where γ is the film tension and h the distance between the dividing surfaces. As the next step, we proceed to the excesses with respect to the phase γ and subtract from (54) the equation

$$dU^\gamma = T\,dS^\gamma - p^\gamma\,d(hA) + \sum_i \mu_i\,dm_i^\gamma\,. \tag{55}$$

Using the definition (53), we obtain

$$d\bar{U}^{\alpha\gamma,\beta\gamma} = T\,d\bar{S}^{\alpha\gamma,\beta\gamma} - \Pi\,d(hA) + \gamma\,dA + \sum_i \mu_i\,d\bar{m}_i^{\alpha\gamma,\beta\gamma}\,, \tag{56}$$

where $\bar{U}^{\alpha\gamma,\beta\gamma}$, $\bar{S}^{\alpha\gamma,\beta\gamma}$, and $\bar{m}_i^{\alpha\gamma,\beta\gamma}$ are the total excesses for both surfaces of energy, entropy, and mass of the component i, respectively.

Equation (56) is valid for arbitrary positions of the dividing surfaces and plays the role of the main fundamental equation for a thin film from which many other thermodynamic relationships may be obtained. In particular, we obtain from (56) the expression

$$\Pi = -\frac{1}{A}\left(\frac{\partial \bar{U}^{\alpha\gamma,\beta\gamma}}{\partial h}\right)_{T,\,A,\,\bar{m}_i^{\alpha\gamma,\beta\gamma}} \tag{57}$$

which may also be considered as a definition of the disjoining pressure.

Two other fundamental equations follow from (56):

$$\bar{U}^{\alpha\gamma,\beta\gamma} = T\bar{S}^{\alpha\gamma,\beta\gamma} - \Pi hA + \gamma A + \sum_i \mu_i \bar{m}_i^{\alpha\gamma,\beta\gamma}\,, \tag{58}$$

$$d\gamma = -(\bar{S}^{\alpha\gamma,\beta\gamma}/A)\,dT + h\,d\Pi - \sum_i \Gamma_i^{\alpha\gamma,\beta\gamma}\,d\mu_i\,. \tag{59}$$

Equation (59) is the analog to the Gibbs adsorption equation (in terms of absolute adsorption). It is not an independent thermodynamic relationship and should be considered together with the fundamental equations for the bulk phases in deriving any physical results.

The above definitions for the disjoining pressure refer only to a flat film. When proceeding to the case of a curved film, the following complications arise: a) the definitions (51)—(53) cease to be equivalent, b) each of these definitions loses its uniqueness. Indeed, if we use the definitions (51) and (52), there will be two disjoining pressures since the pressures p^α and p^β on the opposite sides of the film will be different. The use of definition (53) is even more difficult because the quantity p_N will be a function of the spatial coordinates in the case of a curved film.

3. Thermodynamics of surface separation processes

The processes of separation of substances based on the composition difference between the surface layer and the bulk phases are called surface separation processes. Such processes are widely applied for the separation of complex mixtures. Besides adsorption chromatography, considerable progress has been achieved in some other surface separation methods, in particular, in the foam separation method (see, e.g., [66—69]).

Continuous and equilibrium surface separation processes are analogous to open phase processes (e.g., vaporization) in the course of which the composition of a system changes. The process of once-through surface separation is analogous to a distillation process while the process of multiple surface separation is analogous to a rectification process. For this reason the theory of surface separation processes has many features in common with the theory of open phase processes, which is well developed in modern chemical thermodynamics.

The changes of the composition and the mass of a system in the course of surface separation are related by the equation [70, 71]

$$\frac{dx_i}{d \ln m} = x_i^\sigma - x_i \qquad (i = 1, 2, ...) , \tag{60}$$

where x_i is the mole fraction of component i, m is the total number of moles in the system, the superscript σ refers to the surface layer being separated in the course of the process.

Equations (60) may be written in the form

$$\frac{dx_i}{dx_k} = \frac{x_i^\sigma - x_i}{x_k^\sigma - x_k} \tag{61}$$

and may be considered as a set of differential equations for the surface separation lines along which the composition changes in the state diagram of the system.

The surface tension, σ, also changes in the course of surface separation, and its change is given by the equation [70, 71]

$$a \frac{d\sigma}{d \ln m} \cong - \sum_{i,k=1}^{n-1} (x_i^\sigma - x_i)(x_k^\sigma - x_k) \frac{\partial^2 g}{\partial x_i \, \partial x_k} \tag{62}$$

where a is the ratio of the area to the number of moles in the surface layer, g GIBBS' mole thermodynamic potential, and n the number of components. It follows from the thermodynamic stability conditions that the quadratic form in the right-hand side of (62) is always positive. Hence, the derivative $d\sigma/d \ln m$ is negative. Since the mass of a system always diminishes in the course of surface separation, the surface tension always increases in the course of surface separation.

The analysis of the principles determining the course of the surface separation lines is the content of the thermodynamics of surface separation processes. Both the local relationships determining the shape of the separation lines and general rules for surface separation diagrams are considered. As an example of the latter, we present the formula determining the interrelation between the numbers of singular points of various types (knots and saddles) on the concentration surface separation diagram [72]

$$\sum_{k=1}^{n} 2^k (N_k^+ + C_k^+ - N_k^- - C_k^-) = (-1)^{n-1} + 1 , \tag{63}$$

where N_k^+ and N_k^- are the numbers of the k-component knot-points, and C_k^- and C_k^- are the numbers of the k-component saddle-points with the indices $+I$ and $-I$, respectively.

Formula (63) allows us to classify surface separation diagrams. Since the location of the surface separation lines is uniquely related to the location of the constant surface tension lines. Formula (63) is also the basis for the classification of surface tension diagrams. The number of the possible diagram types grows rapidly with the number of components. For instance, it may be concluded by analogy with distillation line diagrams [73] that already 38 types of diagrams are possible for a three-component system.

There are, among other problems in the thermodynamics of surface separation processes, the theory of multistage processes (such as foam fractionation) [74] and the theory of surface separation in the presence of equilibrium chemical reactions [75].

Conclusion

It is very hard to exhaust the topic of this article, and we have confined ourselves in every section only to examples which demonstrate current progress in the thermodynamics of interfaces. Considerable success has also been achieved in the thermodynamics of adsorption, wetting, nucleation, electrodes processes and in other fields. Moreover, we have barely touched upon non-equilibrium thermodynamics, which is a new trend of increasing importance in interfacial phenomena (see review [76]). One may also mention quasi-thermodynamics (see, e.g., [7]) in which thermodynamic methods are used for the investigation of the structure of surface layers.

It is certain that, together with the development of new trends, Gibbs' theory itself will continue to develop in the future. Detailed monographs devoted to the modern thermodynamic theory of capillarity [20, 36] have already been written. However, it is interesting that none of them completely coincides with Gibbs' original work. In spite of the passage of a century, there are a number of problems which have been discussed by Gibbs alone. For many years Gibbs' work has served as a source of ideas and inspiration for new generations of investigators and sometimes as a source of unexpected finds (the author himself happened to learn a striking and previously unknown fact that the first rigorous proof of the Le Chatelier-Brown reduced principle is to be found in Gibbs' work [77]). It may be said that Gibbs' theory of capillarity has not in the least become outdated. Its significance and range of application continue to widen, and it will for long time to come serve the interests of science and technology.

References

[1] J. W. Gibbs, Trans. Conn. Acad., **3**, 343, 1878
[2] J. W. Gibbs, Proc. Amer. Acad., **16**, 420, 1881
[3] E. B. Wilson, Proc. Nat. Acad. USA, **31**, 34, 1945
[4] E. A. Guggenheim, Trans Faraday Soc., **36**, 397, 1940
[5] J. Rice, in "*A Commentary of the Scientific Writings of J. W. Gibbs*", Vol. I (F. G. Donnan and A. Haas, Eds.), New Haven, 1936, p. 505
[6] S. Kondo, J. Chem. Phys., **25**, 662, 1956
[7] S. Ono, S. Kondo, in "*Handb. d. Physik*", Vol. 10, Springer, Berlin, 1958, p. 237
[8] A. I. Rusanov, V. V. Krotov, in "*Progr. in Surface and Membrane Sci.*" 2, Academic Press, N.Y. (in press).

[9] A. I. RUSANOV, Kolloid. Zh. **28**, 551, 1966
[10] M. VAN DEN TEMPEL, J. LUCASSEN and E. H. LUCASSEN-REYNDERS, J. Phys. Chem., **69**, 1798, 1965
[11] A. SCHELUDKO, "*Colloid Chemistry*", Elsevier, Amsterdam, 1966, p. 254
[12] M. J. ROSEN, J. Colloid Interface Sci., **24**, 279, 1967
[13] E. H. LUCASSEN-REYNDERS, J. LUCASSEN, and M. VAN DEN TEMPEL, J. Colloid Interface Sci., **28**, 339, 1968
[14] V. V. KROTOV, A. I. RUSANOV, Kolloid. Zh. **34**, 81, 1972
[15] J. D. VAN DER WAALS, P. KOHNSTAMM, "*Lehrbuch der Thermostatik*", Teil I, Barth, Leipzig 1927
[16] G. BAKKER, "*Kapillarität und Oberflächenspannung*", "*Handb. d. exper. Phys.*", B. VI, Wien-Harms, Leipzig 1928
[17] J. E. VERSCHAFFELT, Acad. Roy. Belgique, Bull. classe sci., **22**, 373, 390, 402, 1936
[18] J. C. ERIKSSON, Ark. Kemi., **25**, 331, 343; **26**, 49, 117; 1966
[19] A. I. RUSANOV, "*Thermodynamics of Interfacial Phenomena*" (russ.), Leningrad Univ., Leningrad, 1960
[20] A. I. RUSANOV, "*Phase Equilibria and Interfacial Phenomena*" (russ.), Khimia, Leningrad, 1967 (see also German translation: A. I. RUSANOV, "*Phasengleichgewichte und Grenzflächenerscheinungen*", Akademie-Verlag, Berlin, 1978)
[21] F. C. GOODRICH, in "*Surface and Colloid Science*", vol. I (E. MATIJEVIC, Ed.), Wiley, New York, 1969, p. I.
[22] F. P. BUFF, J. Chem. Phys. **19**, 1591, 1951
[23] T. L. HILL, J. Phys. Chem., **56**, 526, 1952
[24] A. I. RUSANOV, Vestnik Leningrad Univ., No. 16, 71, 1959
[25] J. C. ERIKSSON, Surface Sci., **14**, 221, 1969
[26] A. I. RUSANOV, Kolloid. Zh. **39**, 711, 1977
[27] A. I. RUSANOV, J. Colloid Interface Sci., **63**, 330, 1978
[28] A. FRUMKIN, Ergebnisse exakt. Naturw., **7**, 235, 1928
[29] F. O. KOENIG, J. Phys. Chem. **38**, 111, 339, 1934
[30] J. A. V. BUTLER, "*Elektrocapillarity. The Chemistry and Physics of Electrodes and Other Charged Surfaces*", Methuen, London, 1940
[31] R. PARSONS, Canad. J. Chem., **37**, 308, 1959
[32] A. I. RUSANOV, Zh. Fiz. Khim., **36**, 549, 690, 1962
[33] A. SANFELD, "*Introduction to the Thermodynamics of Charged and Polarized Layers*", Wiley, London, 1968
[34] A. I. RUSANOV, Dokl. Akad. Nauk SSSR, **238**, 831, 1978
[35] R. DEFAY, "*Étude Thermodynamique de la Tension Superficielle*", vol. I, Paris, 1934
[36] R. DEFAY, I. PRIGOGINE, "*Tension Superficielle et Adsorption*", Desoer, Liége, 1951 (see also English translation: R. DEFAY, I. PRIGOGINE, A. BELLEMANS, D. H. EVERETT, "*Surface Tension and Adsorption*", Longmans, London, 1966)
[37] A. V. STORONKIN, Vestnik Leningrad Univ., No. 16, 74, 1956
[38] A. V. STORONKIN, A. N. MARINICHEV, V. T. ZHAROV, Zh. Fiz. Khim., **47**, 3016, 1973; **50**, 3048, 1976
[39] A. I. RUSANOV, Kolloid. Zh., **27**, 428, 1965.
[40] A. N. MARINICHEV, S. A. LEVICHEV, A. I. RUSANOV, in "*The Centennial of Gibbs' theory of capillarity*" (russ.) (A. I. RUSANOV, Ed.) Leningrad Univ., Leningrad, 1978, p. 100
[41] A. I. RUSANOV, in "*Progr. in Surface and Membrane Sci.*", vol. 4, (J. F. DANIELLI, M. D. ROSENBERG, D. A. CADENHEAD, Eds.), Academic Press, New York, 1971, p. 57

[42] A. I. RUSANOV, Kolloid. Zh. **28**, 443, 1966
[43] A. I. RUSANOV, in "*Physical Adsorption from Multicomponent Phases*" (russ.) (M. M. DUBININ, Ed.), Nauka, Moscow, 1972, p. 74
[44] A. I. RUSANOV, S. A. LEVICHEV, O. N. MIKHALENKO, Zh. Fiz. Khim. **43**, 2636, 1969
[45] B. V. DERYAGUIN, Kollod. Zh., **17**, 207, 1955
[46] B. V. DERYAGUIN, E. OBUKHOV, Acta physicochim. USSR, **5**, I, 1936
[47] B. V. DERYAGUIN, M. KUSAKOV, Izv. Akad. Nauk SSSR, ser. Khim., 741, 1936; 1119, 1937
[48] B. V. DERYAGUIN, A. S. TITIEVSKAYA, Dokl. Akad. Nauk SSSR, **89**, 1041, 1953
[49] B. V. DERYAGUIN, I. I. ABRIKOSOVA, Dokl. Akad. Nauk SSSR, **108**, 214, 1956; Zh. Fiz. Khim., **32**, 442, 1958
[50] A. I. RUSANOV, F. M. KUNI in "*Research in Surface Forces*" (russ.) (B. V. DERYAGUIN, Ed.), vol. 3, Nauka, Moscow, 1967, p. 129. (see also English translation: "*Consultants Bureau*", New York-London, 1971, p. 111)
[51] A. N. FRUMKIN, Zh. Fiz. Khim., **12**, 337, 1938
[52] B. V. DERYAGUIN, Acta physicochim. URSS, **12**, 181, 1940
[53] G. A. MARTYNOV, B. V. DERYAGIN, Kolloid. Zh., **24**, 480, 1962
[54] B. V. DERYAGUIN, G. A. MARTYNOV, YU. V. GUTOP, Kolloid. Zh., **27**, 357, 1965
[55] B. V. DERYAGUIN, YU. V. GUTOP, Kolloid. Zh., **27**, 674, 1965
[56] B. V. DERYAGUIN, J. Colloid Interface Sci., **24**, 357, 1967
[57] A. I. RUSANOV, Kolloid. Zh., **28**, 718, 1966; **29**, 141, 237, 1967
[58] A. SCHELUDKO, B. RADOEV, T. KOLAROV, Trans. Faraday Soc., **64**, 2213, 1968
[59] A. I. RUSANOV, Dokl. Akad. Nauk SSSR, **203**, 387, 1972
[60] A. I. RUSANOV, J. Colloid Interface Sci., **53**, 20, 1975
[61] B. V. DERYAGUIN, N. V. CHURAEV, Kolloid Zh., **38**, 438, 1976; J. Colloid Interface Sci., **54**, 157, 1976
[62] S. G. ASH, D. H. EVERETT, C. RADKE, J. Chem. Soc., Faraday Trans. Part 2, **69**, 1256, 1973
[63] A. I. RUSANOV, Dokl. Akad. Nauk SSSR, **218**, 882, 1974
[64] B. V. DERYAGUIN, V. M. STAROV, N. V. CHURAEV, Kolloid Zh., **38**, 449, 1976
[65] A. I. RUSANOV, B. G. BELENKY, Zh. Fiz. Khim., **47**, 2046, 1973; J. Chromatogr., **80** 147, 1973
[66] J. J. BIKERMAN, "*Foams*", Springer, New York, 1973
[67] H. G. CASSIDY, in "*Technique of Organic Chemistry*", vol. 10, New York, 1957
[68] H. KISHIMOTO, Kolloid-Z., **192**, 66, 1963
[69] "*Adsorptive Bubble Separation Techniques*" (R. LEMLICH, Ed.), Academic Press, New York, 1971
[70] A. I. RUSANOV, V. T. ZHAROV, S. A. LEVICHEV, Dokl. Akad. Nauk SSSR, 184, 372, 1969
[71] A. V. STORONKIN, A. I. RUSANOV, V. T. ZHAROV, S. A. LEVICHEV, Kolloid. Zh. **31**, 290, 1969
[72] V. T. ZHAROV, A. I. RUSANOV, S. A. LEVICHEV, Kolloid. Zh., **30**, 819, 1968
[73] V. T. ZHAROV, L. A. SERAFIMOV, "*Physical-Chemical Principles of Distillation and Rectification*", Khimia, Leningrad, 1975
[74] V. T. ZHAROV, A. I. RUSANOV, S. A. LEVICHEV, Teor. Osnovy Khim. Tekhnol., **4**, 264, 1970
[75] V. T. ZHAROV, A. I. RUSANOV, S. A. LEVICHEV, Kolloid. Zh., **33**, 679, 1971
[76] R. DEFAY, I. PRIGOGINE, A. SANFELD, J. Colloid Interface Sci., **58**, 498, 1977
[77] A. I. RUSANOV, M. M. SHULTZ, Vestnik Leningrad Univ., No. 4, 60, 1960

Surface viscosity as a capillary excess transport property

F. C. GOODRICH†

(Institute of Colloid and Surface Science, Clarkson College of Technology, Potsdam, N.Y./USA)

Abstract

It is shown that the mathematical methods of GIBBS, devised by him to handle problems in capillary thermodynamics, are equally applicable to problems in capillary hydrodynamics. The introduction of a GIBBS dividing surface into a moving, fluid interface serves to clarify the boundary conditions in current use in the interpretation of experimental results and to interpret the coefficients of surface dilational and of surface shear viscosity as capillary excess transport properties in an entirely analogous way to GIBBS' interpretation of the surface tension as a capillary excess free energy.

1. Introduction

Among the enduring contributions of Willard GIBBS to capillary thermodynamics [1] was to show that the thin but molecularly diffuse interfacial region between two bulk fluid phases may be modeled in such a way that its contribution to the thermodynamic properties of the two phase system is expressed in terms of a set of excess quantities. Thus, if Q be any extensive thermodynamic property of the two phase system and if the two bulk phases be called I and II respectively, then

$$Q = Q^{\mathrm{I}} + Q^{\mathrm{II}} + q \tag{1}$$

in which q represents the contribution of the interface considered as an excess over the sum $Q^{\mathrm{I}} + Q^{\mathrm{II}}$ contributed by the bulk phases.

This approach sounds deceptively simple until it is investigated analytically with the customary rigor which one associates with the name of Willard GIBBS. The fundamental problem which arises is how much of each bulk phase one is to include in the terms Q^{I} and Q^{II}. In Figure 1 the interfacial region is drawn as diffuse. The local physical properties of the system here change rapidly, but not infinitely swiftly; and as the interfacial region is traversed from a point well within phase II to another point well within phase I, the physical properties change continuously. How then is one to interpret Q^{I} and Q^{II}, for they are extensive quantities computed exclusively from bulk phase properties and thus should strictly not include any of the interfacial region? GIBBS solved this problem by introducing into the interfacial region a mathematical surface, the GIBBS dividing surface, and conventionally defining the bulk phases as lying either to one side or the other of this mathematical surface of zero thickness. Q^{I} must therefore be computed from the geometric extent of the conventional volume on the I side of the dividing surface in Figure 1 and similarly for Q^{II}.

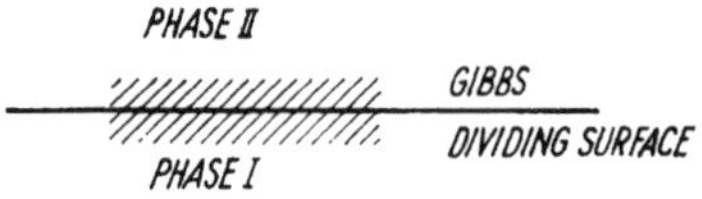

Fig. 1: The diffuse interfacial region between bulk phases I and II.

Neither of Q^{I} or Q^{II} is thus literally the Q value for the indicated region, for each region contains a portion of the diffuse interfacial zone where bulk phase properties do not apply. The sum $Q^{I} + Q^{II}$ cannot thus be the exact amount of Q contained in the two phase system, and the deficiency is supplied by q in equation (1).

Finally, it becomes apparent in attempting to apply this mathematical model to experimental systems that some sort of decision must be adopted as to the precise location of the GIBBS dividing surface. From equation (1) it is evident that the sum $Q^{I} + Q^{II} + q$ is invariant, but that the individual values of the terms will change depending upon the conventional volumes assigned to each of the bulk phases, i.e., upon the location of the dividing surface. This freedom of mathematical choice in the theory was exploited by GIBBS to simplify the resulting thermodynamic equations and thus to ease the task of experimental interpretation of measured values of q. Unfortunately, no single convention regarding the location of the GIBBS dividing surface suffices to simplify all thermodynamic problems connected with fluid interfaces, and the modern investigator has to be keenly aware of the various conventions used if he is to avoid error in his interpretation of experimental results.

2. Capillary excess transport properties

In this article I wish to show that the methods devised by GIBBS for the static interface may be extended to interfaces in motion. Inevitably new macroscopic properties of the two phase system are introduced. Thus a static, bulk fluid may be described thermodynamically by such properties as its density, hydrostatic pressure, internal energy, etc. But if the fluid is in motion, we need quantities which describe the rate of transport of mass, momentum, and energy through it. These appear in the theory of bulk fluids as diffusion, viscosity, and heat transport coefficients respectively.

Similarly if our system consists of two fluid phases, the transport of mass, momentum, and energy can be expected to exhibit peculiarities in the neighborhood of the phase boundary, properties which cannot be predicted from a knowledge of transport coefficients valid only in the interiors of the bulk phases. By analogy with the capillary thermodynamics of GIBBS, we may inquire if these peculiarities can be accounted for by the introduction into the dynamic theory of capillary excess transport coefficients. May we, in short, introduce a GIBBS dividing surface into a moving interface and assign excess transport coefficients to it in the same way that we assign excess thermodynamic properties to a GIBBS dividing surface inserted into a static interface? It is this program which I investigate here in an important special case directed at the transport of momentum only. The result will be an identification of the experimentally measurable surface viscosity coefficients as capillary excess viscosities in precisely the same sense as the familiar interfacial tension γ of a static phase boundary was identified by GIBBS as a capillary excess free energy.

3. The macroscopic hydrodynamic problem

I shall restrict our discussion to problems connected with plane interfaces. Such interfaces are certainly not the only ones of experimental interest, but a more general approach would inevitably require the introduction of curvilinear coordinate systems, and the increase in rigor would be offset by a corresponding complication in the notation and a loss in the reader's intuitive grasp of the physical situation. The interface in Figure 1 is thus not only initially plane, but is to remain plane throughout the course of the motion. Taking the z axis of a Cartesian system as normal to the interfacial region and positive in phase II, this means that the velocity field

$$\vec{v} = (v_x, v_y, v_z)$$

is of such a type that at all times t, $v_z = 0$ at $z = 0$, defined to be the macroscopic location of the phase boundary. By macroscopic I mean the coordinate which an experimentalist would conventionally use to define the position of the optically observed interface. It is true that later in this article we shall have to consider the interface as diffuse and thus define a GIBBS dividing surface at $z = 0$ with some care; but such details are of small concern to the conventional hydrodynamicist to whom matter is treated as a continuum and fluid phase boundaries are infinitely abrupt. Our interface may thus be sheared in the x and y directions parallel to the phase boundary, and it may be extended (dilated) in the plane of the interface, but its curvature remains zero throughout the course of the motion.

In the hydrodynamic analysis of the macroscopic problem [2, 3], the investigator must first solve the Navier-Stokes equations for the velocity fields $\vec{v}^{\mathrm{I}}$, $\vec{v}^{\mathrm{II}}$ in each of the bulk phases and then fit the two solutions together via suitable boundary conditions. For incompressible fluids I and II and for the restricted class of motions described above, one of the boundary conditions expresses the conservation of linear momentum in directions normal and parallel to the interface [4]

$$\nabla_s(\gamma + \varkappa \nabla_s \cdot \vec{v}^0) + \eta \, \nabla_s^2 \vec{v}^0 + \mu^{\mathrm{II}} \frac{\partial \vec{v}^{\mathrm{II}}}{\partial z} - \mu^{\mathrm{I}} \frac{\partial \vec{v}^{\mathrm{I}}}{\partial z} = 0 . \tag{2}$$

Here $\nabla_s = (\partial/\partial x, \partial/\partial y, 0)$ is the surface gradient operator with $\nabla_s^2 = \nabla_s \cdot \nabla_s$; $\vec{v}^{\mathrm{I}}$ and $\vec{v}^{\mathrm{II}}$ are the velocity fields in the interiors of the indicated phases extrapolated to $z = 0$; and μ^{I}, μ^{II} are the respective coefficients of bulk shear viscosity. The velocity $\vec{v}^0 = (v_x^0, v_y^0, 0)$ is the common value achieved by $\vec{v}^{\mathrm{I}}$ and $\vec{v}^{\mathrm{II}}$ at the phase boundary: $\vec{v}^{\mathrm{I}} = \vec{v}^{\mathrm{II}} = \vec{v}^0$ at $z = 0$. Finally, γ is the interfacial tension, and $\varkappa$, η are respectively the coefficients of surface dilational and of surface shear viscosity.

The terms in equation (2) may be identified individually. The quantity $\nabla_s \gamma$ represents momentum transport parallel to the interface due to a gradient in interfacial tension, i.e., a MARANGONI flow. The term $\varkappa \nabla_s \cdot \vec{v}^0$ measures viscous resistance to a local dilation of the interface, and the term $\eta \nabla_s^2 \vec{v}_0$ measures viscous resistance to shearing motion in the plane of the interface. All of these terms result from the transport of momentum along the xy plane of the phase boundary. Finally the terms $\mu^{\mathrm{I}}(\partial \vec{v}^{\mathrm{I}}/\partial z)$ and $\mu^{\mathrm{II}}(\partial \vec{v}^{\mathrm{II}}/\partial z)$ measure transport of momentum in a direction normal to the plane of the interface. Equation (2) thus states that the flux of momentum through and along the phase boundary is locally steady, with time entering only as a parameter. One remarks, in short, that equation (2) lacks an inertial term. Momentum may thus be transported through and along an interface, but it may not accumulate there. It is as though the interface were assigned a zero mass density.

Despite the empirical success of (2) in practice [5—14], the thoughtful hydrodynamicist must be concerned by the absence of an inertial term in equation (2). One could argue heuristically that a zero mass density is a reasonable hypothesis for an interface of zero thickness, but is it not then equally reasonable to assign zero viscosity coefficients $\varkappa$, η to such an interface? Indeed, if all capillary effects are totally ignored, then a sufficient boundary condition would be simply the last two terms of equation (2), and these two terms alone have been successful in the analysis of hydrodynamic problems involving two scrupulously pure fluids in the absence of a temperature induced gradient in the interfacial tension. For such systems one presumes the absence of any momentum transport parallel to the interface, and the two phases communicate only by momentum transport normal to the phase boundary.

Whenever the two phases are contaminated by traces of surface active materials, however, the retention of the capillary terms in equation (1) is essential; and one again returns to the problem of justifying the assignment of a zero mass density to the phase boundary. What I shall attempt in this article is a formal justification of this hypothesis. It will, in fact, turn out not to be an hypothesis at all but a convention related to the proper positioning of a GIBBS dividing surface within the molecularly diffuse interfacial region. I shall identify this convention as simply an extension to the dynamic system of one of GIBBS' most popular conventions used to locate his dividing surface in the static interface: the capillary excess mass density is deliberately chosen to be zero. With this convention it is possible to give a quantitative definition of the surface viscosity coefficients $\varkappa$, η as capillary excess viscosities.

4. The pressure tensor

We begin our investigation by selecting a small wedge of interphase whose mathematical boundaries move with the fluid in the conventional sense of hydrodynamics [15]. Figure 2 should be imagined to be carved mathematically out of the heart of Figure 1. The thickness Δz is small on a laboratory scale, but large enough on a microscopic scale so that the upper surface (hereafter called the "ceiling") lies well within the region where bulk phase II properties pertain, while the lower surface (the "floor") is similarly sufficiently deep inside phase I. Over floor and ceiling the indicated bulk phase densities ϱ and bulk phase shear viscosities μ are valid. Δz is thus large enough

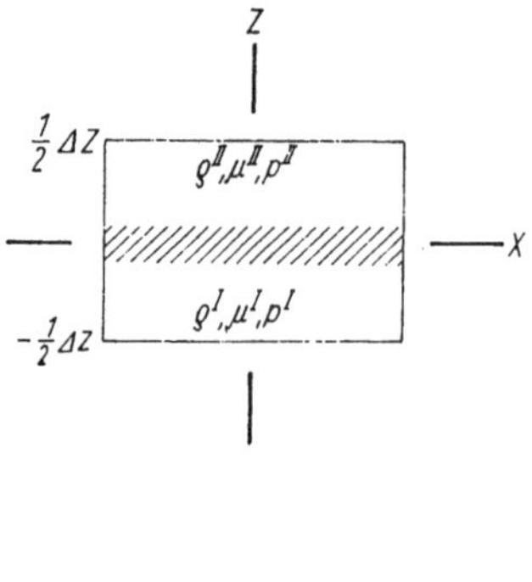

Fig. 2: A sample of plane interphase of dimensions $\Delta x\ \Delta y\ \Delta z$.

The Z axis is chosen in such a way as to diagonalize the static part of the pressure tensor, equation(4). The thickness ΔZ is sufficient so that bulk phase properties may be assigned to the "floor" at $-\frac{1}{2}\Delta Z$ and to the "ceiling" at $+\frac{1}{2}\Delta Z$. The location of the dividing surface at $Z = 0$ is unspecified.

to span all of the anisotropy of the interfacial region. For the moment I make no convention regarding the location of the reference plane $z = 0$ within the region between floor and ceiling.

If the situation is a static one, then it is known that in the interiors of the bulk phases the pressure tensor $\hat{P}$ is isotropic

$$\hat{P} = \begin{pmatrix} -p & 0 & 0 \\ 0 & -p & 0 \\ 0 & 0 & -p \end{pmatrix} . \tag{3}$$

Within the interfacial region, however, we follow BAKKER [16] and assume that the pressure tensor is axially symmetric about z. A general theorem in tensor analysis states that a second order, axially symmetric tensor can have at most two independent components p_T and p_N

$$\hat{P} = \begin{pmatrix} -p_T & 0 & 0 \\ 0 & -p_T & 0 \\ 0 & 0 & -p_N \end{pmatrix} . \tag{4}$$

Because (4) converges to (3) as the interfacial region is traversed into either of the bulk phases, it follows that p_T and p_N are functions of z with $p_N = p_T = p$ at $z = \pm \frac{1}{2} \Delta z$. BAKKER exploited this model to derive his identification of the interfacial tension γ via the integral

$$\gamma = \int_{+\frac{1}{2}\Delta z}^{-\frac{1}{2}\Delta z} (p_N - p_T)\, dz \to \int_{-\infty}^{\infty} (p_N - p_T)\, dz . \tag{5}$$

If the fluids are in motion, then extra terms must be added to the static pressure tensor. I assume here that a NEWTONian model is adequate, so that in the interiors of the bulk phases the new form of the pressure tensor (3) is

$$p_{ij} = -p\delta_{ij} + \sum_m \sum_n c_{ij}^{mn} s_{mn} . \tag{6}$$

Here δ_{ij} is the KRONECKER delta, s_{mn} are the components of the rate of strain tensor, and the quantities c_{ij}^{mn} are the viscosity coefficients of the bulk fluid comprising the components of a tensor of the fourth order. In the interiors of the bulk phases the tensor c_{ij}^{mn} may be assumed to be isotropic, whence general considerations of tensor analysis [17] require that it possess at most two independent components. I take these components to be the LAMÉ coefficients λ, μ, whence for (3) we have explicitly

$$\begin{aligned} p_{xx} &= -p + (\lambda + 2\mu)\, s_{xx} + \lambda s_{yy} + \lambda s_{zz} , \\ p_{yy} &= -p + \lambda s_{xx} + (\lambda + 2\mu)\, s_{yy} + \lambda s_{zz} , \\ p_{zz} &= -p + \lambda s_{xx} + \lambda s_{yy} + (\lambda + 2\mu)\, s_{zz} , \\ p_{xy} &= 2\mu s_{xy}; \qquad p_{xz} = 2\mu s_{xz}; \qquad p_{yz} = 2\mu s_{yz} . \end{aligned} \tag{7}$$

Equations (7) are the constitutive relations for an isotropic, NEWTONian fluid. If they are used to write down a conservation law for the transport of linear momentum, the result is the NAVIER-STOKES equations of fluid dynamics. In the interfacial, region, however, we have good reason to suppose that the local properties of the fluid are no longer isotropic, and generalizing (6), we replace the static part of the pressure tensor by (4) and demand that the tensor c_{ij}^{nm} exhibit axial symmetry, i.e., that it be invariant under all rotations of the coordinate system about the z axis. A general theorem in tensor analysis [17] states that axially symmetric tensors of the fourth order can have at most five independent components, here written λ, λ', μ, μ', μ'', so that explicitly in the interfacial region

$$\begin{aligned} p_{xx} &= -p_T + (\lambda' + 2\mu')\, s_{xx} + \lambda' s_{yy} + \lambda s_{zz}\,, \\ p_{yy} &= -p_T + \lambda' s_{xx} + (\lambda' + 2\mu')\, s_{yy} + \lambda s_{zz}\,, \\ p_{zz} &= -p_N + \lambda s_{xx} + \lambda s_{yy} + (\lambda + 2\mu)\, s_{zz}\,, \\ p_{xy} &= 2\mu' s_{xy}; \qquad p_{xz} = 2\mu'' s_{xz}; \qquad p_{yz} = 2\mu'' s_{yz}\,. \end{aligned} \tag{8}$$

Because (8) must reduce to (7) as the interfacial region is traversed, it is evident that λ, λ', μ, μ', μ'' are all functions of z taking on the asymptotic values λ^{I}, λ^{II}, or μ^{I}, μ^{II} as appropriate on $z = \pm \frac{1}{2} \Delta z$. The five viscosity coefficients thus share a common property with p_T and with ϱ in that each changes rapidly but still continuously throughout the interfacial region.

To date, all experiments in capillary rheology have been performed under conditions such that the bulk phases I and II may be considered incompressible. It is therefore convenient to limit our attention to this important special case by imposing the restriction

$$s_{xx} + s_{yy} + s_{zz} = \frac{\partial v_x}{\partial x} + \frac{\partial v_y}{\partial y} + \frac{\partial v_z}{\partial z} = 0 \tag{9}$$

upon equations (7) and (8). Equations (7) simplify to

$$p_{ij} = -p\delta_{ij} + 2\mu s_{ij} \tag{10}$$

for the isotropic interiors of the bulk phases, while in the anisotropic interfacial region I define from (9) the local rate of areal dilation by

$$\theta = s_{xx} + s_{yy} = -s_{zz} = \nabla_s \cdot \vec{v} \tag{11}$$

and eliminate s_{zz} from equations (8):

$$\begin{aligned} p_{xx} &= -p_T + (\lambda' - \lambda)\, \theta + 2\mu' s_{xx}\,, \\ p_{yy} &= -p_T + (\lambda' - \lambda)\, \theta + 2\mu' s_{yy}\,, \\ p_{zz} &= -p_N - 2\mu\theta\,, \\ p_{xy} &= 2\mu' s_{xy}; \qquad p_{xz} = 2\mu'' s_{xz}; \qquad p_{yz} = 2\mu'' s_{yz}\,. \end{aligned} \tag{12}$$

Because only the difference $\lambda' - \lambda$ appears in (12), the incompressibility restriction (9) has reduced the number of interfacial viscosity coefficients to four.

5. Momentum balance

If we ignore gravity and other body forces, the differential equation of momentum transport at any point in the two phase system is

$$\varrho \frac{d\bar{v}}{dt} = \nabla \cdot \hat{P} \tag{13}$$

in which the operator d/dt stands for the substantial derivative. Because our system is incompressible, we know from the equation of continuity in fluid dynamics that $d\varrho/dt = 0$, and hence ϱ and d/dt in equation (13) may be commuted. I integrate both sides of equation (13) over the small volume sketched in Figure 2, converting the right hand side into a surface integral by GAUSS' theorem. Because the boundaries of this small volume have been specified to move with the fluid, the operator d/dt factors outside of the integral sign on the left and we have

$$\frac{d}{dt}\iiint \varrho\vec{v}\,dx\,dy\,dz = \iint \hat{P}\cdot d\vec{S} \tag{14}$$

in which $d\vec{S}$ is the outward drawn normal surface vector.

Equation (14) is exact, but to make progress towards the derivation of the macroscopic boundary condition (2) appropriate approximations must now be introduced. In the interior of a bulk phase [where (14) is also valid], one would at this point in the analysis expand the integrands on each side of (14) in TAYLOR series in x, y, z about the midpoint of the sample volume. This option is not completely open to us when we have to deal with the interfacial region, for we have already remarked that the local properties change rapidly with z, so that TAYLOR series expansions with respect to z would be slow to converge. On the other hand, the local properties change only slowly with respect to x and y, so that TAYLOR series expansions parallel to the plane of the interface are permissible. Using the leading term of such an expansion for the volume integral on the left in (14),

$$\int\int\int \varrho\vec{v}\,dx\,dy\,dz = \Delta x\,\Delta y \int \varrho\vec{v}\,dz\,. \tag{15}$$

The approximation of the surface integral on the right in (14) is more complicated. Consider first the contribution to it from the ceiling. The momentum flux across the upper surface in Figure 2 is the vector $(p_{xz}^{\mathrm{II}}, p_{yz}^{\mathrm{II}}, p_{zz}^{\mathrm{II}})$ in which the superscript II means that the tensor component is to be evaluated at $z = +\frac{1}{2}\Delta z$. I expand each component p_{iz}^{II} in a TAYLOR series in x and y and retain only the leading term

$$\int\int p_{iz}^{\mathrm{II}}\,dx\,dy = \Delta x\,\Delta y\,p_{iz}^{\mathrm{II}}; \quad i = x, y, z\,. \tag{16}$$

Similarly, the floor contribution from the surface at $z = -\frac{1}{2}\Delta z$ proves to be

$$\int\int p_{iz}^{\mathrm{I}}\,dx\,dy = -\Delta x\,\Delta y\,p_{iz}^{\mathrm{I}}; \quad i = x, y, z\,. \tag{17}$$

The sum of equations (16) plus (17) is the net transport of the three components of momentum along the normal direction

$$\text{net normal transport} = \Delta x\,\Delta y \,\Big|_{\mathrm{I}}^{\mathrm{II}}\, p_{iz}; \quad i = x, y, z\,. \tag{18}$$

In the plane of the interface, the momentum flux along the x axis is the vector (p_{xx}, p_{xy}, p_{xz}). I expand each tensor component in a TAYLOR series in x centered on $x = 0$. On the right hand wall at $x = + \frac{1}{2} \Delta x$ correct to two terms

$$p_{xi} = p_{xi}(z) + \frac{1}{2} \Delta x \frac{\partial p_{xi}(z)}{\partial x} . \tag{19}$$

On the left hand wall at $x = - \frac{1}{2} \Delta x$

$$p_{xi} = p_{xi}(z) - \frac{1}{2} \Delta x \frac{\partial p_{xi}(z)}{\partial x} . \tag{20}$$

The net flux is the difference of equation (19) minus equation (20), whence integrating over the wall surfaces the net tangential transport along x is

$$\int\int p_{xi}\, dy\, dz = \Delta x\, \Delta y \int [\partial p_{xi}(z)/\partial x]\, dz; \qquad i = x, y, z . \tag{21}$$

In equations (19, 20, 21) the reader should remark that the strong dependence of the tensor components and their derivatives on z is explicity recognized in the notation.

The momentum transport along y is similar:

$$\int\int p_{yi}\, dy\, dz = \Delta x\, \Delta y \int [\partial p_{yi}(z)/\partial y]\, dz; \qquad i = x, y, z . \tag{22}$$

Collecting equations (15, 18, 21, 22) we have correct to order $\Delta x\, \Delta y$ the three components of the momentum balance equation (14)

$$\begin{aligned}
\frac{d}{dt}\int \varrho v_x\, dz &= \Big|_{\mathrm{I}}^{\mathrm{II}} p_{xz} + \int \left[\frac{\partial p_{xx}}{\partial x} + \frac{\partial p_{xy}}{\partial y}\right] dz , \\
\frac{d}{dt}\int \varrho v_y\, dz &= \Big|_{\mathrm{I}}^{\mathrm{II}} p_{yz} + \int \left[\frac{\partial p_{xy}}{\partial x} + \frac{\partial p_{yy}}{\partial y}\right] dz , \\
\frac{d}{dt}\int \varrho v_z\, dz &= \Big|_{\mathrm{I}}^{\mathrm{II}} p_{zz} + \int \left[\frac{\partial p_{xz}}{\partial x} + \frac{\partial p_{yz}}{\partial y}\right] dz .
\end{aligned} \tag{23}$$

The next step in the analysis is to substitute the anisotropic form of the pressure tensor (12) into equations (23) and to identify the components of the rate of strain tensor s_{ij} with the symmetric part of the gradient of the velocity field. The details are routine and are reserved for appendix A. The result is the conversion of equations (23) into the form

normal

$$\frac{d}{dt}\int \varrho v_z\, dz = \Big|_{\mathrm{I}}^{\mathrm{II}} p_{zz} + \int \mu'' \frac{\partial \theta}{\partial z} dz + \nabla_s^2 \int \mu'' v_z\, dz; \tag{24}$$

tangential

$$\begin{aligned}
\frac{d}{dt}\int \varrho v_x\, dz &= \Big|_{\mathrm{I}}^{\mathrm{II}} p_{xz} - \frac{\partial}{\partial x}\int p_T\, dz + \frac{\partial}{\partial x}\int (\lambda' - \lambda + \mu')\, \theta\, dz + \nabla_s^2 \int \mu' v_x\, dz , \\
\frac{d}{dt}\int \varrho v_y\, dz &= \Big|_{\mathrm{I}}^{\mathrm{II}} p_{yz} - \frac{\partial}{\partial y}\int p_T\, dz + \frac{\partial}{\partial y}\int (\lambda' - \lambda + \mu')\, \theta\, dz + \nabla_s^2 \int \mu' v_y\, dz
\end{aligned} \tag{25}$$

in which the reader will observe that equation (24) expressing the balance of the normal component of momentum is different in form from equations (25) which express the balance of the two tangential components. This difference arises fundamentally from the anisotropy of the interphase.

6. A model interface

Equations (24, 25) are quite general for any plane, interfacial region between two incompressible, NEWTONian fluids only provided that the condition of axial symmetry be an acceptable description of the local properties. In particular they should then be valid for that favorite model of the macroscopic fluid dynamicist: the infinitely abrupt interface. In the infinitely abrupt interface the local properties change discontinuously when the plane $z = 0$ is traversed. Explicitly I write

$$\begin{aligned}
\lambda' &= \lambda\,,\\
\varrho &= \frac{1}{2}(\varrho^{\mathrm{I}} + \varrho^{\mathrm{II}}) + \frac{1}{2}h(z)(\varrho^{\mathrm{II}} - \varrho^{\mathrm{I}})\,,\\
p_T = p_N &= \frac{1}{2}(p^{\mathrm{I}} + p^{\mathrm{II}}) + \frac{1}{2}h(z)(p^{\mathrm{II}} - p^{\mathrm{I}})\,,\\
\mu = \mu' = \mu'' &= \frac{1}{2}(\mu^{\mathrm{I}} + \mu^{\mathrm{II}}) + \frac{1}{2}h(z)(\mu^{\mathrm{II}} - \mu^{\mathrm{I}})\,,\\
v_x &= v_x^0 + \frac{1}{2}z\left(\frac{\partial v_x^{\mathrm{I}}}{\partial z} + \frac{\partial v_x^{\mathrm{II}}}{\partial z}\right) + \frac{1}{2}\left(\frac{\partial v_x^{\mathrm{II}}}{\partial z} - \frac{\partial v_x^{\mathrm{I}}}{\partial z}\right)\int_0^z h(w)\,dw\,,\\
v_y &= v_y^0 + \frac{1}{2}z\left(\frac{\partial v_y^{\mathrm{I}}}{\partial z} + \frac{\partial v_y^{\mathrm{II}}}{\partial z}\right) + \frac{1}{2}\left(\frac{\partial v_y^{\mathrm{II}}}{\partial z} - \frac{\partial v_y^{\mathrm{I}}}{\partial z}\right)\int_0^z h(w)\,dw\,,\\
v_z &= -z\theta^0 - \frac{1}{4}z^2\left(\frac{\partial \theta^{\mathrm{I}}}{\partial z} + \frac{\partial \theta^{\mathrm{II}}}{\partial z}\right) - \frac{1}{2}\left(\frac{\partial \theta^{\mathrm{II}}}{\partial z} - \frac{\partial \theta^{\mathrm{I}}}{\partial z}\right)\int_0^z (z - w)\,h(w)\,dw\,,\\
\theta &= \theta^0 + \frac{1}{2}z\left(\frac{\partial \theta^{\mathrm{I}}}{\partial z} + \frac{\partial \theta^{\mathrm{II}}}{\partial z}\right) + \frac{1}{2}\left(\frac{\partial \theta^{\mathrm{II}}}{\partial z} - \frac{\partial \theta^{\mathrm{I}}}{\partial z}\right)\int_0^z h(w)\,dw
\end{aligned} \tag{26}$$

in which $h(z)$ is the HEAVISIDE function: $h(z) = -1$ for z negative; $h(z) = 1$ for z positive. From equations (26) and by an examination of Figures 3, 4, the reader will note that the model interface exhibits discontinuous density, viscosity coefficients, and static pressure tensor. The components of the velocity vector are continuous, but the slopes of the tangential components $\partial v_x/\partial z$ and $\partial v_y/\partial z$ are discontinuous, while for the normal component it is the second derivative $\partial^2 v_z/\partial z^2$ which is discontinuous. Finally the quantity θ^0 is the areal dilation rate of the surface of discontinuity,

$$\theta^0 = \frac{\partial v_x^0}{\partial x} + \frac{\partial v_y^0}{\partial y} = \nabla_s \cdot \vec{v}^0$$

and the condition of incompressibility (9) is satisfied.

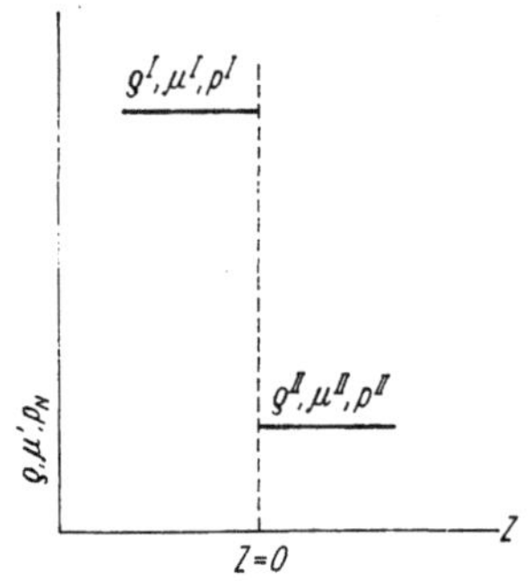

Fig. 3: A model interface.

The density ϱ, shear visocity μ, and hydrostatic pressure p are discontinuous at $z = 0$.

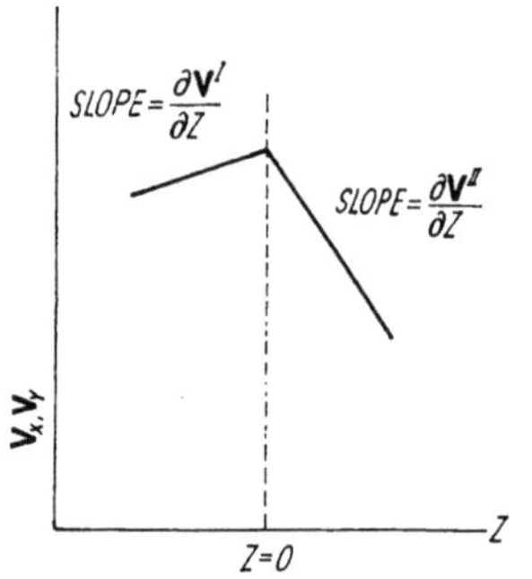

Fig. 4: A model interface.

The velocity vector $\overleftrightarrow{v}$ is continuous but undergoes a discontinuous change in slope at $z = 0$.

By the substitution of equations (26) into (24, 25) all of the integrations may formally be completed for the model interface. The interested reader may follow through one of these integrations in appendix B, but we shall have no real need for the explicit results. It is sufficient for our purposes to accept the fact that all of these integrals are of $O(\Delta z)$ or less, which is to say they are the products of bulk phase quantities such as ϱ^{I}, μ^{II}, etc. times the distance Δz. Our real purpose in constructing the model interface was to compare the situation in the real interface with the model. This we do by adding and subtracting the results for the model integrals to equations (24, 25).

As an example of one of these operations, consider the integral $\int p_T\, dz$ which appears among equations (25). I write

$$\int p_T\, dz = \int [p_T - \tfrac{1}{2}(p^{\mathrm{I}} + p^{\mathrm{II}}) - \tfrac{1}{2} h(z)(p^{\mathrm{II}} - p^{\mathrm{I}})]\, dz$$
$$+ \int [\tfrac{1}{2}(p^{\mathrm{I}} + p^{\mathrm{II}}) + \tfrac{1}{2} h(z)(p^{\mathrm{II}} - p^{\mathrm{I}})]\, dz$$
$$= -\gamma + O(\Delta z)$$

in which

$$\gamma = \int [\tfrac{1}{2}(p^{\mathrm{I}} + p^{\mathrm{II}}) + \tfrac{1}{2} h(z)(p^{\mathrm{II}} - p^{\mathrm{I}}) - p_T]\, dz \qquad (27)$$

is evidently a capillary excess property closely related to the surface tension defined by BAKKER's equation (5) for the static interface. I have, indeed, used the same symbol γ for both, but I postpone a reconciliation of equations (5) and (27) until section 7.

In an analogous way one can develop the integrals $\int \varrho v_x \, dz$ and $\int \mu' v_x \, dz$ into the forms

$$\int \varrho v_x \, dz = \int \{\varrho v_x - v_x^0 [\tfrac{1}{2} (\varrho^{\mathrm{I}} + \varrho^{\mathrm{II}}) + \tfrac{1}{2} h(z) (\varrho^{\mathrm{II}} - \varrho^{\mathrm{I}})]\} \, dz + O(\Delta z) , \quad (28)$$

$$\int \mu' v_x \, dz = \int \{\mu' v_x - v_x^0 [\tfrac{1}{2} (\mu^{\mathrm{I}} + \mu^{\mathrm{II}}) + \tfrac{1}{2} h(z) (\mu^{\mathrm{II}} - \mu^{\mathrm{I}})]\} \, dz + O(\Delta z) . \quad (29)$$

To make a further simplification I argue at this point that the velocity field in the experimental system changes very little across the width Δz of our sample of interphase. To be sure, the derivatives $\partial v_x/\partial z$ and $\partial v_y/\partial z$ change significantly across this distance, but I presume nevertheless that Δz may be taken so small that to a good approximation $\vec{v} \approx \vec{v}^{\mathrm{I}} \approx \vec{v}^{\mathrm{II}}$ everywhere within the interfacial region. Were this not the case, the experimental investigator would observe changes in the velocity field on either side of an interface so large that he would be forced to deduce the presence in his system of a slip plane coincident with the optically observed interface. There is no experimental evidence for the existence of such slip planes, and I therefore choose v_x^0 in equations (28, 29) to be the x component of the experimentally observed interfacial velocity vector and set $v_x = v_x^0$ throughout the range of the integral. Factoring v_x^0 outside of the integral sign, equations (28, 29) become

$$\int \varrho v_x \, dz = \Gamma v_x^0 + O(\Delta z) , \qquad \int \mu' v_x \, dz = \eta v_x^0 + O(\Delta z)$$

in which

$$\Gamma = \int [\varrho - \tfrac{1}{2} (\varrho^{\mathrm{I}} + \varrho^{\mathrm{II}}) - \tfrac{1}{2} h(z) (\varrho^{\mathrm{II}} - \varrho^{\mathrm{I}})] \, dz , \quad (30)$$

$$\eta = \int [\mu' - \tfrac{1}{2} (\mu^{\mathrm{I}} + \mu^{\mathrm{II}}) - \tfrac{1}{2} h(z) (\mu^{\mathrm{II}} - \mu^{\mathrm{I}})] \, dz \quad (31)$$

are respectively a capillary excess mass density and a capillary excess shear viscosity.

Equations (30, 31) are central to the type of development of capillary rheology which is the objective of this article. Possible forms for the integrands are sketched in Figures 5, 6. Without the aid of a rigorous statistical mechanical investigation, the analytic form of the integrands is unknown, but the shapes selected are at least consistent with the experimentally known properties of real fluids. We shall return to this point in section 7. For the moment, it is sufficient to identify the integrals (30, 31) with the shaded areas in Figures 5, 6. It is also obvious that the values of these integrals will depend upon the location of the dividing surface $z = 0$, which in our model is the surface of discontinuity of the Heaviside function $h(z)$.

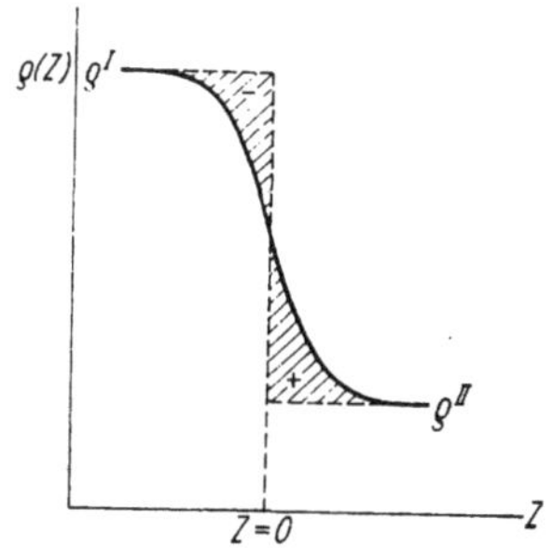

Fig. 5: A comparison of the diffuse, experimental interface with the model of Figs. 3, 4.

The density ϱ changes monotonically through the anisotropic region. The capillary excess mass density Γ defined by equation (30) is the sum of the shaded areas. By a proper choice of the dividing surface $Z = 0$, Γ can be made to vanish.

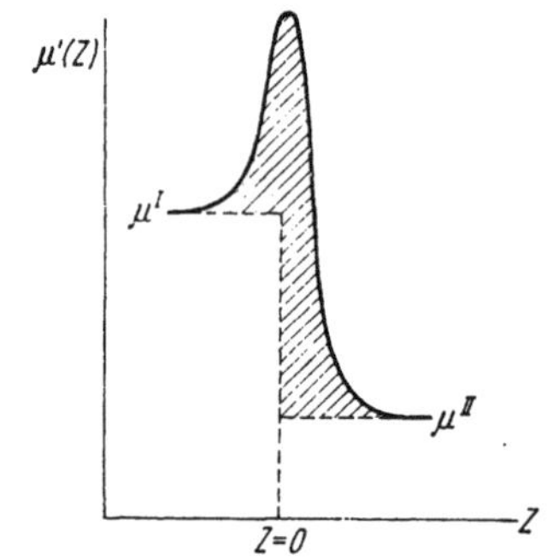

Fig. 6: A comparison of the diffuse, experimental interface with the model of Figs. 3, 4

The shear viscosity μ' is drawn with a pronounced maximum. The surface shear viscosity η of equation (31) is the shaded area. For the dividing surface $Z = 0$ as drawn, η will be positive, but η could be made to vanish by a sufficiently large displacement to the right of the plane $Z = 0$.

A consistent application of these methods to the equations of momentum balance (24, 25) leads to their transformation into the form

normal

$$0 = \Big|_{\mathrm{I}}^{\mathrm{II}} p_{zz} + O(\Delta z); \tag{32}$$

tangential

$$\begin{aligned} \frac{d}{dt}\, \Gamma v_x^0 &= \Big|_{\mathrm{I}}^{\mathrm{II}} p_{xz} + \frac{\partial \gamma}{\partial x} + \varkappa \frac{\partial \theta_0}{\partial x} + \eta \nabla_s^2 v_x^0 + O(\Delta z)\,; \\ \frac{d}{dt}\, \Gamma v_y^0 &= \Big|_{\mathrm{I}}^{\mathrm{II}} p_{yz} + \frac{\partial \gamma}{\partial y} + \varkappa \frac{\partial \theta_0}{\partial y} + \eta \nabla_s^2 v_y^0 + O(\Delta z)\,. \end{aligned} \tag{33}$$

γ, Γ, and η are defined by equations (27, 30, 31) respectively, and

$$\varkappa = \int [\lambda' - \lambda + \mu' - \tfrac{1}{2}(\mu^{\mathrm{I}} + \mu^{\mathrm{II}}) - \tfrac{1}{2} h(z)\,(\mu^{\mathrm{II}} - \mu^{\mathrm{I}})]\, dz \tag{34}$$

is a capillary excess dilational viscosity. In common with γ, Γ, and η; $\varkappa$ is not defined precisely until a choice is made for the location of the dividing surface $z = 0$.

7. A dividing surface convention

Equations (32, 33) are valid for arbitrary location of the mathematical surface $z = 0$. The integrals γ, Γ, η, $\varkappa$ all shift with each choice of the dividing surface in such a way as to maintain the validity of (32, 33). It is thus a matter of pure convenience as to how this surface shall be located; and following GIBBS, we wish to make our choice in such a way as to simplify the form of the macroscopic boundary condition. One strong influence in guiding such a choice is that $z = 0$ should pass through the zone of maximum anisotropy of the interphase, for then Δz may be taken small ($\Delta z \sim 10^{-7}$ to 10^{-6} cm), and the quantities $O(\Delta z)$ in equations (32, 33) may be neglected. Such a choice would also aid in the interpretation of such quantities as

$$\Big|_{\mathrm{I}}^{\mathrm{II}} p_{zz} = p_{zz}^{\mathrm{II}} - p_{zz}^{\mathrm{I}}$$

as calculable from bulk phase properties extrapolated to either side of the optically observable interface. If in addition to passing through the anisotropic region we can choose $z = 0$ in such a way as to make one of the four integrals γ, Γ, η, $\varkappa$ equal to zero, then equations (32, 33) will undergo even further simplification.

To illustrate this approach, consider the integral γ defined by equation (27). The conjectured form of the integrand is sketched in Figure 7 in which the dividing surface $z = 0$ is chosen to lie near the region of maximum anisotropy of the interface. A reasonable experimental value for a static surface tension would be 50 erg cm^{-2}, and from an examination of BAKKER's integral (5) it follows that for $\Delta z \sim 5 \times 10^{-7}$ cm the departure of p_N from p_T in the interfacial region is of the order 100 atm. The difference between the two curves p_N and p_T in the zone of anisotropy is thus very great. We might nevertheless choose $z = 0$ in such a way that the integral (27) vanishes if we displace the discontinuity in the HEAVISIDE function sufficiently far to the left, deep inside phase I. The extent to which such a displacement is necessary depends upon the magnitude of the HEAVISIDE discontinuity sketched in Fig. 7. For the static plane interface it is known that $p^{I} = p^{II}$, so that the HEAVISIDE discontinuity is zero; and it follows that for this interface it is impossible to choose a dividing surface which will set $\gamma = 0$. For the dynamic interface, $p^{II} - p^{I}$ should be of order $\mu^{II}\theta^{II} - \mu^{I}\theta^{I} \sim$ $\sim 10^{-8}$ atm, which is to say that the leftward displacement of the dividing surface in Figure 7 would have to be of the order 10^{10} cm to secure a zero value for γ! Such a model is palpably absurd.

The convention of choice emerges from heuristic considerations of the properties of bulk matter. Under the most extreme conditions of pressure and temperature accessible in terrestrial laboratories, the densities of condensed fluids vary over only about 1 order of magnitude, from $\varrho \sim 1$ g/ml for hydrocarbons to $\varrho \sim 10$ g/ml for liquid metals. Under the same conditions of temperature and pressure, the shear viscosities of bulk fluids vary over many orders of magnitude from $\mu \sim 10^{-2}$ poise for ligth hydrocarbons to $\mu \sim 10^{7}$ poise for materials such as asphalt. One is forced to conclude that the viscosity of a substance is far more sensitive to details of internal structure than is the density, and that whatever the physical conditions in an interfacial region, they cannot be so extreme that ϱ will pass through sharp maxima or minima as the interface is traversed. Theoretical attempts [18] to describe the density of interfacial matter bear out this heuristic argument, and it is for this reason that the curve ϱ in Figure 5 is shown as monotonic. Even in the presence of adsorbed films of surface active species, the function ϱ, while no longer necessarily monotonic, cannot be expected to exhibit pronounced irregularities in the zone of anisotropy. By the same token we may reasonably expect in some interfaces, particularly those at which adsorbed films are present, that the local organization will be such as to make the viscosities λ, λ', μ, μ', μ'' depart widely from bulk phase values. It is for this reason

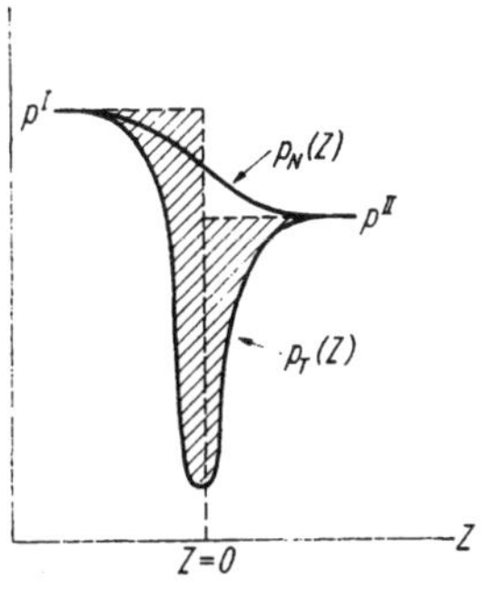

Fig. 7: A comparison of the diffuse, experimental interface with the model of Figs. 3, 4.

The integral γ of equation (27) is the shaded area. For the dividing surface $Z = 0$ as drawn, γ is positive, but it could be made to vanish by a sufficiently large displacement to the left of the plane $Z = 0$.

that the curve μ' drawn in Figure 6 is displayed with a sharp maximum. In principle the investigator could still choose the dividing surface $z = 0$ in such a way as to make either of the integrals η (31) or $\varkappa$ (34) vanish, but these conventions would be inconvenient for the same reason that it would be inconvenient to make the integral γ (27) vanish.

We are thus left with the only remaining practicable convention of choosing $z = 0$ in such a way as to make the integral Γ of equation (30) vanish. Not only will this choice set equal to zero the inertial terms on the left of equations (32, 33), but in addition the postulated monotonic nature of the interfacial density function ϱ drawn in Figure 5 guarantees that the dividing surface will pass through the optically observed interface of the experimentalist. Δz may thus be taken so small that the quantities $O(\Delta z)$ may be dropped from equations (32, 33). The result is a set of boundary conditions suitable for use in hydrodynamic problems involving a plane interface: normal

$$\Big|_{\mathrm{I}}^{\mathrm{II}} p_{zz} = -(p^{\mathrm{II}} - p^{\mathrm{I}}) - 2(\mu^{\mathrm{II}} - \mu^{\mathrm{I}})\,\theta_0 = 0; \tag{35}$$

tangential

$$\Big|_{\mathrm{I}}^{\mathrm{II}} p_{xz} + \frac{\partial\gamma}{\partial x} + \varkappa\,\frac{\partial\theta^0}{\partial x} + \eta\nabla_s^2 v_x^0 = 0\,, \qquad \Big|_{\mathrm{I}}^{\mathrm{II}} p_{yz} + \frac{\partial\gamma}{\partial y} + \varkappa\,\frac{\partial\theta^0}{\partial y} + \eta\nabla_s^2 v_y^0 = 0\,. \tag{36}$$

Equations (36) are identical with equation (2) if γ as defined by equation (27) is experimentally indistinguishable from γ defined by BAKKER's integral (5). I argue that a further convenience of the dividing surface convention $\Gamma = 0$ is to remove any practical distinction between the two definitions of surface tension. From equation (35) the difference $p^{\mathrm{II}} - p^{\mathrm{I}} = -2(\mu^{\mathrm{II}} - \mu^{\mathrm{I}})\,\theta_0$ for any dilation rate θ^0 and bulk phase viscosity μ^{I}, μ^{II} of practical interest in interfacial rheology is of the order 10^{-10} to 10^{-6} atm. Given that $p_N - p_T$ in the interfacial region has already been shown above to be of the order 100 atm, the step discontinuity produced by the HEAVISIDE function $h(z)$ in Figure 7 is very small indeed, and so long as this discontinuity is placed in the region of maximum interfacial anisotropy, then we may safely identify the integrals (5) and (27) as equivalent definitions of the surface tension.

8. Summary and Conclusions

I have shown in these pages that the mathematical methods of Willard GIBBS, applied by him so successfully to static, fluid interfaces, are equally applicable to fluid interfaces in motion. In particular, by modeling an interface as a mathematical plane and by choosing the location of that plane in such a way as to set equal to zero the capillary excess mass density, not only thermodynamic quantities such as adsorption concentrations of surface active species are then defined, but so also are the interfacial rheological coefficients $\varkappa$ and η, respectively the coefficient of surface dilational viscosity and the coefficient of surface shear viscosity. Through use of the integrals (30, 31, 34) the path is now open to a theoretical estimation of these parameters via the methods of non equilibrium statistical mechanics.

Acknowledgement

This work was supported by the National Science Foundatation, Grant No. CHE-7505240.

Appendix A

We wish to convert equations (23) into the form (24, 25). As an example, substitute from equations (12) into the first of (23). We have

$$\frac{\partial p_{xx}}{\partial x} + \frac{\partial p_{xy}}{\partial y} = -\frac{\partial p_T}{\partial x} + (\lambda' - \lambda)\frac{\partial \theta}{\partial x} + 2\mu'\frac{\partial s_{xx}}{\partial x} + 2\mu'\frac{\partial s_{xy}}{\partial y}.$$

But

$$s_{xx} = \frac{\partial v_x}{\partial x},$$

$$s_{xy} = \frac{1}{2}\left[\frac{\partial v_x}{\partial y} + \frac{\partial v_y}{\partial x}\right],$$

$$\theta = \frac{\partial v_x}{\partial x} + \frac{\partial v_y}{\partial y},$$

whence

$$\frac{\partial s_{xx}}{\partial x} + \frac{\partial s_{xy}}{\partial y} = \frac{1}{2}\left[\frac{\partial \theta}{\partial x} + \frac{\partial^2 v_x}{\partial x^2} + \frac{\partial^2 v_x}{\partial y^2}\right] = \frac{1}{2}\left[\frac{\partial \theta}{\partial x} + \nabla_s^2 v_x\right].$$

It follows that

$$\frac{\partial p_{xx}}{\partial x} + \frac{\partial p_{xy}}{\partial y} = -\frac{\partial p_T}{\partial x} + (\lambda' - \lambda + \mu')\frac{\partial \theta}{\partial x} + \mu'\nabla_s^2 v_y$$

as is written in the first of equations (25). The transformation of the remainder of equations (23) into the form (24, 25) is analogous.

Appendix B

Consider the integral $\int p_T\, dz$ which occurs among equations (25).

Substituting from equations (26),

$$\int p_T\, dz = \int_{-\frac{1}{2}\Delta z}^{+\frac{1}{2}\Delta z} \tfrac{1}{2}\left[(p^{\mathrm{I}} + p^{\mathrm{II}}) + \tfrac{1}{2} h(z)\,(p^{\mathrm{II}} - p^{\mathrm{I}})\right] dz$$

$$= \tfrac{1}{2}\,(p^{\mathrm{I}} + p^{\mathrm{II}})\,\Delta z = O(\Delta z).$$

The remaining integrations are similar. One has only to identify

$$\int h(z)\, dz = 0, \qquad \int z h(z)\, dz = \tfrac{1}{4}\,(\Delta z)^2 \quad \text{etc.}$$

List of symbols

c_{ij}^{mn}	= components of the fourth order viscosity tensor.
$h(z)$	= HEAVISIDE function: $h(z) = -1$ for $z < 0$; $h(z) = 1$ for $z > 0$.
p	= hydrostatic pressure in the interior of a bulk phase.
p_{ij}	= components of the pressure tensor $\hat{P}$.
p_T	= tangential component of the pressure tensor in the interfacial region.
p_N	= normal component of the pressure tensor in the interfacial region.
$p^{\mathrm{I}}, p^{\mathrm{II}}$	= hydrostatic pressure in bulk phases I and II extrapolated to the interfacial region.

Q = any extensive thermodynamic property.

q = a capillary excess extensive thermodynamic property.

s_{ij} = components of the rate of strain tensor.

v_x, v_y, v_z = CARTESIAN components of the fluid velocity vector $\vec{v}$.

$v_x^0, v_y^0, 0$ = CARTESIAN components of the fluid velocity vector $\vec{v}^0$ at the interface.

x, y, z = CARTESIAN coordinates. The z axis is oriented normal to the interface and positive in phase II.

γ = surface or interfacial tension.

δ_{ij} = the KRONECKER delta.

η = coefficient of surface shear viscosity.

$\theta = \dfrac{\partial v_x}{\partial x} + \dfrac{\partial v_y}{\partial y}$ = areal dilation rate of an xy plane.

θ^0 = areal dilation rate of the interface $z = 0$.

$\varkappa$ = coefficient of surface dilational viscosity.

$\lambda, \lambda', \mu, \mu', \mu''$ = the five distinct viscosity coefficients of the interphase.

$\mu^{\mathrm{I}}, \mu^{\mathrm{II}}$ = bulk phase shear viscosity coefficients.

ϱ = local denstiy in the interfacial region.

$\varrho^{\mathrm{I}}, \varrho^{\mathrm{II}}$ = bulk phase densities.

∇_s = surface gradient operator $(\partial/\partial x,\ \partial/\partial y,\ 0)$.

∇_s^2 = surface LAPLACIAN operator $(\partial^2/\partial x,\ \partial^2/\partial y^2,\ 0)$.

References

[1] J. W. GIBBS, "*The Scientific Papers*", vol. I, Dover, New York (1961)

[2] L. D. LANDAU, E. M. LIFSHITZ, "*Fluid Mechanics*", Pergamon Press, New York (1959), chapter 7

[3] V. G. LEVICH, "*Physicochemical Hydrodynamics*", Prentice-Hall, Inc., Englewood Cliffs, N.J. (1962), chapters 7, 9

[4] L. E. SCRIVEN, Chem. Eng. Sci. **12**, 98 (1960)

[5] R. J. MANNHEIMER, R. S. SCHECHTER, J. Colloid and Interface Sci. **25**, 434 (1967); ibid. **27**, 324 (1968)

[6] R. J. MANNHEIMER, R. A. BURTON, J. Colloid and Interface Sci. **32**, 73 (1970)

[7] R. J. MANNHEIMER, R. S. SCHECHTER, J. Colloid and Interface Sci. **32**, 195, 212, 225 (1970)

[8] D. T. WASAN, L. GUPTA, M. K. VORA, AIChE J. **17**, 1287 (1971)

[9] L. GUPTA, D. T. WASAN, Ind. Eng. Chem. Fundamentals **13**, 26 (1974)

[10] E. H. LUCASSEN-REYNDERS, J. LUCASSEN, Advances in Colloid and Interface Sci. **2**, 347 (1969)

[11] J. LUCASSEN, D. GILES, J. Chem. Soc., Faraday Trans. I **75**, 26 (1974)

[12] F. C. GOODRICH, L. H. ALLEN, A. POSKANZER, J. Colloid and Interface Sci. **52**, 201 (1975)

[13] A. POSKANZER, F. C. GOODRICH, J. Colloid and Interface Sci. **52**, 213 (1975)

[14] A. POSKANZER, F. C. GOODRICH, J. Phys. Chem. **79**, 2122 (1975)

[15] L. D. LANDAU, E. M. LIFSHITZ, ibid., chapter 1

[16] G. BAKKER, "*Handbuch der Experimental-Physik*", vol. 6, Wien-Harms ([illegible])

[17] I. N. SNEDDON, D. S. BERRY, „*Handbuch der Physik*", vol. 6, Springer Verlag, Berlin (1958)

[18] A. BRIN, Doctoral Thesis, Université de Paris (1956)

Some applications of GIBBS idea to the statistical thermodynamics of surface phenomena

B. V. DERJAGUIN

(Department of Surface Phenomena, Institute of Physical Chemistry of the Academy of Sciences of the USSR, Moscow/USSR)

Introduction

During the time elapsed since the publication of GIBBS' works, the wide scope of phenomena covered by them have become completely clear, as well as the depth and power of the methods of interpreting as developed by him. In this case, especially wonderful is the general character of those methods that allow developing on their basis not only those fields of the theory of capillarity which were treated by GIBBS himself, but also new trends.

In the present paper, we would like to show, first of all, how the GIBBS' thermodynamic method is applied to treating thin layers, no part of which, in distinction to the cases studied by GIBBS, possesses the properties of a bulk phase. It has been shown that in such a case, owing to the overlapping of interphase transition layers there arises a new world of phenomena determining the basic properties of disperse systems. Secondly, the paper is indicative, on one hand, of the application of GIBBS Grand Ensemble and, on the other hand, of the work of formation of the critical nucleus of a new phase to the strict derivation of the new phase formation probability.

1. Disjoining pressure of thin films

In the thermodynamics of GIBBS of heterogeneous systems all the extensive values, e.g., energy, entropy, mass of components, are determined by summing up the terms proportional to the volumes of phases and the interfaces areas. This approach remains true under condition that nowhere in the system the transition layers belonging to different interfaces or to different parts of an interface overlap one another (see Fig. 1). No such overlapping will be observed only in the case where, starting from one interface, one cannot reach the other without intersecting the areas that are homogeneous with regard to intensive properties, i.e., those possessing the properties of a phase. It is just in view of this that GIBBS had studied only such thin films whose medium parts possessed the phase properties.

Yet for the science of colloids and disperse systems, of special importance are situations, in which the approaching of two particles causes the overlapping of their surface zones.

Let us consider the effects that are new in comparison with those treated in the work of GIBBS and his direct successors, and that do appear with such an overlapping in the case of a thin plane-parallel interlayer of liquid (or gas). First of all, at such an overlapping, if it occurs in the isothermal-isobaric process involving the bulk masses of two phases approaching each other, as a rule, the GIBBS free energy of the system (i.e. the free enthalpy) must change. Hence, strong interactions depending on the

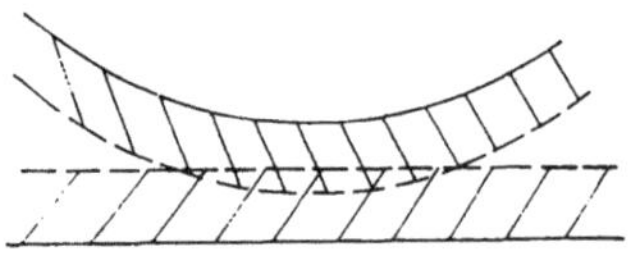

Fig. 1

extent of overlapping must arise. In the case where the overlapping occurs within the region of the plane-parallel interlayer, the interaction is unambiguously characterized by a variation in the pressure of this interlayer as compared with that of the bulk phase of which the interlayer is a part.

This excess pressure was introduced and defined by us as disjoining pressure [1]. In this case, however, a clarification is required, which is connected with the following fact: if overlapping does occur, there disappears those parts of the interlayer that preserve the isotropy of properties, including the isotropic character of hydrostatic pressure. Therefore, the pressure in it becomes everywhere anisotropic, and is characterized by a pressure tensor. In the general case, this tensor changes its components depending on the location of a point, i.e. on its coordinate oriented across the interlayer.

A substantial simplification in interpreting the equilibrium of such interlayers is achieved, if the forces dependent on distance are disregarded, by changing those to the near-range forces that are equivalent to them. It is the MAXWELL electrostatic field stress tensor which is used to reduce the electrostatic interaction forces that are connected with the volume charges of ionic origin, to the near-range forces.

The molecular attraction forces may be replaced by the near-range forces by using the theory developed by LIFSHITZ [2], which reduces those forces to the electromagnetic field fluctuations. By including in the pressure tensor, a tensor corresponding to Coulomb forces, and a tensor corresponding to those fluctuations, too, we exclude the long-range forces that are applied to the elements the interlayer is made of.

From the condition of equilibrium between the latter follows the equation

$$P_{zz} = \text{const},$$

where P_{zz} is the component of the pressure tensor, which is normal to the surfaces confining the interlayer (liquid or gas) [3].

Then the disjoining pressure $\Pi(h)$ will be defined as follows:

$$\Pi(h) = P_{zz} - P_0, \tag{1}$$

where P_0 is the pressure in the bulk phase, whose thinning out produced an interlayer of thickness h. Thus, the disjoining pressure $\Pi(h)$ is the main thermodynamic characteristic of thin interlayers, which is indispensably required for developing the thermodynamics of GIBBS' heterogeneous systems in a new direction.

As is well known, to calculate the values of thermodynamic functions determining the equilibrium of heterogeneous systems requires resorting to statistical methods. These must be applied, for example, to deriving the adsorption isotherm, or to deriving the interphase tension versus components concentration relationship. In a similar way, a problem arises involving the application of statistical methods to calculating the disjoining pressure as a function of the interphase interlayer thickness. The approach to solving that problem depends on the nature of those surface forces that determine the overlapping effect of interphase transient layers.

In the general case, that proves to be a very complicated problem. However, there is a possibility of solving it in a simpler way as compared with the case of classical GIBBS' surface phenomena. Such a possibility presents itself if the film thickness considerably exceeds the size of molecules, so that its state in overlapping may be interpreted basically in a macroscopic way. As an example there may be mentioned calculations of the dispersion component of the disjoining pressure, carried out on the basis of LIFSHITZ' macroscopic approach. Another example is the calculation of the electrostatic component arising if diffuse ionic atmospheres [4] are overlapped or if diffuse adsorption layers are overlapped in the binary solutions of non-electrolytes [5].

Let us consider the first example, for the purpose of examining what part the statistical and the macroscopical methods play in these calculations. For maximum simplicity, let us avail ourselves of the GIBBS-DUHEM equation, by supplementing the expression for work done in the equilibrium isothermal-isobaric processes with two terms:

$$-S\Pi\, dh + S(\psi_1\, d\sigma_1 + \psi_2\, d\sigma_2)\,, \qquad (2)$$

where $\psi_{1,2}$ is the potential of one of the two interphase surfaces (Fig. 2) as compared with that of an infinitely remote electrode, $\sigma_{1,2}$ is its charge density; S is the area of a plane-parallel interlayer of a thickness h; Π is the disjoining pressure of the interlayer.

The first term is the usual expression for work which is to be done against the disjoining pressure as the interlayer thickness decreases. The second term expresses the work of transfer of charges $d\sigma_{1,2}$ from infinity onto the surface of plates. Let us subtract the sum $S\, d(\psi_1\sigma_1 + \psi_2\sigma_2)$ from both members of the GIBBS-DUHEM equation. Thus we obtain for the differential of the supplemented GIBBS free energy, $G_{T,P}$

$$dG_{T,P} = -S\Pi\, dh - S(\sigma_1\, d\psi_1 + \sigma_2\, d\psi_2)\,. \qquad (3)$$

As $dG_{T,P}$ is the total differential, we may write:

$$\left|\frac{\partial \Pi}{\partial \psi_1}\right|_{h,\psi_2} = \left|\frac{\partial \sigma_1}{\partial h}\right|_{\psi_1,\psi_2}. \qquad (4)$$

Integrating POISSON-BOLTZMANN's equation for an electrolyte interlayer, we obtain:

$$\left(\frac{d\psi}{dx}\right)^2 = \frac{a}{b}\left[\frac{e^{z_1 b\psi}}{z_1} + \frac{e^{-z_2 b\psi}}{z_2} - C\right], \qquad (5)$$

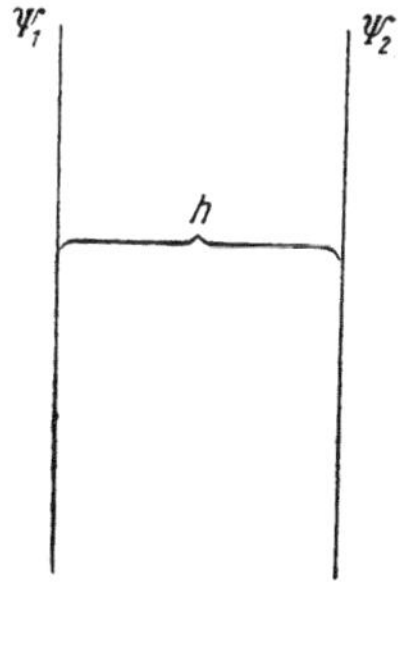

Fig. 2

where the potential ψ is a function of the coordinate χ measured from the left surface of the interlayer, normal to it:

$$b = \frac{e}{kT}, \qquad a = \frac{8\pi}{\varepsilon} e c_1 z_1 = \frac{8\pi}{\varepsilon} e c_2 z_2 ,$$

where e is the electron charge, k is BOLTZMANN's constant, T is KELVIN temperature, ε is permittivity, c_1, c_2 are the concentrations of cations and anions, respectively; z_1, z_2 are their charges, respectively. In equation (5), C is the integration constant. At $x = 0$ we obtain:

$$\psi = \psi_1 , \tag{6}$$

$$\sigma_1 = -\frac{D}{4\pi}\left(\frac{\partial\psi}{\partial x}\right)_{x=0} . \tag{7}$$

Hence, it follows:

$$\frac{a}{b} C = \frac{a}{b}\left[\frac{e^{z_1 b\psi_1} - 1}{z_1} + \frac{e^{-z_2 b\psi_1} - 1}{z_2}\right] - \left(\frac{\partial\psi}{\partial x}\right)^2_{x=0} . \tag{8}$$

Thus, it is obvious that as $h \to \infty$

$$C = C_\infty = 0 . \tag{8¹}$$

From equations (5), (6), and (7), it follows:

$$\sigma_1^2 = \frac{\varepsilon^2}{16\pi^2}\frac{a}{b}\left[\frac{e^{z_1 b\psi_1}}{z_1} + \frac{e^{-z_2 b\psi_1}}{z_2} - C\right], \tag{9}$$

$$-\sigma_1\left(\frac{\partial\sigma_1}{\partial h}\right)_{\psi_1,\psi_2} = \frac{\varepsilon^2 a}{32\pi^2 b}\left(\frac{\partial C}{\partial h}\right)_{\psi_1,\psi_2} . \tag{10}$$

Using the generally known identity

$$\left(\frac{\partial C}{\partial h}\right)_{\psi_1}\left(\frac{\partial h}{\partial \psi_1}\right)_C\left(\frac{\partial \psi_1}{\partial C}\right)_h = -1 \tag{11}$$

and the obvious (see Fig. 2) relationship

$$\left(\frac{\partial\psi_1}{\partial h}\right)_{c,\psi_2} = \frac{4\pi\sigma_1}{\varepsilon}, \tag{12}$$

from equation (10) we obtain:

$$\left(\frac{\partial\sigma}{\partial h}\right)_{\psi_1} = \frac{\varepsilon}{8\pi}\frac{a}{b}\left(\frac{\partial C}{\partial\psi_1}\right)_h . \tag{13}$$

From this expression and from equations (4) and (8), by integrating those and taking into consideration that as $h \to \infty$ the equality (8¹) is valid, we derive a well-known expression:

$$\Pi = \frac{\varepsilon}{8\pi}\frac{a}{b} C = bT\left[c_1(e^{z_1 b\psi_1} - 1) + c_2(e^{-z_2 b\psi_1} - 1)\right] - \frac{\varepsilon}{8\pi}E_1^2 . \tag{14}$$

The first term expresses the excess hydrostatic pressure; the second term, the Maxwell electrostatic tension.

2. Examination of overlapping of thin interlayers

Examination of overlapping of thin interlayers of liquid or gas leads to other differences from GIBBS thermodynamics of heterogeneous systems. One of the essential differences is revealed for a particular case where an interlayer of phase 3 separates two identical phases 1 and 2. In such a case the following approach is possible: instead of two geometrical interfaces whose localization is arbitrary, one quasi-interface is considered, which is disposed exactly in the symmetry plane of an interlayer. This eliminates all indeterminacy, when introducing excesses of the masses of components, energy, and entropy, relative to that plane, whose excesses characterize the excess content of these extensive values in the interlayer. Thus, if, for example, the concentration of the ith component in phases 1 and 2 is c_i, and if the volume of the system is V then the excess of the ith component in the interlaeyr is equal to:

$$\Delta m_i = m_i - c_i V ,$$

where m_i is the total mass of the ith component.

It is supposed that the volume of the system is determined on the basis of the location of interfaces with the external phase or the walls of a vessel, the location of which corresponds to zero adsorption of the ith component. Thus the value of Δm is determined independently of the interlayer thickness, h, which is indeterminate at small values of h. Of course, in passing over to statistical methods, the third dimension and some equivalent of the interlayer thickness will have to be used.

By using a single interface between two similar phases, one succeeded [7] in developing the thermodynamics of symmetrical films and finding the conditions of equilibrium and of the stability of equilibrium for systems made up of 3 and 4 components [8]. In this case, owing to the effect of overlapping, the tension of such a film is not equal to the double surface tension σ_{13} at the boundary of phases 1—3. If it proves to be smaller than that sum (the ordinary case), a quasi-contact angle arises between the bulk phase 3 and its interlayer (Fig. 3).

This quasi-contact angle was first introduced and examined by Martynov, Gutop and me [9]. The essential difference of this contact angle from the ordinary one consists in that, distinct from the latter, it owes its very existence to the overlapping effect. In the contrary case, when the interlayer tension exceeds the double value of σ_{13}, there does not arise any contact angle different from zero. Nonetheless, the interlayer may be in equilibrium coexistence with the bulk phase 3 when the interlayer has a certain thickness. This was discussed in more detail in study [10]. It will have to be noted that the conditions of stability of the equilibrium of free films in four-component systems, which were first examined in [8], are substantially more complicated than with a smaller number of components.

In the case where the interlayer of phase 3 is found between two dissimilar fluid phases 1 and 2, the contact angles α and β (see Fig. 4) may be determined from NEUMANN's triangle. One of the sides of this triangle is the interlayer tension. Nevertheless,

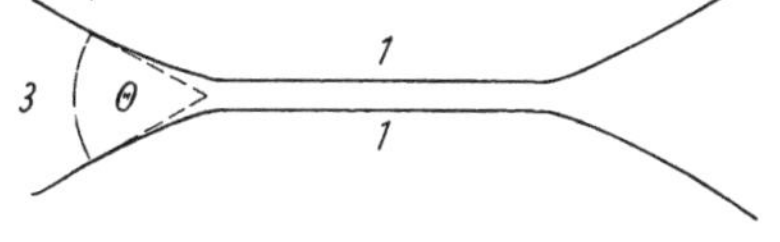

Fig. 3

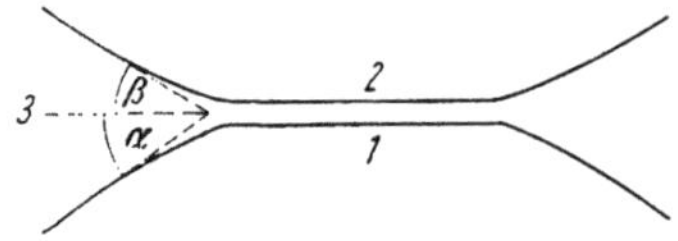

Fig. 4

this case is also not trivial, because the difference of angles α and β from zero also stems completely from the overlapping effect.

Now, contrary to the symmetrical case if is impossible to determine the excesses of extensive values without taking into consideration the interlayer thickness. But it would be possible to determine those values by resorting to another method of general applicability.

We have already discussed the excess Gibbs free energy. It is equal to a change in Gibbs free energy of the system, which occurs when a thin interlayer forms as a result of the overlapping effect. It may be calculated from the formula $\Delta G = \int_h^\infty \Pi \, dh$. In a similar way, by having measured the heat of formation of a thin interlayer, one may determine its excess enthalpy; hence, knowing also the excess free enthalpy, it would be possible to determine the excess energy.

Knowing the volumes of bulk phases 1, 2 and 3 and the changes in the concentrations of the ith component in these phases, as caused by the formation of a thin interlayer and the overlapping effect, it would be possible to determine the excess content of the ith component.

All these excesses must be determined as a function of the interlayer thickness, h. This is the basic complication, whether from the point of view of the experimental procedure or the essential points: owing to an indefiniteness of the concept of the thickness of films, which approach molecular size. This situation is a feature that characterizes the present approach in distinction from the classical work of Gibbs, although it represents its natural development. Now, this development enables one to pass from the purely thermodynamic treatment of heterogeneous systems over to the treatment using statistical and molecular-physical methods for calculation of surface forces in heterogeneous and disperse systems.

3. The second classical work of Gibbs covers the fundamentals of statistical mechanics

The further development of that science proceeded along the way indicated by Gibbs, yet many aspects of that work of Gibbs still await further use and development. Perhaps the most interesting example is the theory of formation of a new phase. This is essentially a kinetic problem. Its solution combines both thermodynamic and molecular-statistical calculations. The former are required, as has been by Gibbs, for calculating the work of formation of a critical nucleus — the concept that was also introduced by Gibbs [11]. The latter are required for calculating the probability of attaining and passing over the activation barrier corresponding to the critical nucleus. Calculation of that probability presents the main difficulty. Thus, for example, in the case of condensation the complete statistical calculation of the processes of after-growth of molecular complexes until the critical nucleus is obtained, generally presents an extremely difficult problem, whether in view of the mathematical difficulties involved or of the necessity of knowing numerous and difficult-to-determine constants that are characteristic of various stages of the processes of aggregation and disaggregation of molecular complexes. The aggregate growth processes essentially represent a complex-ramified chain reaction. It would be still more difficult to carry out the strict calculation of the process of birth of a new phase inside a liquid, for example, when it is starting to boil.

The initial stage of that process even from the qualitative point of view is rather a matter for speculation. In any case, it is determined by collective interactions, whose calculation presents one of the most complicated problems of statistical physics. Therefore it is reasonable, prior to trying to give a solution of the general case, which would be of low precision, to proceed first to a more readily accessible strict solution of a particular case. The problem of calculation of the nucleation probability is substantially facilitated in the limit case of very small supersaturations. A critical nucleus of an essentially macroscopic size corresponds to that case. In view of that, determination of its size and formation work may be carried out in the strictest manner using GIBBS thermodynamics as the basis.

The second simplification consists in a possibility of applying the pass-over method to that case. As it is known, its essence consists in the following: a differential equation of the growth of subcritical nuclei is taken in the form of KRAMERS [12], ZELDOVICH [13], or FOCKER-PLANCK, and it is applied only to the range of sizes which are close to that of the critical nucleus. This may be justified only in the case where the critical nucleus size is rather great.

In this case, however, two difficulties arise. First, when integrating, for example, KRAYERS' equation for the pass-over region, there appears an indefinite cofactor. It may be determined for the case of rarefied vapours, as FRENKEL had shown [41].

As it was shown in [16, 17], for more general cases, and especially for the case of nucleation in a liquid phase, the known attempts for determining that cofactor [13, 14] are unsubstantiated and lead to erroneous results. Simultaneously, cases are referred to [17], when the knowing of that cofactor is more important than that of the exponential factor.

Secondly, there is considered, as a rule, the steady-state solution of kinetic equations of nuclei growth and their passing over the "barrier". In order to obtain that solution, a limit condition is imposed, at which the concentration of supercritical nuclei is equal to zero.

This means that the grown supercritical nuclei, in spite of their guaranteed possibility of a further growth, are removed from the system, and their substance is converted to the initial phase. This maintains the supersaturation of the latter at a constant level. Simultaneously, when integrating the kinetic equation, an equilibrium function of the distribution of nuclei sizes is considered. That such a consideration is justified is difficult to prove, especially as the consideration relates, as a rule, to the medium in an unlimited volume.

All these difficulties and uncertainties, however, are excluded, if the GIBBS Grand Ensemble is applied to the solution of a nucleation problem [16 — 18]. For this purpose, it would be sufficient to consider nucleation within a limited volume V, which is separated from the remaining part of the system of the prescribed volume W and energy E by a partition allowing exchange of molecules, but not that of near critical nuclei. Volume V, in which the possibility of new phase formation is considered, is limited. In view of this fact, when building the GIBBS Grand Ensemble for volume, we can examine in the strictest manner the equilibrium statistical distribution of the states of the ensemble, that differ, in particular, in the size of the new phase nuclei. Under these circumstances, the return of the new phase to its initial state is considered to be possible, even though an extremely long time is needed, the return being caused by a fluctuation of an extremely low yet finite probability.

To determine the probability of the direct passage over the critical nucleus size from the initial, supersaturated phase it would be sufficient to divide the flux of nuclei over the pass, which is expressed on the basis of counting the states of GIBBS Grand Ensemble, by the number of states of this ensemble corresponding to the initial metastable state. With such a method of counting, the similar numbers of states in the numerator and denominater are mutually reduced, and the nucleation probability is obtained expressed in terms of the critical nucleus formation work. The earlier methods did not use GIBBS Grand Ensemble. In distinction from those methods, the formula thus derived contains a strict and unambiguously determined pre-exponential factor.

To derive this formula does not require knowing and taking into account the initial stages of growing nuclei. It would be sufficient to consider the kinetics of growth in the near critical region.

It is especially important to be able to apply that method to calculating the probability of nucleation in a liquid, for example, the probability of boiling up of a liquid of arbitrary volatility and viscosity [18]. In this case, the problem is complicated by the fact that the fluctuation growth of a nucleus must be characterized by two parameters, the size and the number of molecules of vapour in it, or by the volume and pressure whose fluctuations are mutually independent.

The method and the results of solving this problem have been published in [18], we will limit ourselves to the following. It is the application of GIBBS Grand Ensemble in conjunction with the KRAMERS-ZELDOVICH equation that allows for the case where the critical nucleus is large, a strict formula to be derived for calculating ν, i.e. probability of supercritical nucleus formation, without taking into consideration the initial, microscopic stage of bubble growth. The general formula includes a dimensionless parameter

$$\omega = \frac{P_c R_c}{\beta v \eta},$$

where P_c and R_c are the pressure and the radius of the critical nucleus, v is the thermal velocity of molecules, η is the viscosity of the liquid, β is the coefficient of condensation.

If $\omega \gg 1$, then at $\frac{2\sigma}{P_c R_c} \gg 3$ where σ is the surface tension of liquid, the formula assumes the form

$$\nu = \frac{V \varrho_l^2 R_c}{2\eta} \sqrt{\theta \sigma}\, e^{-\psi_c/kT}, \tag{15}$$

where V is the volume of the medium, ϱ_l is the density of the liquid, ψ_c is the critical nucleus formation work. At $\frac{2\sigma}{P_c R_c} \ll 3$:

$$\nu = \frac{\varrho_l^2}{\varrho_v^c} \frac{V\beta v}{2} \sqrt{\frac{\sigma}{kT}}\, e^{-\psi_c/kT}\,, \tag{16}$$

where ϱ_v^c is the density of vapour in the critical nucleus.

This formula differs from formula (36) of reference [15] by the factor ϱ_l/ϱ_v^c.

At $\omega \ll 1$, the general formula converts to

$$\nu = \frac{V\varrho_l^2 R_c}{2\eta} \sqrt{\theta\sigma}\, e^{-\psi_c/kT}\,. \tag{17}$$

This formula coincides with that derived earlier [17] and with formula (15).

References

[1] B. V. DERJAGUIN and M. M. KUSAKOV, Izvestiya Akad. Nauk SSSR, Ser. khim., No. 5, 741, 1936; No. 5, 1119; 1937; Acta Phys. Chim. B. V. DERJAGUIN, Koll. zh., 17, 207, 1955. URSS, 10, 25, 153, 1939

[2] E. M. LIFSHITZ, JETF, **29**, 94, 1955

[3] B. V. DERJAGUIN and N. V. CHURAEV, Koll. zh., **38**, No. 3, 1976

[4] A. I. RUSANOV, "*Phase Equilibria and Surface Phenomena*", published in Russian by "Khimiya", Leningrad, 1967

[5] B. V. DERJAGUIN and N. V. CHURAEV, Koll. zh., **37**, No. 6, 1975
B. V. DERJAGUIN, V. M. STAROV and N. V. CHURAEV, Koll. zh., **38**, No. 3, 1976

[6] I. LANGMUIR, J. Chem. Phys., **6**, 893, 1938

[7] B. V. DERJAGUIN, G. A. MARTYNOV, YU. V. GUTOP, Papers in the Collection "*Researches in the Field of Surface Forces*", volume 2, published in Russian by "Nauka", Moscow, 1964. Res. in Surface Forces. Vol. 2, Consult. Bureau Press, N.Y., 1966. B. V. DERJAGUIN, YU. V. GUTOP, Doklady AN SSSR, **171**, 652, 1966; Collection "*Researches in the Field of Surface Forces*", volume 3, p. 141, 1967, published in Russian by "Nauka", Moscow

[8] B. V. DERJAGUIN and YU. V. GUTOP, Koll. zh., **30**, 19, 1968

[9] B. V. DERJAGUIN, G. A. MARTYNOV, Yu. V. GUTOP, "*Research in the Field of Surface Forces*", Vol. 2, p. 9, Consult Bureau Press, N.Y., 1966; B. V. DERJAGUIN, G. A. MARTYNOV, YU. V. GUTOP; Koll. zh., **27**, 357, 1965

[10] G. A. MARTYNOV, I. B. IVANOV, B. V. TOSHEV, Koll. zh., **38**, 474, 1976

[11] W. GIBBS, "*The Scientific Papers*", v. 1, N.Y., 1961

[12] K. KRAMERS, Physika, **7**, 284, 1940

[13] YA. B. ZELDOVICH, JETF, **12**, 525, 1942

[14] YA. I. FRENKEL, "*Kinetic Theory of Liquids*", (russ.) Chapter 7, Moscow, published in Russianby "Nauka", 1975

[15] YU. M. KAGAN, Zhurn. Fis. Khim., **34**, 92, 1960

[16] B. V. DERJAGUIN, Doklady AN SSSR, **193**, 1096, 1970; J. Colloid. a. Interf. Sci., **38**, 517, 1972

[17] B. V. DERJAGUIN, Zhurn. eksper. teor. fisiki (JETF), **65**, 2261, 1973; J. Chem. Phys. **61**, 3665, 1974

[18] B. V. DERJAGUIN, A. V. PROKHOROV, N. N. TUNITSKIJ, Zhurn. eksper. teor. fisiki (JETF), **73**, 11 (5), 1977

Excess interfacial entropy and energy in binary liquid-liquid systems

R. J. Good

(Department of Chemical Engineering, State University of New York at Buffalo, Buffalo, New York/USA)

and

F. P. Buff

(Department of Chemistry, University of Rochester, Rochester, New York/USA)

Abstract

The rigorous, general theory of the excess interfacial energy and entropy for a binary, liquid-liquid system is developed. Explicit attention is given to the selection of the appropriate Gibbsian dividing surfaces, and to the fact that energy and entropy are determined only to within additive constants. The latter consideration leads to the choice of the pure liquids as reference states, for systems in which mutual solubilities are low. With this choice, it is found that simple expressions for interfacial entropy and energy arise, which are capable of being interpreted in terms of interfacial structure, concentration gradients and intermolecular forces. The result in the special case when mutual solubilities are low, is:

$$S_s(\text{rel}) = -\left(\frac{\partial \gamma}{\partial T}\right)_P \qquad U_s(\text{rel}) = \left[\frac{\partial(\gamma/T)}{\partial(1/T)}\right]_P,$$

where S_s(rel) und U_s(rel) are the excess interfacial entropy and energy relative to the pure liquids as reference states.

In addition to S_s(rel) and U_s(rel), there is a second component of S_s, and of U_s, directly associated with interfacial excess masses, I_1 and I_2. These entropy and energy components are not invariant with respect to the choice of the dividing surface, and are of much less physical interest than S_s(rel) and U_s(rel).

It is proposed that the functions S_s(rel) and U_s(rel) be used in the "geometric mean" relations connecting total energy of adhesion to the total energies of cohesion of the separate phases, and the entropy of adhesion to the entropies of cohesion of the separate phases.

The general theory of the surface excess thermodynamic functions is appropriate for the investigation of critical phenomena. The special theory is applied to the benzene-water system, and a qualitative interpretation of the results is presented.

1. Introduction

The publication of the final installment of Gibbs' great paper, "*On the Equilibrium of Heterogeneous Substances*" [1] took place one hundred years ago. With penetrating physical insight, Gibbs formulated a phenomenological theory of matter in thermal equilibrium, which has withstood the passage of time.

Gibbs' thermodynamic theory of capillarity laid the foundation for a vast number of studies, both theoretical and experimental, which have gone far towards elucidating the structure of interfaces. A scientifically important fraction of these studies have been on binary, liquid-liquid systems. In such systems, it is possible to measure interfacial tension and its temperature and pressure derivatives, and to study the diffuseness of the interface by optical methods. Theoretical interpretations of these results, using statistical mechanical models of varying degrees of sophistication, have been developed by a number of writers; we will mention some of these below. It is important to note that all such studies must turn to thermodynamics — i.e., to the methods of GIBBS — when it comes to relating theoretical models to the observations that can be made in the laboratory.

GIBBS himself was deeply interested in the question of interfacial structure, even though he was aware that detailed speculations as to structure, by him, would be premature.

GIBBS' grand synthesis of the work in general thermodynamics, of his great predecessors and contemporaries, had three main features. First, from GAUSS [2], GIBBS drew a variational principle, which he used to establish the distribution and stability of fluids and interfaces in an external field. (He implicitly acknowledged this inspiration in the last chapter of his monograph, "Elementary Principles of Statistical Mechanics" [3], where he gave the grand potential, Ω, in GAUSS' notation.) GIBBS freed GAUSS' principle from its modellistic origins. Second, he recognized that the use of the First and Second Laws of thermodynamics (which he quoted in the summary form given by CLAUSIUS [4]) would put the variational principle on a solid thermal basis. Third, he saw the importance of intensive variables such as chemical potential, as well as extensive variables, in characterizing the states of matter in bulk and in interfacial regions.

There is a fourth feature to GIBBS' program, which enabled him to apply rigorous thermodynamics to interfaces, and particularly to fluid interfaces. This was the "dividing surface" concept. In his discussion of interfacial phenomena, GIBBS carried his rigorous treatment of the intrinsic inhomogeneity of the interfacial region as far as was possible at that time.

He understood that, for all ordinary interfaces (i.e., far from the critical region) the gradient of concentration is steep, compared to the dimensions that are observable in the laboratory. At the same time, he understood that the gradient must exist over regions that are no smaller than the distance between molecules, and that in general, the thickness of the interfacial region will be greater than that distance.

Before GIBBS, Thomas YOUNG [5], the discoverer of many of the concepts and formulas of capillarity, had not only recognized the existence of a macroscopically sharp but microscopically diffuse interfacial profile, but also emphasized the optically anisotropic nature of the interfacial region. GIBBS was aware of the studies using polarized light, which by that time had already led to the detection of less-than-infinite concentration gradients at the liquid-vapor surface of pure liquids. Subsequent studies with elliptically polarized light, and more recently with laser light scattering, have revealed the correctness of GIBBS' intuition in regard to concentration gradients. Recent theoretical work [6, 7] on the capillary wave theory of interfacial thickness [12, 13] (the so-called BLS theory) has been confirmed by the optical measurements of WEBB, et al. [8], with regard to the extrapolation of the BLS treatment to the critical region.

One of the early implementations of GIBBS' ideas on interfaces was due to VAN DER WAALS. Building on MAXWELL's rudimentary density gradient theory of interfaces [9], by 1888 VAN DER WAALS was in possession of a treatment [10, 11] which has maintained its power up to present times. Two exact, alternative formulations of the VAN DER WAALS density gradient theory, first presented in 1971 [7, 12, 13], provide the most powerful modern approach to statistical mechanical theories of surface tension, its curvature dependence, and related properties.

Since the second portion, or about 100 pages, of Gibbs' paper, Ref. [1], is devoted to the subject of interfacial phenomena, it is fitting that our contribution should be an elaboration of Gibbs' work on two-phase binary systems. We derive, in Section 2, the general theory of the excess superficial energy and entropy, in a general form appropriate for liquid-liquid systems. (Some algebraic details of this derivation will be found in the Appendix). We develop it in Section III, in a form particularly suited to applications to systems with very low mutual solubility. And we apply the theory in Section IV to the water-benzene system, for which experimental data are available. Although some of our final results formally resemble Gibbs' Eqs. (580)—(581), the interpretation that we give to the terms is both appreciably different, and more explicit.

With regard to the use of dividing surfaces, it may be noted that Gibbs employed two alternative approaches. In the footnote, Ref. [1] p. 267, Gibbs presented a treatment which was intended primarily for the extensive state, that avoided the explicit use of his dividing surface (and hence, his "imaginary", reference system) entirely. Gibbs dismissed this approach for two reasons. The first is of a mathematical nature; he remarked that *"Such a ... method could not readily be extended so as to treat cases of more than two components with complete generality"*. The second and overriding reason is the inhomogeneous nature of the interface itself. Gibbs pointed this out in the abstract of his paper [14] and elsewhere in Ref. [1]; it is clear that this inhomogeneity justified the employment of the dividing surface.

We now introduce our treatment of interfacial thermodynamics, with the facts noted above: that though a typical interface is sharp on a macroscopic scale, it is, in general, diffuse on a microscopic scale. As a consequence of the fact that the concentrations of the components vary spatially, there must also be a corresponding spatial variation of energy and entropy densities. On account of the molecular nature of matter and the fact that entropy and intermolecular energy are not properties attributable to single molecules, the functions that represent entropy and energy densities are not unique. These functions must, however, conform to the requirement that they trend to the bulk values, far from the interface; and in addition, their structure should exhibit maximum mathematical simplicity.

We consider a columnar section having unit area, and extending perpendicular to the (planar) interface. When the energy and entropy densities are integrated over the whole system, the unique entropy S^s, energy U^s, and numbers of moles, N_i^s of the entire system (all per unit area) are obtained. The functions U^s, and S^s correspond to the symbols Gibbs used, ref. [1] p. 267: E_s for extensive (or total) energy and H_s [a Greek capital eta] for extensive (or total) entropy.

We now proceed with the first step in an iterative, subtractive procedure, which was formulated in 1955 [15]. In Ref. [15], it was also explicitly recognized that Gibbsian surface thermodynamics is based on an asymptotic estimate of free energy. That is to say, for example, higher order terms in the grand potential, Ω, involve ratios of the range of intermolecular forces to the dimensions of the system.

We consider a spatial reference system having the same volume and shape as the realsystem, and the same bulk properties for each of the respective phases, up to the selected dividing surface. (Under deformation of the real system, the volume of the spatial reference system remains the same as that of the real system.) We subtract from U^s the bulk energies, $U^{s\alpha}$, with α = I or II, for the two phases, I and II, computed for the spatial reference system. The difference corresponds to the excess superficial energy, U_s (GIBBS' ε_s). The corresponding procedure provides an evaluation of the excess superficial entropy, S_s, and the excess superficial densities, Γ_1 and Γ_2, with respect to this dividing surface.

We introduce, at this point, a notation [16] for a dividing surface specified by an arbitrary criterion: a superscript (b). Thus $S_s^{(b)}$, $U_s^{(b)}$ and $\Gamma_i^{(b)}$ are, respectively, the surface excess entropy, energy and number of moles of i per unit area, with respect to the dividing surface specified by the criterion, b. With this notation, we generalize the formalism of GUGGENHEIM and ADAM [17]. (See Ref. [16] for a critique of GUGGENHEIM's formalism). We, following GIBBS, envision a family of spatial reference systems ("imaginary systems") and for each, a specified dividing surface. Cf. BUFF [18], who found it of value to place a dividing surface, for a one-component system, such that $U_s = 0$ and, alternatively, so that $S_s = 0$. Any arbitrary criterion may be used, of course, e.g. that $\Gamma_1^{(b)} = -\Gamma_2^{(b)}$. This notation is consistent with previous formulations of the GIBBS equation; the use of the superscript (b) does not restrict the treatment, as against the GIBBS equation without superscripts.

As GIBBS recognized [1], the dividing surfaces that are of physical interest, in a binary liquid-liquid system, are generally quite close to the physical "surface of discontinuity", i.e. where the concentration gradient is steepest. (GIBBS also pointed out that, for a second component that is at nearly the same concentration in the two phases, the appropriate dividing surface may *"fail to coincide sensibly with the surface of physical discontinuity"*. (See Ref. 1, p. 234.)

Linear combinations of surface thermodynamic functions such as are found in the GIBBS equation, which correspond to macroscopically observable properties, must be invariant with respect to the location of the dividing surface. Thus the GIBBS equation may be written,

$$-d\gamma = S_s^{(b)}\, dT + \sum \Gamma_i^{(b)}\, d\mu_i\,. \tag{1.1}$$

The experimental functions, γ and its derivatives with respect to temperature and pressure, are, of course, independent of the choice of b. It is not, however, directly clear from Eq. (1.1) just what physical significance is to be attached to $(\partial\gamma/\partial T)_P$.

The existence of a family of dividing surfaces leads to the concept of a family of thermodynamic distance parameters, with respect to pairs of surfaces. These parameters are of two kinds. The first is purely thermodynamic in character. (See GIBBS, Ref. [1], p. 236 and 266.) GIBBS pointed out that while $(\partial\gamma/\partial P)_T$ represents a "diminution of volume", associated with unit extension of the surface, it also represents the distance between two dividing surfaces, namely the $\Gamma_1 = 0$ and $\Gamma_2 = 0$ surfaces. This idea has been examined in more detail recently [19], with regard to a two-phase system containing a surfactant as a third component.

The second type of thermodynamic distance requires extra-thermodynamic assumptions that arise from molecular theory [18, 20, 21]. Examples of this type of distance parameter have been presented previously, for one-component systems [18, 21]. The application of this second type of distance parameter to binary systems, based on the equations of the present paper, is reserved for a subsequent communication.

We now turn to two important ingredients in our formulation. The first is the choice of suitable intensive variables for the bulk phases. For a two-phase, binary system, two degrees of freedom are available. A previous summary of Gibbsian surface thermodynamics [22] utilized temperature and the mole fraction of a component in a specified phase. However, in the present application to two liquid phases, temperature and pressure are more appropriate.

The second essential ingredient involves a basic aspect of the nature of energy and entropy. It is commonly assumed in surface thermodynamics that interfacial entropy, S_s, and energy, U_s, are not subject to the ambiguity intrinsic in the energy and entropy functions, of an unknown, additive constant. S_s and U_s are excess functions, and are, by definition, related to the bulk entropy and energy, respectively. The "absolute" nature of U_s and S_s has been commented upon explicitly by Pippard [23].

On a more detailed examination of this question, it may be seen that it is only in a very limited sense, that U_s and S_s are absolute. For a one-component system, the independence of arbitrary reference states and unknown additive constants is valid only for $U_s^{(1)}$ and $S_s^{(1)}$, the surface functions relative to the $\Gamma_1 = 0$ dividing surface. With respect to any other dividing surface, $\Gamma_s^{(b)} \neq 0$; and the energy and the entropy intrinsically associated with the mass $\Gamma_1^{(b)}$ do contain the unknown, additive constants. Hence the general surface energy, $U_s^{(b)}$, and entropy, $S_s^{(b)}$ are not absolute.

Since the reference system is defined to have the same volume as the real system, but a different number of moles of the two components, there is not in general any dividing surface with respect to which both Γ_1 and Γ_2 are zero.

Hence $U_s^{(b)}$ and $S_s^{(b)}$ will, in general, be subject to the ambiguity of the additive constants. This ambiguity cannot be removed by the device of selecting the $\Gamma_1 = 0$ surface or the $\Gamma_2 = 0$ surface. For example, it will in general be true that, for the interface between phases 1 and 2,

$$S_s^{(1)} \neq S_s^{(2)}, \qquad U_s^{(1)} \neq U_s^{(2)}. \tag{1.2}$$

We will, in sections II and III, show how this dilemma may be resolved. We will introduce the reference-state energy, and entropy explicitly. When this is done and when the appropriate reference state is identified, it will be seen that expressions for interfacial entropy and energy are obtained in a simple form. The values of U_s and S_s that can be calculated with these expressions will be found to admit of physical interpretation in terms of interfacial structure.

There is a motivation for this study, that should be mentioned here. In an earlier paper [24], an equation was proposed relating the free energy of adhesion, ΔG_{12}^a, to the free energies of cohesion of phase 1 and of phase 2, ΔG_1^c and ΔG_2^c, respectively:

$$\frac{\Delta G_{12}^a}{\sqrt{\Delta G_1^c \, \Delta G_2^c}} = \Phi_{12}\,. \tag{1.3}$$

Φ_{12} may be estimated from molecular properties of substances 1 and 2. It has been shown [25] that the calculated value of Φ_{12} is in good agreement with experiment. It was also shown in the original study of this subject [24] that a similar relation should hold for the total energy of adhesion, ΔU_{12}^a, and the total energies of cohesion, ΔU_1^c and ΔU_2^c:

$$\frac{\Delta U_{12}^a}{\sqrt{\Delta U_1^c \, \Delta U_2^c}} = \Phi_{12}^*\,. \tag{1.4}$$

A related equation in terms of entropy has also been suggested recently [26], which we may put in the form,

$$\frac{\Delta S_{12}^{a}}{\sqrt{\Delta S_1^c \, \Delta S_2^c}} = \Psi_{12} \,. \tag{1.5}$$

Since (unlike energy) surface entropy is a function of disorder, e.g. of molecular orientation as well as concentration gradients in the interface, it would be expected that, in general,

$$\Psi_{12} \neq \Phi_{12} \,. \tag{1.6}$$

In order to test Equations 1.4, 1.5 and 1.6, it is necessary to compute S_{s12} and U_{s12} from experimental data. But it is not clear how this should be done. Cf. Eq. (1.2), above. The problem of the additive constants in the liquid-liquid interfacial total energy and entropy must be solved before this test can be made. The results that will be given in Sections 2 and 3 will lead to the needed solution of this problem.

2. General theory

We now turn to the determination of the excess interfacial entropy, S_s, and excess interfacial energy, U_s, within the framework of GIBBSian thermodynamics. Since the ultimate application of the formalism is to binary, liquid-liquid systems, it is appropriate (as already noted) to select temperature, T, and pressure, P, as the independent thermodynamic variables. The system of equations [1] which interrelate the pertinent intensive variables consists of the GIBBS-DUHEM equations for the bulk phases, I and II and the GIBBS adsorption equation which accounts for the presence of the interface:

$$dp = S_v^{\mathrm{I}}\, dT + c_1^{\mathrm{I}}\, d\mu_1 + c_2^{\mathrm{I}}\, d\mu_2 \,, \tag{2.1}$$

$$dp = S_v^{\mathrm{II}}\, dT + c_1^{\mathrm{II}}\, d\mu_1 + c_2^{\mathrm{II}}\, d\mu_2 \,, \tag{2.2}$$

$$-d\gamma = S_s^{(b)}\, dT + \Gamma_1^{(b)}\, d\mu_1 + \Gamma_2^{(b)}\, d\mu_2 \,, \tag{2.3}$$

where S_v^α (with $\alpha = \mathrm{I}$ or II) represents the entropy density of phase α, μ_1 and μ_2 are the chemical potentials, and c_i^α is the concentration of component i in phase α. The set of equations, (2.1) to (2.3), describing entropy properties, has a counterpart for the description of energy (enthalpy) properties [22], whose form is motivated by the molecular formulation of thermodynamically open systems [15]. The entropy terms may be eliminated with use of the classical free energy relations for the bulk phases and the interface,

$$H_v^\alpha = TS_v^\alpha + \sum_{i=1}^{2} c_i^\alpha \mu_i \,, \qquad \alpha = \mathrm{I}, \mathrm{II} \,, \tag{2.4}$$

$$U_s^{(b)} = TS_s^{(b)} + \gamma + \sum_{i=1}^{2} \Gamma_i^{(b)} \mu_i \,, \tag{2.5}$$

where H_v^α is the enthalpy per unit volume of phase α

When solved for S_v^α and $S_s^{(b)}$, Eqs. (2.4) and (2.5) take the form

$$S_v^\alpha = \frac{H_v^\alpha}{T} - \sum_{i=1}^{2} \frac{c_i^\alpha \mu_i}{T}, \qquad \alpha = \mathrm{I, II}, \tag{2.6}$$

$$S_s^{(b)} = \frac{U_s^{(b)}}{T} - \frac{\gamma}{T} - \sum_{i=1}^{2} \frac{\Gamma_i^{(b)} \mu_i}{T}. \tag{2.7}$$

Substitution of Eqs. (2.6) and (2.7) into Eqs. (2.1) to (2.3), followed by division by T, leads to the desired relations

$$\frac{1}{T} dp = -H_v^{\mathrm{I}}\, d(1/T) + c_1^{\mathrm{I}}\, d(\mu_1/T) + c_2^{\mathrm{I}}\, d(\mu_2/T), \tag{2.8}$$

$$\frac{1}{T} dp = -H_v^{\mathrm{II}}\, d(1/T) + c_1^{\mathrm{II}}\, d(\mu_1/T) + c_2^{\mathrm{II}}\, d(\mu_2/T), \tag{2.9}$$

$$-d(\gamma/T) = -U_s^{(b)}\, d(1/T) + \Gamma_1^{(b)}\, d(\mu_1/T) + \Gamma_2^{(b)}\, d(\mu_2/T). \tag{2.10}$$

It should be noted that for planar interfaces, $U_s^{(b)} = H_s^{(b)}$. The physical origin of this equality of the excess energy and enthalpy arises from the fact that the pressure normal to the interface is constant, in a weak gravitational field.

The evaluation of $[\partial\gamma/\partial T]_P$ and $[\partial\gamma/\partial P]_T$ from Eqs. (2.1) to (2.3) is attained by elimination of the chemical potential terms. The Appendix contains this mathematical manipulation, as well as the operations which explicitly refer each thermodynamic function to the component reference states, $\bar{S}_1^0(\mathrm{ref})$ and $\bar{S}_2^0(\mathrm{ref})$, whose values must be selected on physical grounds. The results may be expressed in the form

$$-\left(\frac{\partial\gamma}{\partial T}\right)_P = S_s^{(b)}(\mathrm{rel}) - \frac{\Gamma_{\mathrm{I}}^{(b)}}{c_{\mathrm{I}}} S_v^{\mathrm{I}}(\mathrm{rel}) - \frac{\Gamma_{\mathrm{II}}^{(b)}}{c_{\mathrm{II}}} S_v^{\mathrm{II}}(\mathrm{rel}), \tag{2.11}$$

$$-\left(\frac{\partial\gamma}{\partial P}\right)_T = \frac{\Gamma_{\mathrm{I}}^{(b)}}{c_{\mathrm{I}}} + \frac{\Gamma_{\mathrm{II}}^{(b)}}{c_{\mathrm{II}}}, \tag{2.12}$$

where

$$S_s^{(b)}(\mathrm{rel}) \equiv S_s^{(b)} - \Gamma_1^{(b)}\bar{S}_1^0(\mathrm{ref}) - \Gamma_2^{(b)}\bar{S}_2^0(\mathrm{ref}), \tag{2.13}$$

$$S_v^\alpha(\mathrm{rel}) \equiv c_1^\alpha[\bar{S}_1^\alpha - \bar{S}_1^0(\mathrm{ref})] + c_2^\alpha[\bar{S}_2^\alpha - \bar{S}_2^0(\mathrm{ref})] \tag{2.14}$$

and $\bar{S}_i^\alpha$ is the partial molar entropy of component i in phase α. The superficial densities of matter, Γ_α, and bulk densities, c_α, formally associated with phase α, are given explicitly by

$$\Gamma_{\mathrm{I}}^{(b)} \equiv \Gamma_1^{(b)} - \frac{c_1^{\mathrm{II}}\Gamma_2^{(b)}}{c_2^{\mathrm{II}}}, \qquad \Gamma_{\mathrm{II}}^{(b)} \equiv \Gamma_2^{(b)} - \frac{c_2^{\mathrm{I}}\Gamma_1^{(b)}}{c_1^{\mathrm{I}}}, \tag{2.15}$$

$$c_{\mathrm{I}} \equiv c_1^{\mathrm{I}} - \frac{c_2^{\mathrm{I}}c_1^{\mathrm{II}}}{c_2^{\mathrm{II}}}, \qquad c_{\mathrm{II}} \equiv c_2^{\mathrm{II}} - \frac{c_2^{\mathrm{I}}c_1^{\mathrm{II}}}{c_1^{\mathrm{I}}}. \tag{2.16}$$

The concentrations, c_{I} and c_{II}, are the composition variables in this formulation. They are related to the mole fractions, x_1^{I} and x_2^{II}, by

$$x_1^{\mathrm{I}} \equiv \frac{x_{\mathrm{I}}}{1 - (1 - x_{\mathrm{I}})(1 - x_{\mathrm{II}})}, \qquad x_2^{\mathrm{II}} \equiv \frac{x_{\mathrm{II}}}{1 - (1 - x_{\mathrm{I}})(1 - x_{\mathrm{II}})} \tag{2.17}$$

with

$$c_{\rm I} \equiv (c_1^{\rm I} + c_2^{\rm I})\, x_{\rm I}\,, \qquad c_{\rm II} \equiv (c_1^{\rm II} + c_2^{\rm II})\, x_{\rm II}\,.$$

The right hand side of Eqs. (2.11) and (2.12) must be invariant with respect to placement of the dividing surface. The choices of $\Gamma_{\rm I} = 0$ with associated $\Gamma_{\rm II}^{(\rm I)}$, or of $\Gamma_{\rm II} = 0$ with associated $\Gamma_{\rm I}^{(\rm II)}$, are appropriate. With respect to these dividing surfaces, Eq. (2.12) takes the alternative form

$$-\left(\frac{\partial \gamma}{\partial p}\right)_T = \frac{\Gamma_{\rm I}^{(\rm II)}}{c_{\rm I}} = \frac{\Gamma_{\rm II}^{(\rm I)}}{c_{\rm II}}\,. \tag{2.18}$$

In this formalism, $(\partial\gamma/\partial P)_T$ represents the distance between the $\Gamma_{\rm I} = 0$ and $\Gamma_{\rm II} = 0$ dividing surfaces, in addition to being (as already noted) the diminution of volume associated with unit extension of the interface.

Substitution of Eq. (2.13) into Eq. (2.11) and specializing to the $\Gamma_{\rm II} = 0$ or $\Gamma_{\rm I} = 0$ dividing surfaces leads to one of the main results of this section:

$$S_s^{(\rm II)}({\rm rel}) = -\left(\frac{\partial\gamma}{\partial T}\right)_P - S_v^{\rm I}({\rm rel})\left(\frac{\partial\gamma}{\partial P}\right)_T \tag{2.19a}$$

or

$$S_s^{(\rm I)}({\rm rel}) = -\left(\frac{\partial\gamma}{\partial T}\right)_P - S_v^{\rm II}({\rm rel})\left(\frac{\partial\gamma}{\partial P}\right)_T\,. \tag{2.19b}$$

$S_s^{(\rm II)}({\rm rel})$ and $S_s^{(\rm I)}({\rm rel})$ are the excess interfacial entropy computed relative to specified reference states (as defined by Eqs. 2.13 and 2.14) and referred to the dividing surfaces $\Gamma_{\rm II} = 0$ and $\Gamma_{\rm I} = 0$ respectively.

The analogy between Eqs. (2.1) to (2.3) and Eqs. (2.8) to (2.10) is utilized in the Appendix. We define the energy analogs of the quantities given in Eqs. (2.13) and (2.14):

$$U_s^{(b)}({\rm rel}) \equiv U_s^{(b)} - \Gamma_1^{(b)}\overline{H}_1^0({\rm ref}) - \Gamma_2^{(b)}\overline{H}_2^0({\rm ref})\,, \tag{2.20}$$

$$H_v^\alpha({\rm rel}) \equiv c_1^\alpha[\overline{H}_1^\alpha - \overline{H}_1^0({\rm ref})] + c_2^\alpha[\overline{H}_2^\alpha - \overline{H}_2^0({\rm ref})]\,, \quad \alpha = {\rm I, II}\,, \tag{2.21}$$

where $\overline{H}_i^\alpha$ is the partial molal enthalpy of component i in phase α. The final result for energy, analogous to the entropy result, Eq. (2.19), thus leads to another principal result of this section:

$$U_s^{(\rm II)}({\rm rel}) = \left[\frac{\partial(\gamma/T)}{\partial(1/T)}\right]_P - H_v^{\rm I}({\rm rel})\left(\frac{\partial\gamma}{\partial P}\right)_T\,, \tag{2.22a}$$

$$U_s^{(\rm I)}({\rm rel}) = \left[\frac{\partial(\gamma/T)}{\partial(1/T)}\right]_P - H_v^{\rm II}({\rm rel})\left(\frac{\partial\gamma}{\partial P}\right)_T\,. \tag{2.22b}$$

It is finally observed that, while Eqs. (2.18) to (2.20) apply to the case of arbitrary miscibility, they are in a form which reduces to the formulae applicable to the important case of very small mutual solubility.

3. Theory of systems with small miscibility

We now undertake a direct examination of the important case of a binary, liquid-liquid system in which mutual solubilities are negligibly small. This approximation allows us to simplify the concentration variables defined in Equations 2.15 to 2.17 and used in the final equations 2.19 to 2.22.

We take component 1 to be the principal constituent of phase I, and component 2 to be the principal substance present in phase II; $c_1^{\mathrm{I}} \gg c_2^{\mathrm{I}}$, $c_2^{\mathrm{II}} \gg c_1^{\mathrm{II}}$. In this case, Eqs. 2.15 and 2.16 reduce to

$$\Gamma_{\mathrm{I}}^{(b)} = \Gamma_1^{(b)}\,, \qquad \Gamma_{\mathrm{II}}^{(b)} = \Gamma_2^{(b)}\,, \tag{3.1a}$$

$$c_{\mathrm{I}} = c_1^{\mathrm{I}}\,, \qquad c_{\mathrm{II}} = c_2^{\mathrm{II}}\,. \tag{3.1b}$$

This amounts to a return to more familiar concentration and surface excess functions.

As an alternative to the above approach, we may proceed by a simpler route, to the identical final equations. It has been shown [16] that the distance, λ_{21}, between the $\Gamma_1 = 0$ surface and the $\Gamma_2 = 0$ surface is given by

$$\lambda_{21} = \left(\frac{\partial \gamma}{\partial P}\right)_T = -\lambda_{12}\,. \tag{3.2}$$

The negative sign before λ_{12}, as opposed to λ_{21}, is a consequence of the choice of directions in the coordinate system. For systems with low miscibility, we may also write [16],

$$\lambda_{21} = \Gamma_2^{(1)}/c_2^{\mathrm{II}} = \Gamma_1^{(2)}/c_1^{\mathrm{I}}\,, \tag{3.3}$$

$$= \Gamma_2^{(b)}/c_2^{\mathrm{II}} + \Gamma_1^{(b)}/c_1^{\mathrm{I}}\,. \tag{3.4}$$

When $c_2^{\mathrm{I}} \ll c_1^{\mathrm{I}}$ and $c_1^{\mathrm{II}} \ll c_2^{\mathrm{II}}$, it may be shown that the terms $c_2^{\mathrm{I}}\, d\mu_2$ and $c_1^{\mathrm{II}}\, d\mu_1$ can be neglected in Equations 2.1 and 2.2 (See Equations A.4 and A.5, in the Appendix, which can be used to provide a basis for this approximation.) Then the system of equations, 2.1 to 2.3, becomes

$$\mathrm{d}P = S_v^{\mathrm{I}}\, dT + c_1^{\mathrm{I}}\, d\mu_1\,, \tag{3.5}$$

$$dP = S_v^{\mathrm{II}}\, dT + c_2^{\mathrm{II}}\, d\mu_2\,, \tag{3.6}$$

$$-d\gamma = S_s^{(b)}\, dT + \Gamma_1^{(b)}\, d\mu_1 + \Gamma_2^{(b)}\, d\mu_2\,, \tag{3.7}$$

where now, S_v^{I} and S_v^{II} are, respectively, the entropy densities of the pure components 1 and 2, and $S_v^{\mathrm{I}} = c_1^{\mathrm{I}}\bar{S}_1(\text{pure})$, $S_v^{\mathrm{II}} = c_2^{\mathrm{II}}\bar{S}_2(\text{pure})$. Equations 3.5 and 3.6 can be used directly, to eliminate $d\mu_1$ and $d\mu_2$ from Equation 3.7.

Now, as already mentioned, operationally, energy is not an absolute quantity; it is ambiguous to the extent of containing an unknown, additive constant. We can treat total energy rigorously only by expressing it as a difference function, i.e. as being relative to the energy of a reference state. Mathematically, we may say the same with regard to entropy. We can express this relative nature explicitly by writing, for example,

$$S_v^{\mathrm{I}}(\text{rel}) \equiv S_v^{\mathrm{I}}(\text{pure}) - c_1^{\mathrm{I}}\bar{S}_1^0(\text{ref})\,, \tag{3.8a}$$

$$S_v^{\mathrm{II}}(\text{rel}) \equiv S_v^{\mathrm{II}}(\text{pure}) - c_2^{\mathrm{II}}\bar{S}_2^0(\text{ref})\,. \tag{3.8b}$$

(Cf. Eq. 2.14, above, and in the Appendix, Eqs. A9a and A9b).

Here, $\bar{S}_1^0(\text{ref})$ and $\bar{S}_2^0(\text{ref})$ are reference state entropies per mole of pure liquid, the choice of the reference states not being dependent on the mathematical formalism.

The relative excess surface entropy is given by

$$S_s^{(b)}(\text{rel}) \equiv S_s^{(b)} - \Gamma_1^{(b)}\bar{S}_1^0(\text{ref}) - \Gamma_2^{(b)}\bar{S}_2^0(\text{ref}) \,. \tag{3.9}$$

We can now use Equations 3.8 and 3.9 together with 3.5 and 3.6, to write Equation 3.7 in the form,

$$-d\gamma = \{S_s^{(b)}(\text{rel}) - \Gamma_1^{(b)}[\bar{S}_1(\text{pure}) - \bar{S}_1^0(\text{ref})] - \Gamma_2^{(b)}[\bar{S}_2(\text{pure}) - \bar{S}_2^0(\text{ref})]\}\, dT$$
$$- \left[\frac{\Gamma_1^{(b)}}{c_{\text{I}}^1} + \frac{\Gamma_2^{(b)}}{c_2^{\text{II}}}\right] dP \,. \tag{3.10}$$

This is the desired solution to the system of Equations 3.5—3.7. The coefficient of dP in Eq. 3.10 is identical with the right side of Eq. 3.4.

We can next write the expression for the relative excess surface entropy, by setting $dP = 0$ in Eq. 3.10 and rearranging:

$$S_s^{(b)}(\text{rel}) = -\left(\frac{\partial\gamma}{\partial T}\right)_P - \Gamma_1^{(b)}[\bar{S}_1(\text{pure}) - \bar{S}_1^0(\text{ref})] - \Gamma_2^{(b)}[\bar{S}_2(\text{pure}) - \bar{S}_2^0(\text{ref})] \,, \tag{3.11}$$

$$S_s^{(i)}(\text{rel}) = -\left(\frac{\partial\gamma}{\partial T}\right)_P - \Gamma_j^{(i)}[\bar{S}(\text{pure}\, j) - \bar{S}_j^0(\text{ref})] \,, \tag{3.12}$$

where $i = 1$ or 2 and $j = 2$ or 1. If we let $\beta = \text{II}$ or I in correspondence with j, then on combining with Eq. 3.1, we obtain

$$S_s^{(i)}(\text{rel}) = -\left(\frac{\partial\gamma}{\partial T}\right)_P - \left(\frac{\partial\gamma}{\partial P}\right)_T [S_{vj}^{\beta} - S_{vj}^0(\text{ref})] \,. \tag{3.13}$$

The interfacial energy can be obtained by an exactly analogous method, starting with the system consisting of the Gibbs equation and the Gibbs-Duhem equation for the two bulk phases in the modified forms, that was derived in section II. We employ definitions of $H_v^{\text{I}}(\text{rel})$ and $H_v^{\text{II}}(\text{rel})$ and of $U_s^{(b)}(\text{rel})$ that are exactly analogous to the entropy definitions, Equations 3.8a, 3.8b and 3.9. The resulting equations are,

$$U_s^{(b)}(\text{rel}) = \left[\frac{\partial(\gamma/T)}{\partial(1/T)}\right]_P - \Gamma_1^{(b)}[\bar{H}_1(\text{pure}) - \bar{H}_1^0(\text{ref})] - \Gamma_2^{(b)}[\bar{H}_2(\text{pure}) - \bar{H}_2^0\,(\text{ref})] \,, \tag{3.14}$$

$$U_i^{(i)}(\text{rel}) = \left[\frac{\partial(\gamma/T)}{\partial(1/T)}\right]_P - \Gamma_j^{(i)}[(\bar{H}(\text{pure}\, j) - \bar{H}_j^0(\text{ref})] \tag{3.15}$$

$$= \left[\frac{\partial(\gamma/T)}{\partial(1/T)}\right]_P - \left(\frac{\partial\gamma}{\partial P}\right)_T [(H_{vj}^{\beta} - H_{vj}^0(\text{ref})] \,. \tag{3.16}$$

It is evident at this point that there are two kinds of entropy components present in $S_s^{(b)}$; and the corresponding statement can be made about $U_s^{(b)}$. The last two terms in Eq. (3.10) correspond to the excess entropy in the real system over that in a spatial reference system with dividing surface b, which is present by virtue of the excess masses $\Gamma_1^{(b)}$ and $\Gamma_2^{(b)}$. The reference state bulk entropies, $\bar{S}_1^0$ and $\bar{S}_2^0$, are, formally, present in this component of $S_s^{(b)}$. And the value of the sum of these two terms will vary with the placement of the dividing surface. As we have indicated, an exactly analogous statement with respect to the corresponding components of surface energy, $U_s^{(b)}$, can be made in discussing Eq. (3.14). There is an element of arbitrariness in this result with respect both to U_s and S_s which is quite undesirable. The ambiguity of the energy and entropy functions, mentioned above, has manifested itself. Therefore these terms will be of relatively minor physical interest.

The components of interfacial entropy and of interfacial energy in Eqs. 3.11 to 3.16 which have by far the greatest physical importance are the terms, $(\partial\gamma/\partial T)_P$ and $[\partial(\gamma/T)/\partial(1/T)]_P$. To examine these terms, we first take note of the flexibility (that is allowed in choosing reference states) for the two components. Now, in the spatial reference systems used for developing the dividing surface concept (see above) the bulk properties for the reference system are those of the two bulk phases, I and II. This suggests a "natural" physical basis, according to which we should choose the two pure bulk liquids for the thermodynamic reference states, as well as for the spatial reference system. When this choice is made, the resulting superficial entropy and energy terms are, of course, relative to the values in those reference states; and so we will designate them S_s(rel) and U_s(rel).

The result of this choice is that the coefficients of Γ_i, Γ_j and $(\partial\gamma/\partial P)_T$ in Eqs. 3.11 to 3.16 vanish, i.e.,

$$\bar{S}_1(\text{pure}) - \bar{S}_1^0 = 0\,, \qquad \bar{S}_2(\text{pure}) - \bar{S}_2^0 = 0\,, \tag{3.17}$$

$$\bar{H}_1(\text{pure}) - \bar{H}_1^0 = 0\,, \qquad \bar{H}_2(\text{pure}) - \bar{H}_2^0 = 0\,. \tag{3.18}$$

We obtain, as a consequence,

$$S_s(\text{rel}) = -\left(\frac{\partial\gamma}{\partial T}\right)_P , \tag{3.19}$$

$$U_s(\text{rel}) = \left[\frac{\partial(\gamma/T)}{\partial(1/T)}\right]_P . \tag{3.20}$$

It will be noted that U_s(rel) and S_s(rel), Eqs. (3.19) and (3.20), carry no superscripts such as would designate, e.g., the $\Gamma_1 = 0$ or $\Gamma_2 = 0$ surfaces. This is because, when the reference state is the pure liquid,

$$S_s^{(1)}(\text{rel}) = S_s^{(2)}(\text{rel}) = S_s^{(b)}(\text{rel})\,,$$

and

$$U_s^{(1)}(\text{rel}) = U_s^{(2)}(\text{rel}) = U_s^{(b)}(\text{rel})\,.$$

That is to say, these two functions are independent of the placement of the dividing surface, when the solubilities are very small.

In view of the discussion following Eq. (3.16), above, we can now state the physical interpretation of S_s(rel) and U_s(rel). These may be regarded as the excess entropy and energy that would be present in the system if the surface excess masses, $\Gamma_1^{(b)}$ and $\Gamma_2^{(b)}$, made no contribution to the excess entropy and energy of the system.

Hence S_s(rel) and U_s (rel) are also the excess entropy and energy that must be due to all other structural properties or features of the interface per se, in addition to the excess masses of substance 1 and 2. This means that they must be the manifestations of (a) the composition gradients in the interfacial region; (b) the forces between the molecules that are present in the region where composition gradients exist; (c) the entropy of mixing of molecules in this region; and (d) the degree of orientation, for unsymmetrical molecules in this region.

In the development above, we have taken the reference states to be the pure liquids. However, for completeness, we should discuss what happens when other reference states are chosen. For example, in view of the Third Law of Thermodynamics, there is intuitive appeal to basing the reference states on the perfect crystals of the pure components, 1 and 2, at 0 K.

To indicate that we are choosing any reference state other than pure liquid, we will employ an asterisk; as we must retain the superscript (b) designating the dividing surface. Then, instead of using Eqs. (3.17) and (3.18), we use the definitions,

$$\bar{S}_i^{\alpha *}(\text{ref}) \equiv \bar{S}_i^{\alpha} - \bar{S}_i^{0*}\,, \tag{3.21}$$

$$\bar{H}_i^{\alpha *}(\text{ref}) \equiv \bar{H}_i^{\alpha} - \bar{H}_i^{0*}\,, \tag{3.22}$$

where if $i = 1$, $\alpha = \text{I}$ and if $i = 2$, $\alpha = \text{II}$. Eqs. (3.11) and (3.12) then become, respectively,

$$-S_s^{(b)*}(\text{rel}) = \left(\frac{\partial \gamma}{\partial T}\right)_P + \Gamma_1^{(b)}\bar{S}_1^{\text{I}*} + \Gamma_2^{(b)}\bar{S}_2^{\text{II}*}\,, \tag{3.23}$$

$$U_s^{(b)*}(\text{rel}) = \left[\frac{\partial(\gamma/T)}{\partial(1/T)}\right]_P - \Gamma_1^{(b)}\bar{H}_1^{\text{I}*} - \Gamma_2^{(b)}\bar{H}_2^{\text{II}*}\,. \tag{3.24}$$

There is no direct physical reason to compute U_s using as a reference state the perfect crystals at 0 K. We will, however, carry out a computation of surface entropy, in section IV, which will put this kind of choice of reference state into concrete terms. It will be found, for example, that the value of the computed $S_s^{(i)*}(\text{rel})$ is very sensitive to the spatial location of the dividing surface.

4. Application

(The benzene-water system)

It is desirable to illustrate the theory with a physical system. The binary system for which, probably, the best values of γ are available, is benzene vs. water. The relevant data are given in Table 1.

Table 1: Measured interfacial thermodynamic data for benzene vs. water

		Source or Reference
γ, 20°	33.7 ± 0.2 erg/cm	[27]
$\left(\frac{\partial \gamma}{\partial T}\right)_P$, 20—30°, 1 atm	−0.038 ± 0.005 erg/cm²/deg	[27]
$\left(\frac{\partial \gamma}{\partial P}\right)_T$, 25°, 1—50 atm	−0.0007 erg/cm²/bar = 0.07	[28—31]
$\Gamma_w^{(B)}$	4×10^{-11} mol/cm²	[16]
$\Gamma_B^{(W)}$	0.8×10^{-11} mol/cm²	[16]

The value of γ in Table 1 differs slightly from that reported by HARKINS and BROWN [32]. However, the Harkins value was obtained by a relative method (drop weight) of measuring interfacial tension, while the authors in Ref. [27] used absolute methods of measuring γ. Hence these results supersede those of HARKINS and BROWN.[1]

Table 2a) shows the values of S_s(rel) and U_s(rel), computed with the aid of Eqs. (3.19) and (3.20), for this system. Table 2b) shows the values of $S_s^{(W)*}$(rel) and $S_s^{(B)*}$(rel) computed using Eqs. (3.23) and (3.24), with the reference state being based on the Third Law. The absolute molar entropy of benzene at 25° was taken to be 64.34 cal/mol [33] and that of water, 45.11 cal/mol [34].

Table 2: Interfacial entropy and total energy for C_6H_6 vs. H_2O

a) Relative excess entropy and energy, with pure liquids as reference states:		
S_s(rel)	0.038 ± 0.005 erg/cm²/deg	Eq. 3.19
U_s(rel)	45 ± 2 erg/cm²	Eq. 3.20
b) Excess entropy, using the Third Law as a basis for the reference states:		
$S_s^{(W)*}$(rel)	0.059 ± 0.010 erg/cm²/deg	Eq. 3.23
$S_s^{(B)*}$(rel)	0.111 ± 0.025 erg/cm²/deg	Eq. 3.23

The accuracy of the value, 0.0007 erg²/cm/bar in Table 1, for $(\partial\gamma/\partial P)_T$, is hard to determine. From an examination of MICHAELS' and HAUSERS data, one might estimate the limits, very roughly, as ± 0.0002 erg/cm²/bar. The error limits in Table 2 are computed on the basis of these limits, together with the limits indicated for $(\partial\gamma/\partial T)_P$ in Table 1.

Discussion

An examination of Table 2b) shows, first, that the computed values of S_s based on the Third Law are sensitive to errors in $(\partial\gamma/\partial P)_T$- particularly in the case of $S_s^{(B)*}$(rel). The physical reason for this may be seen from the "diminution of volume" interpretation [1] of the pressure coefficient. An uncertainty of ± 0.0002 in $(\partial\gamma/\partial P)_T$ means an uncertainty of $\pm 2 \times 10^{-10}$ cm³ in the volume enclosed between unit areas of the $\Gamma_W = 0$ and $\Gamma_B = 0$ surfaces. This volume, times the entropy per unit volume of the pure component, leads to the uncertainty in the thermodynamic quantities computed.

Table 2b) also shows that if S_s is computed using reference states based on the Third Law, it is very sensitive to the choice of the dividing surface. The physical

[1] HARKINS and BROWN determined an empirical calibration curve for drop-weight interfacial tension measurements, using as standard, values of liquid-vapor interfacial tension that had been measured independently. For a liquid-liquid system, intertial and viscous effects are appreciable, so HARKINS' "principle of similitude" cannot be relied upon, as between liquid-vapor and liquid-liquid interfaces. Hence HARKINS' result for γ of benzene vs. water cannot be considered as absolute. WOSILAIT's measurements were by the pendant drop method; and PIETRAS measured interfacial tension by the WILHELMY plat method, with detachment of the plate from the interface.

reason for this sensitivity is the same as that given above: The fact that the volume between unit areas of the two dividing surfaces cannot be neglected.

The degree of physical interest in computations in which the reference state is any state other than the pure liquid, can now be examined. For example, the fact that $S_s^{(B)*}(\text{rel}) - S_s^{(W)*}(\text{rel}) = 0.052$ erg/cm²/deg means only that the entropy is associated with the superficial excess masses in the interfacial region. A corresponding estimate of difference in interfacial energy, with respect to the $\Gamma_W = 0$ vs. the $\Gamma_B = 0$ surface, could be computed. These numbers tell us nothing new physically, beyond the information contained in $\Gamma_B^{(W)}$ and $\Gamma_W^{(B)}$. Therefore they contribute nothing to the growth of physical understanding of interfacial structure; there is no advantage in trying to find additional structural interpretations of quantities such as $S_s^{(i)}$ and $U_s^{(i)}$ beyond interpreting $S_s(\text{rel})$, $U_s(\text{rel})$, $\Gamma_1^{(2)}$ and $\Gamma_2^{(1)}$.

The values of $S_s(\text{rel})$ and $U_s(\text{rel})$, on the other hand, can be given interpretations which reveal a great deal about interfacial structure. They are physically suitable for comparison to the surface energy and entropy of the two pure components. Table 3 lists the appropriate data for pure water and benzene.

Table 3: Surface Thermodynamic Properties of Water and Benzene

	Water	Benzene
γ, dyne/cm	72.0	28.15
$U_s^{(1)}$, erg/cm²	115.2	68.3
$S_s^{(1)}$, erg/cm²/deg	0.145	0.135

The first observation that presents itself, in a comparison of Tables 2 and 3, is that while the benzene-water interfacial tension lies between the surface tensions of water and of benzene, the value of $U_s(\text{rel})$ in Table 2 is considerably below the total surface energy of benzene as well as that of water. And the values of $S_s(\text{rel})$ are considerably less than the surface entropy of benzene and of water.

The low interfacial entropy can only mean much greater molecular order in the liquid-liquid interface than at the liquid-vapor interface. If, as suggested by Good and Elbing [25], there is a hydrogen bond between the water hydroxyls and the π electrons of benzene, that would contribute an orientation entropy, in the benzene, of the order of $R \ln 2$ or 5.8 joule/deg/mol of surface molecules [35]; this is about 0.035 erg/cm²/deg. About one-third of the water molecules in the layer adjoining the benzene could participate in hydrogen bonds (because of the smaller area of a water molecule) so orientation of water molecules would contribute, at most, about 0.012 erg/cm²/deg of negative entropy. The rest of the deficit for the liquid-liquid interface vs. the two liquid-vapor interfaces no doubt comes from the diffuse nature of liquid vapor interfaces; the interfacial thickness is about two or three molecular layers [13] for liquids such as benzene and water. The benzene-water interface should be sharper than the liquid-vapor interfaces, and there should be little intermixing of water and benzene molecules beyond the first layer.

The low value of $U_s(\text{rel})$, 45 erg/cm² in Table 2, presents no qualitative problem for its explanation. The explanation formally resembles that given some time ago by

Harkins [36] with respect to surface free energy. The coordination number of a molecule at a liquid-vapor interface is lower than that for a molecule in the bulk. But at a liquid-liquid interface the coordination number is close to that in the bulk, for each component, when unlike species are included. So it is to be expected that, for the interface, U_s(rel) will be lower than U_s for either of the pure liquids.

We may note that the interpretation of the difference between S_s(rel) for the interface and $S_s^{(1)}$ for the pure liquids is, qualitatively, quite different from that of the difference between U_s(rel) and $U_s^{(1)}$. Therefore Harkins' explanation in terms of free energy, referred to above, must be inadequate, even though it formally resembles the explanation in terms of total energy which we have given.

Finally, we may make the computations called for by Eqs. 1.3, to 1.5. The appropriate thermodynamic functions for adhesion (designated by superscript a) and cohesion (designated by superscript c) are given by

$$\Delta G_{12}^a = \gamma_{12} - \gamma_1 - \gamma_2 \,, \tag{4.1}$$

$$\Delta G_\alpha^c = -2\gamma_\alpha \,, \qquad \alpha = 1 \text{ or } 2 \,, \tag{4.2}$$

$$\Delta U_{12}^a = U_{s12}(\text{rel}) - U_s^{(1)}(\text{pure } 1) - U_s^{(2)}(\text{pure } 2) \,, \tag{4.3}$$

$$\Delta U_\alpha^c = -2U_{s\alpha}^{(\alpha)} \,, \qquad \alpha = 1 \text{ or } 2 \,, \tag{4.4}$$

$$\Delta S_{12}^a = S_{s12}(\text{rel}) - S_s^{(1)}(\text{pure } 1) - S_2^{(2)}(\text{pure } 2) \,, \tag{4.5}$$

$$\Delta S_\alpha^c = -2S_{s\alpha}^{(\alpha)} \,, \qquad \alpha = 1 \text{ or } 2 \,. \tag{4.6}$$

Using the data in Tables 1, 2a) and 3, it is found that

$$\Phi = \frac{\Delta G_{BW}^a}{\sqrt{\Delta G_B^c \, \Delta G_W^c}} = 0.77 \,,$$

$$\Phi^* = \frac{\Delta U_{BW}^a}{\sqrt{\Delta U_B^c \, \Delta U_W^c}} = 0.78 \,,$$

$$\Psi = \frac{\Delta S_{BW}^a}{\sqrt{\Delta S_B^c \, \Delta S_W^c}} = 0.86 \,.$$

It is clear that the expectations indicated in Section 1 are fulfilled, for benzene vs. water: Φ is approximately equal to Φ^*. And the value of Ψ_{22} is unequal to Φ_{22}, the difference being well outside the range of experimental error.

Conclusion

The relative interfacial entropy, S_s(rel), and energy, U_s(rel), for benzene vs. water, can easily be interpreted in terms of structural features that are physically quite reasonable. This cannot be said of the interfacial entropy and energy computed on the basis of reference states other than the pure liquids, e.g. based on the Third Law. In this latter case, the entropy and energy associated with the interfacial excess masses, $\Gamma_1^{(b)}$ and $\Gamma_2^{(b)}$, mask those components of interfacial entropy and energy from which structural information can be deduced.

Thus we have produced a physical interpretation of the thermodynamic invariants, $(\partial\gamma/\partial T)_P$ and $[\partial(\gamma/T)/\partial(1/T)]_P$. These quantities give promise of leading to useful conclusions regarding interfacial structure in many liquid-liquid systems, both for cases exhibiting low mutual solubility, and (using the theory presented in Section 2) for systems in which solubilities are appreciable.

Appendix

It is the purpose of the Appendix to derive the working equations described in the text, by a general method. The starting point is the GIBBS-DUHEM equation in intensive form, for the bulk phases I and II, and the GIBBS adsorption equation for the interfacial region, as a system of simultaneous equations:

$$dp = S_v^{\mathrm{I}}\, dT + c_1^{\mathrm{I}}\, d\mu_1 + c_2^{\mathrm{I}}\, d\mu_2\,, \tag{A1}$$

$$dp = S_v^{\mathrm{II}}\, dT + c_1^{\mathrm{II}}\, d\mu_1 + c_2^{\mathrm{II}}\, d\mu_2\,, \tag{A2}$$

$$-d\gamma = S_s^{(b)}\, dT + \Gamma_1^{(b)}\, d\mu_1 + \Gamma_2^{(b)}\, d\mu_2\,. \tag{A3}$$

Upon solving Eqs. (A1) and (A2) for $d\mu_1$ and $d\mu_2$, we obtain

$$\begin{vmatrix} c_1^{\mathrm{I}} & c_2^{\mathrm{I}} \\ c_1^{\mathrm{II}} & c_2^{\mathrm{II}} \end{vmatrix} d\mu_2 = (c_2^{\mathrm{II}} - c_2^{\mathrm{I}})\, dp - \begin{vmatrix} S_v^{\mathrm{I}} & c_2^{\mathrm{I}} \\ S_v^{\mathrm{II}} & c_2^{\mathrm{II}} \end{vmatrix} dT\,, \tag{A4}$$

$$\begin{vmatrix} c_1^{\mathrm{I}} & c_2^{\mathrm{I}} \\ c_1^{\mathrm{II}} & c_2^{\mathrm{II}} \end{vmatrix} \mathrm{d}\mu_2 = (c_1^{\mathrm{I}} - c_1^{\mathrm{II}})\, dp + \begin{vmatrix} S_v^{\mathrm{I}} & c_1^{\mathrm{I}} \\ S_v^{\mathrm{II}} & c_1^{\mathrm{II}} \end{vmatrix} dT\,. \tag{A5}$$

Substituting into Eq. (A3) and rearranging, we obtain

$$d\gamma = \frac{-\begin{vmatrix} S_s^{(b)} & \Gamma_1^{(b)} & \Gamma_2^{(b)} \\ S_v^{\mathrm{I}} & c_1^{\mathrm{I}} & c_2^{\mathrm{I}} \\ S_v^{\mathrm{II}} & c_1^{\mathrm{II}} & c_2^{\mathrm{II}} \end{vmatrix} dT - \begin{vmatrix} \Gamma_1^{(b)} & \Gamma_2^{(b)} \\ (c_1^{\mathrm{II}} - c_1^{\mathrm{I}}) & (c_2^{\mathrm{II}} - c_2^{\mathrm{I}}) \end{vmatrix} dP}{c_1^{\mathrm{I}} c_2^{\mathrm{II}} - c_2^{\mathrm{I}} c_1^{\mathrm{II}}}\,. \tag{A6}$$

We may, without change in the value of the 3×3 determinant, multiply column 2 by $\bar{S}_1^0(\mathrm{ref})$ and column 3 by $\bar{S}_2^0(\mathrm{ref})$, and subtract from column 1. The result is

$$\begin{vmatrix} [S_s^{(b)} - \Gamma_1^{(b)}\bar{S}_1^0(\mathrm{ref}) - \Gamma_2^{(b)}\bar{S}_2^0(\mathrm{ref})] & \Gamma_1^{(b)} & \Gamma_2^{(b)} \\ [S_v^{\mathrm{I}} - c_1^{\mathrm{I}}\bar{S}_1^0(\mathrm{ref}) - c_2^{\mathrm{I}}\bar{S}_2^0(\mathrm{ref})] & c_1^{\mathrm{I}} & c_2^{\mathrm{I}} \\ [S_v^{\mathrm{II}} - c_1^{\mathrm{II}}\bar{S}_1^0(\mathrm{ref}) - c_2^{\mathrm{II}}\bar{S}_2^0(\mathrm{ref})] & c_1^{\mathrm{II}} & c_2^{\mathrm{II}} \end{vmatrix}\,. \tag{A7}$$

We introduce the definitions, Eqs. (2.13) and (2.14),

$$S_s^{(b)}(\mathrm{rel}) \equiv S_s^{(b)} - \Gamma_1^{(b)}\bar{S}_1^0(\mathrm{ref}) - \Gamma_2^{(b)}\bar{S}_2^0(\mathrm{ref})\,, \tag{A8}$$

$$S_v^{\mathrm{I}}(\mathrm{rel}) \equiv S_v^{\mathrm{I}} - c_1^{\mathrm{I}}\bar{S}_1^0(\mathrm{ref}) - c_2^{\mathrm{I}}\bar{S}_2^0(\mathrm{ref})\,, \tag{A9a}$$

$$S_v^{\mathrm{II}}(\mathrm{rel}) \equiv S_v^{\mathrm{II}} - c_1^{\mathrm{II}}\bar{S}_1^0(\mathrm{ref}) - c_2^{\mathrm{II}}\bar{S}_2^0(\mathrm{ref})\,. \tag{A9b}$$

The physical significance of these definitions is discussed in the main text.

On combining Eqs. (A7) to (A9) with (A6), and expanding the second determinant, we obtain

$$d\gamma = \frac{-\begin{vmatrix} S_s^{(b)}(\mathrm{rel}) & \Gamma_1^{(b)} & \Gamma_2^{(b)} \\ S_v^{\mathrm{I}}(\mathrm{rel}) & c_1^{\mathrm{I}} & c_2^{\mathrm{I}} \\ S_v^{\mathrm{II}}(\mathrm{rel}) & c_1^{\mathrm{II}} & c_2^{\mathrm{II}} \end{vmatrix} dT - \begin{vmatrix} \Gamma_1^{(b)} & \Gamma_2^{(b)} \\ c_1^{\mathrm{II}} & c_2^{\mathrm{II}} \end{vmatrix} dP + \begin{vmatrix} \Gamma_1^{(b)} & \Gamma_2^{(b)} \\ c_2^{\mathrm{I}} & c_1^{\mathrm{I}} \end{vmatrix} dP}{c_1^{\mathrm{I}}c_2^{\mathrm{II}} - c_2^{\mathrm{I}}c_1^{\mathrm{II}}}\,. \tag{A10}$$

Eq. (A10) may be expanded, and the expressions given in Eqs. 2.15 to 2.17 for the concentration variables employed, with the result:

$$d\gamma = -\left[S_s^{(b)}(\mathrm{rel}) - \frac{\Gamma_{\mathrm{I}}^{(b)}}{c_{\mathrm{I}}} S_v^{\mathrm{I}}(\mathrm{rel}) - \frac{\Gamma_{\mathrm{II}}^{(b)}}{c_{\mathrm{II}}} S_v^{\mathrm{II}}(\mathrm{rel})\right] dT - \left[\frac{\Gamma_{\mathrm{I}}^{(b)}}{c_{\mathrm{I}}} + \frac{\Gamma_2^{(b)}}{c_{\mathrm{II}}}\right] dP\,. \tag{A11}$$

When the entropy terms in Eq. (A1) to (A3) are eliminated with use of the respective intensive free energy properties for the bulk phases and the interfacial region (see Section 2), the result may be put in the form

$$\frac{1}{T}\,dp = -H_v^{\mathrm{I}}\,d\left(\frac{1}{T}\right) + c_1^{\mathrm{I}}\,d\left(\frac{\mu_1}{T}\right) + c_2^{\mathrm{I}}\,d\left(\frac{\mu_2}{T}\right), \tag{A12}$$

$$\frac{1}{T}\,dp = -H_v^{\mathrm{II}}\,d\left(\frac{1}{T}\right) + c_1^{\mathrm{II}}\,d\left(\frac{\mu_1}{T}\right) + c_2^{\mathrm{II}}\,d\left(\frac{\mu_2}{T}\right), \tag{A13}$$

$$-d\left(\frac{\gamma}{T}\right) = -U_s^{(b)}\,d\left(\frac{1}{T}\right) + \Gamma_1^{(b)}\,d\left(\frac{\mu_1}{T}\right) + \Gamma_2^{(b)}\,d\left(\frac{\mu_2}{T}\right). \tag{A14}$$

(As mentioned in Section 2, the excess interfacial energy $U_s^{(b)}$, for a planar interface, is equal to $H_s^{(b)}$, the excess interfacial enthalpy.)

Comparison of Eqs. (A12) to (A14) with Eqs. (A1) to (A3), yields, by simple matching of corresponding terms, the analogue of Eq. (A11):

$$d\left(\frac{\gamma}{T}\right) = \left[U_s^{(b)}(\mathrm{rel}) - \frac{\Gamma_1^{(b)}}{c_{\mathrm{I}}} H_v^{\mathrm{I}}(\mathrm{rel}) - \frac{\Gamma_{\mathrm{II}}^{(b)}}{c_{\mathrm{II}}} H_v^{\mathrm{II}}(\mathrm{rel})\right] d\left(\frac{1}{T}\right) - \left[\frac{\Gamma_1^{(b)}}{c_{\mathrm{II}}} + \frac{\Gamma_{\mathrm{II}}^{(b)}}{c_{\mathrm{II}}}\right]\frac{dP}{T}\,. \tag{A15}$$

References

[1] J. W.GIBBS, "*Collected Works*", vol. I, pp. 55—353 and pp. 354—371, Longmans, Green and Co., London, 1906; Dover, N.Y. (1961)
[2] C. F. GAUSS, "*Comment. Soc. Reg. Sci.*", Göttingen, 7, (1930)
[3] J. W. GIBBS, "*Collected Works*", vol. II, Longmans, Green and Co., London, (1906); Yale Univ. Press, New Haven, 1948
[4] R. CLAUSIUS, Pogg. Ann. **75**, 400 (1865)
[5] T. YOUNG, "*Miscellaneous Work*", vol. I, p. 365, J. Murray, London, (1855)
[6] F. P. BUFF, R. A. LOVETT and F. H. STILLINGER, jr., Phys. Rev. Letters, **15**, 621 (1965)
[7] F. P. BUFF, "*Abstracts of 45th National Colloid Symposium*", Atlanta, Georgia, (1971) p. 2
[8] G. H. GILMER, W. GILMORE, J. HUANG and W. W. WEBB, Phys. Rev. Letters, **14**, 491 (1965)
[9] J. C. MAXWELL, "*Capillary Action*", *Encyclopedia Britannica*, 9th ed., A. and C. Black, Edinburgh (1875)
[10] J. D. VAN DER WAALS, "*Vers. en Meded Afd. Natuurk.*", 3rd series, Part V, p.171 (1899)
[11] J. D. VAN DER WAALS, Z. phys. Chem., **13**, 657 (1894)
[12] D. G. TRIEZENBERG and R. ZWANZIG, Phys. Rev. Letters **28**, 1183 (1972)
[13] R. A. LOVETT, P. W. DE HAVEN, J. J. VIECELI and F. P. BUFF, J. Chem. Phys. **58**, 1880 (1973)
[14] J. W. GIBBS, Ref. [1], p. 365
[15] F. P. BUFF, J. Chem. Phys. **23**, 419 (1955)
[16] R. J. GOOD, Pure Appl. Chem., **48**, 427 (1976)
[17] E. A. GUGGENHEIM and N. K. ADAM, Proc. Roy Soc. (London) **A139**, 218 (1933)
[18] F. P. BUFF, 38th National Colloid Symposium, Austin, Texas, 1964
[19] R. J. GOOD (in press)
[20] F. P. BUFF, J. Chem. Phys., **19**, 1591 (1951)
[21] F. P. BUFF and R. A. LOVETT, in "*Simple, Dense Fluids*", H. L. FRISCH and Z. W. SALTZBURG, eds., Academic Press, N.Y., 1968, p. 22
[22] F. P. BUFF, in "*Handbuch der Physik*", S. FLUGGE, ed., v. ol. X, Springer, Berlin (1960)
[23] A. B. PIPPARD, "*Elements of Classical Thermodynamics*", Cambridge University Press, 1957, p. 87
[24] R. J. GOOD and L. A. GIRIFALCO, J. Phys. Chem. **61**, 904 (1957)
[25] R. J. GOOD and E. ELBING, Ind. Eng. Chem. **62**, 54 (1970)
[26] R. J. GOOD, J. Coll. Interface Sci., **59**, 398 (1977)
[27] (a) R. J. GOOD and A. A. WOSILAIT, unpublished results, M. S. Thesis of A. A. WOSILAIT, State University of New York at Buffalo, 1967
(b) R. J. GOOD and C. S. PIETRAS, unpublished results, M. S. Thesis of C. S. PIETRAS, State University of New York at Buffalo, 1969
[28] A. S. MICHAELS and E. A. HAUSER, J. Phys. Chem., **55**, 408 (1951)
[29] M. E. HASSAN, R. F. NIELSEN, and J. C. CALHOUN, Trans. AIME, **192**, 57 (1951)
[30] R. R. HARVEY, J. Phys. Chem. **62**, 322 (1958)
[31] H. Y. JENNINGS, J. Coll. Interface Sci., **24**, 323 (1967)

[32] W. D. Harkins and F. E. Brown, J. Am. Chem. Soc. **41**, 499 (1919)
[33] Rossini et al., API Project 44, U.S. National Bureau of Standards Circular 461, Washington, D.C. (1947)
[34] Rossini et al., U.S. National Bureau of Standards Circular 500, Washington, D.C. (1952)
[35] R. J. Good, J. Phys. Chem. **61**, 810 (1957)
[36] W. D. Harkins, "*The Physical Chemistry of Surface Films*", Reinhold, N.Y., 1952

Capillary phenomena and information on porous structure of adsorbens

M. M. Dubinin

(Institute of Physical Chemistry of the Academy of Sciences of the USSR, Moscow/USSR)

Abstract

The foundations of the theoretical concepts of capillary evaporation from the mesopores of porous bodies are considered. The flaw in the classical treatment based on the Kelvin equation consisted in neglecting the effect of the adsorption field, induced by the mesopore walls, on the thicknesses of the adsorption layers and the curvature of the concave menisci of the liquid in the mesopores. The fundamentals of the Derjaguin-de Boer-Broeckhoff theory are presented; the theory makes an approximate allowance for the effect of the adsorption field on the equilibrium capillary evaporation. But it contains a simplifying assumption about the constancy of the surface tension of a condensed sorbate. The author makes an attempt at approximate determination of the effect of the curvature of the liquid menisci on the surface tension.

The above-mentioned three theoretical versions are applicable for calculating the volume distribution of the mesopores of an aluminosilicate catalyst, a large-pore silica gel, and microporous carbonaceous adsorbents, from experimental data on capillary evaporation of nitrogen at 78 K and benzene at 293 K. On transition from the classical method to that of de Boer and Broeckhoff and its improved version the distribution curves shift towards the larger radii. These shifts are rather substantial and correspond to considerable changes in the nature of porosity, which cannot be neglected even in a crude approximation.

Investigations into capillary phenomena, such as capillary condensation of vapours and mercury penetration into the pores are widely used for characterizing or describing the structure of adsorbent and catalyst pores. The fields of application of these methods depend on the varieties of the adsorbent pores. According to the classification based on the mechanisms of adsorption and capillary phenomena taking place in pores, the adsorbent pores are divided into the following varieties: micropores $r < 6-7$ Å, supermicropores $6-7 < r < 15-16$ Å, mesopores $15-16 < r < 1000-2000$ Å, and macropores $r < 1000 - 2000$ Å. The symbol r denotes the equivalent pore radii, which are equal to the double ratios of the areas of the normal pore sections to their perimeters [1, 2].

The finest absorbent pores — micropores — are commensurate with the molecules adsorbed. Physical adsorption in micropores is characterized by volume filling of the adsorption space, i.e. of the adsorption force field. Adsorption in the mesopores amounts to the formation of successive adsorption layers on the pore surface and terminates with their filling by the capillary condensation mechanism.

The supermicropores constitute the intermediate range of porosity between the micro- and mesopores. Within this range the micropore properties gradually degenerate, and the mesopore properties increasingly manifest themselves. In pores with equivalent radii exceeding 1000—2000 Å their filling by the capillary condensation mechanism practically cannot be realized. Therefore larger pores belong to the macropore variety. Thus, capillary condensation and capillary evaporation can take place only in mesopores, i.e. in the range of equivalent pore radii from 15—16 to 1000—2000 Å.

Calculation of the volume and surface distribution of the adsorbent mesopores according to the sizes is based on two sets of equations. The first set expresses the thickness of the adsorption layer on the surface of mesopores within the size range under consideration and the dependence of the equilibrium vapour pressure on the average radius of curvature of the liquid menisci in the mesopores. The second set of equations relates the reduction in sorption volume as a result of evaporation of a certain amount of liquid from the mesopores to the corresponding changes in the volume and surface of the mesopores proper.

Both sets of equations require an assumption as to the geometrical shape of the pores. As a result, one calculates the distribution functions of the mesopore volume and surface for the equivalent model of an adsorbent with the adopted pore shape, rather than for the real adsorbent with its complex porous structure [3, 4]. In this approach the real adsorbent and its equivalent model are characterized by coinciding capillary evaporation isotherms.

The above assumptions introduce some conventionality and render an approximate character to the following theoretical conceptions of capillary evaporation and of methods for calculating the volume and surface distribution of the adsorbent mesopores.

Fundamental theoretical conceptions of capillary evaporation from mesopores

As distinct from a convex liquid surface, which is readily realized in aerosols in the shape of spherical particles of different sizes, a concave liquid surface usually cannot be obtained without the presence of solid walls. Therefore, in the general case the adsorption field created by the walls of the adsorbent mesopores affects the curvature of the equilibrium concave liquid meniscus as well as the thickness of the adsorption layer. The theory of this phenomenon was published by DERJAGUIN in 1940; almost 30 years later, in 1967, DE BOER and BROEKHOFF approximately calculated the effect of the adsorption field of the pore walls on the chemical potential of the sorbed substance when deriving an improved Kelvin equation [5—8]. The present author made an attempt at approximate development of the DERJAGUIN-BROEKHOFF-DE BOER method by additional inclusion of the dependence of the surface tension on the average curvature radius of the liquid meniscus [9, 10]. Capillary evaporation is considered for an equivalent model adsorbent with cylindrical pores.

1. The most commonly applied classical KELVIN equation is based on neglecting the effect of the adsorption field in the pore both on the thickness of the adsorption layer and on the curvature of the spherical liquid meniscus formed in the space between the adsorption layers. Therefore the average statistical thickness of the adsorption layer t at a constant temperature is a function only of the equilibrium relative pressure p/p_s and is independent of the mesopore surface curvature. In this case the thickness of the adsorption layer can be expressed by the equation of the standard t-curve,

i.e. actually by the equation of the isotherm of adsorption on the surface of a non-porous adsorbent of identical chemical nature. This equation is usually taken in the form of the modified HARKINS and JURA formula

$$\lg (p_s/p) = \frac{C}{t^2} - D(t), \tag{1}$$

where C is a constant parameter, and the correction term $D(t)$ is either a constant, or a function weakly dependent on t, according to the nature of the adsorption system. In calculating the adsorption layer thickness from experiments on adsorption on non-porous adsorbents, the adsorbate density is assumed to be equal to that of the corresponding bulk liquid phase at the temperature under consideration.

The KELVIN equation for the cylindrical pore model is expressed by Eq. (2)

$$r - t = \frac{2\sigma v}{RT \ln (p_s/p)}, \tag{2}$$

where r is the radius of the cylindrical pore, σ is the surface tension, and v is the molar volume of the liquid.

2. BROEKHOFF and DE BOER made an attempt to take into account the effect of adsorption on capillary evaporation from cylindrical pores. In this case the chemical potential of the adsorption phase μ_a is determined by the distance t of the adsorbate-vapour interface from the pore walls. Neglecting the dependence of the potential on the geometrical shape of the pores, the authors obtained Eq. (3)

$$\mu_a - \mu_L = F(t), \tag{3}$$

where μ_L is the chemical potential of the bulk liquid, and $F(t)$ is the function

$$F(t) = 2.303RT\left[\frac{C}{t^2} - D(t)\right], \tag{4}$$

which can be determined from the equation of the t-curve (1). Note that the use of the t-curve to take into account adsorption interactions opens up the way for practical application of the method described.

Let us consider the equilibrium between the adsorption layer in a cylindrical pore of radius r and the gas phase. From the condition that the increments of GIBBS' free energy during transfer of dN moles of substance from the gas to the adsorption phase is equal to zero follows Eq. (5)

$$RT \ln (p_s/p) - F(t) = \frac{\sigma v}{r - t} \quad \text{for} \quad t = t_e. \tag{5}$$

It contains the adsorption layer thickness t_e changed under the effect of the adsorption field. From Eqs. (4) and (5) we obtain the desired equation (6), in which the adtorption layer thickness t_e is a function of the equilibrium relative pressure p/p_s and of the radius r of the cylindrical pore[1]:

$$\lg (p_s/p) - \frac{C}{t_e^2} + D(t_e) - \frac{\sigma v}{2.303RT(r - t_e)} = 0. \tag{6}$$

[1] We use logarithms to the base 10 to correspond with the numerical values of the constants published in [7, 8].

A somewhat more complicated thermodynamic consideration of equilibrium between the liquid meniscus and the gas phase in a homogeneous cylindrical pore leads to Eq. (7)

$$r - t_e = \frac{2\sigma v}{RT \ln (p_s/p)} + \frac{2}{RT \ln (p_s/p)} \int_{t_e}^{r} \frac{r - t}{r - t_e} F(t) \, dt \,. \tag{7}$$

The two equations (6) and (7) obtained contain the three interrelated variables p/p_s, r, and t_e. In solving these equations numerically with the aid of an electronic computer one can, for instance, for a given equilibrium relative pressure, determine the adsorption layer thickness t_e and the pore radius r at which the pore will be emptied as a result of capillary evaporation.

If we restrict ourselves to the first term in Eq. (7), we obtain the KELVIN equation, i.e. the solution of the classical problem, and therefore the second term of Eq. (7) in the approximation considered plays the part of a correction term. Note that the integrand includes the function $F(t)$, which is determined, according to Eq. (4), only by the parameters of the equation of the t-curve, i.e. of the adsorption isotherm of the same vapour on a nonporous adsorbent of identical chemical nature.

3. BROEKHOFF and DE BOER's conceptions discussed here contained a simplifying assumption of the constancy of the surface tension of the adsorbate. In actuality it essentially depends on the curvature radius of the liquid meniscus. On the basis of a quasithermodynamic approach TOLMAN calculated the dependence of the surface tension on the interfacial curvature [11]. Later, AHN and others, proceeding from the statistical theory of liquids, derived the TOLMAN equation and found concrete values of the coefficient appearing in it for a number of liquids [12]. As applied to the problem at hand, the TOLMAN equations have the following form:

$$\sigma = \sigma_0 \bigg/ \left(1 - \frac{2\omega}{\varrho}\right), \tag{8}$$

$$\sigma = \sigma_0 \bigg/ \left(1 - \frac{\omega}{\varrho}\right), \tag{9}$$

$$\varrho = r - t_e \,. \tag{10}$$

Equation (8) corresponds to a concave spherical meniscus, and Eq. (9) to a concave cylindrical meniscus, where ϱ is the absolute value of the curvature radius. The symbol σ_0 denotes the surface tension for a plane liquid surface.

To make an additional approximate allowance for the effect of the curvature of the liquid menisci on the surface tension we assume, in accordance with Eq. (10), that the average effective radius of the meniscus curvature in a cylindrical pore ϱ is equal to the difference between the pore radius r and the adsorption layer thickness t_e. From the general equations (6) and (7) obtained by BROEKHOFF and DE BOER it follows that the surface tension does not appear in the integrand. Therefore taking into account the change in σ amounts to substitution of Eqs. (9) and (10) into (6) and (7). The equations obtained are represented by the expressions (11) and (12)

$$\lg (p_s/p) - \frac{C}{t_e^2} + D(t) - \frac{\sigma_0 v}{2.30RT(r - t_e - \omega)} = 0\,, \tag{11}$$

$$r - t_e = \frac{2\sigma_0 v(r - t_e)}{RT(r - t_e - 2\omega)\ln (p_s/p)} + \frac{2}{RT\ln (p_s/p)} \int_{t_e}^{r} \frac{r - t}{r - t_e} F(t)\, dt\,. \tag{12}$$

They make it possible, in principle, to calculate the adsorption layer thickness t_e and the radii of the emptied pores r of the equivalent adsorbent for each experimental point of the capillary evaporation or desorption isotherm from the values of equilibrium relative pressures.

Basic patterns for calculating volume and surface distribution of adsorbent mesopores

The first approximate pattern for calculating the distribution of the pore volume according to the sizes from the capillary evaporation isotherm, based on the KELVIN equation, was published by WHEELER in 1945 [13]. Since then many versions of the methods have been proposed, mostly for adsorbents with cylindrical pores. DE BOER's school has made an attempt to take into account the pore geometry on the basis of the nature of the hysteresis region in capillary condensation [14]. This would allow one, in principle, to adopt in each particular case a porous-structure model nearest to the real structure. But this did not yield an unambiguous solution of the problem. Calculations for adsorbents with cylindrical pores are used most widely. From the above-described standpoint of the author, in such cases the volume and surface distribution of the pores is calculated for the equivalent model of the adsorbent with the adopted pore shape, rather than for the real adsorbent.

Since there is no universal mathematical method for describing the pore volume distribution according to the sizes suitable for adsorbents of different nature, it is customary to use the method of stepwise analysis of capillary evaporation. For each step, i.e. for a small reduction in equilibrium relative pressure, the volume of the evaporating substance consists of its own capillary evaporation, i.e. the emptying of a certain volume of the sorption space of the mesopores enclosed between the adsorption layers, and the reduction of the adsorption layer thicknesses in the pores emptied in the preceding stages of desorption.

The most perfect and hence most complicated method for calculating the volume and surface distribution of the pores for an adsorbent with cylindrical pores was proposed by DOLLIMORE and HEAL in 1964 [15]. An example of an approximate technique, which is 2—2.5 times less laborious, computationally, is the authors' method for a similar adsorbent model published in 1956 [3, 4, 9, 10].

We first consider the last method for a slightly conical pore as an illustration. Suppose an amount of vapour ΔV_n, expressed in terms of the volume of a bulk liquid and shaded in the scheme of Fig. 1, is evaporated in the nth stage of desorption. To calculate the volume of the pore proper ΔV_n^*, corresponding to ΔV_n and enclosed between the liquid menisci and the adsorption layers at the beginning and at the end

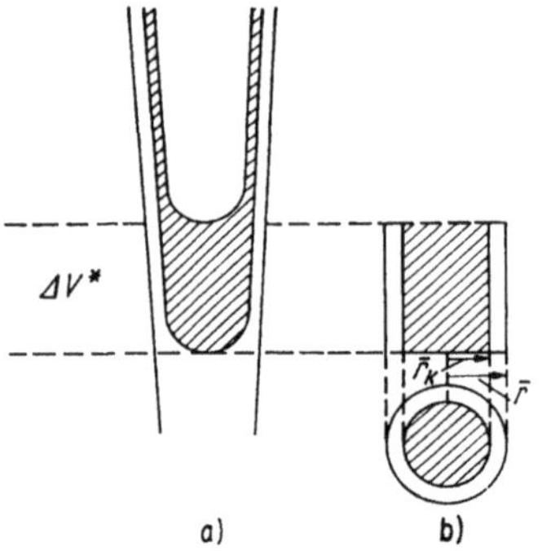

Fig. 1: Scheme of capillary evaporation.

of the stage under review, it is necessary to add to ΔV_n the volume of the adsorption layer and deduct the volume of the vapour desorbed due to the reduction in adsorption layer thicknesses in the previously emptied pores. To a first approximation, ΔV_n^* can be expressed by Eq. (13) in accordance with the above.[1]

$$\Delta V_n^* = \Delta V_n + \bar{t}_n \cdot 2\Delta V_n/(\bar{r}_n - \bar{t}_n) - \Delta t_n \sum_{i=1}^{n-1} 2\Delta V_i/(\bar{r}_i - \bar{t}_i) , \qquad (13)$$

$$\Delta V^*/\Delta r = f(\bar{r}) , \qquad (14)$$

$$V^* = \sum_{i=1}^{n} \Delta V_i^* , \qquad (15)$$

$$S^* = \sum_{i=1}^{n} \Delta S_i^* . \qquad (16)$$

The obtained values of ΔV^* are used for calculating the differential pore volumes $\Delta V^*/\Delta r$ and plotting a distribution curve of the form (14). By summing the values of ΔV^* and the corresponding pore surfaces ΔS^* it is possible to obtain the total mesopore volume V^* and the specific surface area S^* of the equivalent model adsorbent by Eqs. (15) and (16). A more rigorous and accurate calculation of ΔV_n^* for ideal cylindrical pores after DOLLIMORE and HEAL expressed by Eq. (17) is performed similarly, in principle.

$$\Delta V_n^* = \left[\Delta V_n - \Delta t_n \sum_{i=1}^{n-1} \frac{2\Delta V_i^*}{\bar{r}_i} + \Delta t_n \bar{t}_n \sum_{i=1}^{n-1} \frac{2\Delta V_i^*}{\bar{r}_i^2}\right] \left(\frac{\bar{r}_n}{\bar{r}_n - \bar{t}_n + \Delta t_n}\right)^2 . \qquad (17)$$

Experimental analysis of methods for calculating volume and surface distribution of adsorbent pores

The fundamental difference between the above methods for calculating the volume and surface distribution of adsorbent pores according to the sizes lies in their theoretical basis. The use of the more accurate method of DOLLIMORE and HEAL [15] or the approximate method of DUBININ [9, 10] for computing the distribution proper may only result in some divergency between the numerical values due to the adopted

[1] For simplicity we drop e from the symbols t_e.

simplifications. Therefore we will conventionally denote by method I calculations based on the classical Kelvin equation, by method II calculation after Derjaguin, Broekhoff and de Boer, and by method III the further development of method II by including the dependence of the surface tension on the curvature of the liquid meniscus.

A number of sorption systems, for which capillary evaporation was studied in detail, were used for theoretical analysis. These are low-temperature evaporation of nitrogen from an aluminosilicate catalyst [16], of benzene at 25 °C from a large-pore silica gel KCK-2 according to Isirikyan's experiments [17], and of nitrogen at −195 °C and benzene at 20 °C from three speciments of active carbons with developed mesoporosity according to investigations of Dubinin and Zhukovskaya [18]. Table 1 lists numerical values of the parameters appearing in the above-considered equations for adsorption layer thicknesses and equilibria of the liquid menisci with the gas phase.

Table 1: Parameters of Eqs. (1) and (4) for adsorption systems under consideration

Adsorbents	Vapour	C	$D(t)$	Range of applicability p/p_s	ω
Mineral and carbonaceous	N_2, 78 K	16.11	$0.1682 \exp[-0.1137 \cdot t]$	0.45–0.95	3.31
Mineral and carbonaceous		13.99	0.034	0.08–0.45	
Silica gel	C_6H_6, 293 K	5.924	−0.0156	0.20–0.92	4.64
Carbonaceous		22.23	−0.0035	0.08–0.75	

According to van Aken's experiments the t-curves for low-temperature adsorption of nitrogen practically do not differ for aluminosilicate and carbonaceous adsorbents [19]. We learned this from our own experiments. Therefore Broekhoff and de Boer's t-curve equations were adopted for these adsorbents. The t-curve equation for benzene on silica gel was obtained by us from Isirikyan's experiments with nonporous quartz [17]. For adsorption of benzene on active carbons we used the van Aken equation (see Eq. (1) and Table 1).

Let us consider the results of experiments on capillary evaporation of nitrogen from the mesopores of the aluminosilicate catalyst studied by Oulton [16]. As noted above, with the aid of an electronic computer it is possible to calculate with the chosen method the adsorption layer thicknesses r_e and the radii of the emptied pores r for each experimental point of the desorption isotherm. But then methods II and III become nearly inaccessible. Our own experience indicates the expedience of preliminarily computerized calculation of the table of initial data, i.e. the adsorption layer thicknesses, and the radii of the emptied pores for the entire range of equilibrium relative pressures, which change by small values, for instance. 0.02.

The initial data for the adsorption system at hand will be found in Table 2. They have been calculated for all the three methods from the above equations, whose parameters and ranges of applicability are indicated in Table 1. After calculating the integral in the second term of Eq. (7) the computational formulae (18)—(28) acquire the final form. They are given in the Appendix. In these formulae the adsorption layer thicknesses and the pore radii are expressed in Å. The lower boundary of calculations by all the methods corresponded to a pore radius of about 16 Å, i.e. to the lower boundary of the mesopores.

Table 2: Initial data of capillary evaporation of nitrogen at 78 K from mesopores of mineral and carbonaceous adsorbents

p/p_s	Method I		Method II		Method III	
	r, Å	t, Å	r, Å	t_e, Å	r, Å	t_e, Å
0.98	500.04	38.45	565.94	56.32	572.06	56.17
0.96	251.79	23.34	297.53	34.98	303.49	34.77
0.94	169.10	18.38	203.47	25.52	209.35	25.33
0.92	127.73	15.88	154.91	20.99	160.76	20.82
0.90	102.80	14.29	125.20	18.41	131.06	18.25
0.88	86.09	13.14	105.13	16.69	110.99	16.52
0.86	74.09	12.26	90.65	15.43	96.52	15.26
0.84	65.02	11.54	79.70	14.44	85.58	14.27
0.82	57.93	10.94	71.12	13.64	77.02	13.45
0.80	52.21	10.42	64.22	12.96	70.11	12.77
0.78	47.50	9.97	58.53	12.37	64.43	12.18
0.76	43.55	9.57	53.77	11.86	59.67	11.66
0.74	40.18	9.21	49.72	11.40	55.61	11.19
0.72	37.28	8.89	46.22	10.99	52.11	10.77
0.70	34.74	8.59	43.18	10.61	49.05	10.39
0.68	32.50	8.32	40.50	10.27	46.36	10.04
0.66	31.08	8.14	38.80	10.04	44.66	9.81
0.64	28.72	7.83	35.98	9.65	41.83	9.41
0.62	27.11	7.61	34.05	9.37	39.89	9.13
0.60	25.65	7.40	32.31	9.11	38.14	8.86
0.58	24.32	7.20	30.72	8.86	36.53	8.61
0.56	23.10	7.01	29.25	8.62	35.06	8.37
0.54	21.97	6.83	27.91	8.40	33.70	8.14
0.52	20.92	6.66	26.66	8.18	32.44	7.92
0.50	19.95	6.50	25.50	7.97	31.27	7.71
0.48	19.05	6.34	24.41	7.77	30.17	7.51
0.46	18.20	6.19	23.20	7.61	28.89	7.34
0.44	17.40	6.04	22.12	7.44	27.74	7.15
0.42	16.65	5.90	21.28	7.25	26.89	6.96
0.40	15.94	5.76	20.47	7.06	26.08	6.77
0.38	—	—	19.69	6.87	25.30	6.58
0.36	—	—	18.95	6.69	24.56	6.40
0.34	—	—	18.23	6.51	23.84	6.23
0.32	—	—	17.54	6.34	23.14	6.05

Table 2 (Continued)

p/p_s	Method I		Method II		Method III	
	r, Å	t, Å	r, Å	t_e, Å	r, Å	t_e, Å
0.30	—	—	16.86	6.17	22.47	5.88
0.28	—	—	16.21	6.00	21.81	5.71
0.26	—	—	15.57	5.83	21.17	5.55
0.24	—	—	14.94	—	20.55	5.38
0.22	—	—	14.32	—	19.93	5.22
0.20	—	—	—	—	19.32	5.05
0.18	—	—	—	—	18.72	4.88
0.16	—	—	—	—	18.12	4.71
0.14	—	—	—	—	17.52	4.54
0.12	—	—	—	—	16.90	4.36
0.10	—	—	—	—	16.27	4.17
0.08	—	—	—	—	15.60	3.97
0.06	—	—	—	—	14.88	3.75

Such a table makes it possible to determine the layer thicknesses and radii of the emptied pores, necessary for calculating the pore volume distribution according to the sizes, for each experimental point, i.e. the value of p/p_s, with a sufficient accuracy by the linear interpolation method. It follows from our experience, however, that it is more useful to plot, on a sufficiently large scale, the graph of the desorption branch of the isotherm and determine from it, by interpolation in the tabulated initial data of the relative pressures, the values of the adsorption volumes V, for which the adsorption layer thicknesses and the pore radii have already been calculated beforehand for each of the three methods in question. Besides, this procedure enables one to smooth out accidental deviations of the experimental points.

If, for calculating by the first method, the point of beginning of hysteresis in capillary condensation is associated with a pore radius of about 16 Å, then for the second, and particularly for the third method these values of the radius are reached at relative pressures of 0.25 and, respectively, 0.10, which already correspond to the reversible part of the desorption curve. This is not unusual, since the point of beginning of the hysteresis indicates only the appearance of irreversible capillary condensation, and not its beginning.

Knowing the adsorption volumes for different relative pressures with the corresponding values of the adsorption layer thicknesses and pore radii, we have at our disposal all the necessary data for computing the mesopore volume distribution according to the sizes, their total volume, and the specific surface area of the mesopores. The graph of Fig. 2 compares the differential curves for mesopore volume distribution according to the sizes calculated after Dollimore and Heal for the three methods under study for equivalent model adsorbents with cylindrical pores. It follows from the graph that inclusion of the adsorption effect on capillary evaporation results in a shift of distribution curve II toward the higher radii. Additional inclusion of the change in surface tension with the menisci curvature is manifested in further shifts of curve III in the same direction. The positions of the maxima on the curves determined by the average

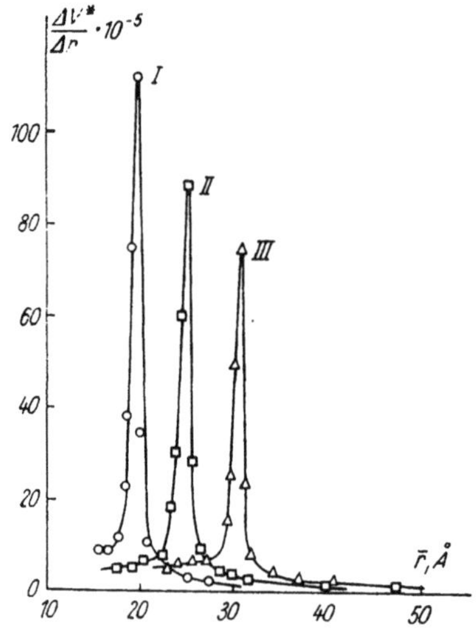

Fig. 2: Mesopore volume distribution according to size for alumosilicate catalyst.

values of the radii $\bar{r}$, are, respectively, 19.7 Å for curve I, computed from the Kelvin equation, 25.2 Å for curve II obtained after DERJAGUIN, BROEKHOFF and DE BOER, and 30.9 Å for curve III, where the approximate effect of the meniscus curvature on the surface tension is included. These shifts are extremely important and result in considerable changes in the porosity characteristics, which cannot be neglected even to a crude approximation.

Table 3 compares the results of computations of the differential mesopore volumes for different radii computed by method III using the DOLLIMORE-HEAL method and

Table 3: Comparison of various methods for calculating differential volumes of mesopores

$\bar{r}$ Å	$(\Delta V^*/\Delta r) \cdot 10^{-5}$ cm³/g		$\bar{r}$ Å	$(\Delta V^*/\Delta r) \cdot 10^{-5}$ cm³/g	
	DOLLIMORE HEAL	DUBININ		DOLLIMORE HEAL	DUBININ
154	0.10	0.10	41.7	2.64	3.05
108	0.68	0.68	35.2	2.45	2.94
97.9	1.60	1.60	30.9	3.23	3.62
93.7	3.14	3.14	27.3	6.47	6.49
90.0	13.0	13.0	25.8	7.04	7.05
84.4	20.3	20.2	24.8	7.21	7.22
81.1	29.1	29.1	23.8	7.63	7.69
78.1	39.0	38.9	22.8	7.90	7.96
76.1	28.5	38.4	21.8	8.11	8.20
73.9	36.1	36.1	19.1	8.84	9.00
69.8	27.1	27.4	17.8	8.26	8.36
64.1	15.4	15.7	16.9	7.57	7.66
55.7	7.38	7.93			
47.2	3.42	3.80			

DUBININ's approximate method. It should be noted that the results obtained by the two methods practically coincide.

Table 4 gives the cumulative characteristics of the mesoporosity of the aluminosilicate catalyst, i.e. the mesopore volume V^* and the specific surface area S^* computed by the three methods in question. Due to the fact that the aluminosilicate catalyst contains micropores, the specific surface area of 339 m²/g determined by the BET method from low-temperature adsorption of nitrogen is a formal value, which considerably exceeds the mesopore specific surface area of about 200 m²/g estimated from the capillary evaporation of nitrogen for the equivalent model adsorbent with cylindrical pores.

Table 4: Aluminosilicate catalyst, Nitrogen, 78 K

Method	Calculation up to:					
	Point of beginning of hysteresis $p/p_s \approx 0.45$			Pore radius $r \approx 16$ Å		
	r, Å	V^*, cm³/g	S^*, m²/g	p/p_s	V^*, cm³/g	S^*, m²/g
I	15.2	0.245	218	0.39	0.245	218
II	22.7	0.220	148	0.25	0.264	195
III	28.4	0.192	105	0.10	0.271	180

Note that calculations of the mesopore volume distribution according to the sizes, proper, the total mesopore volume, and the specific surface area of the equivalent model adsorbent with cylindrical pores both by the exact method of DOLLIMORE and HEAL and by DUBININ's approximate method lead to practically identical results. In computerized calculations the choice of the method does not matter, as in our case. But in computing with the aid of an electronic calculator our method is much less laborious. In the following text we will use the DOLLIMORE and HEAL method, since the results obtained by us do not involve manual computation.

Let us consider the results of experiments on capillary evaporation of benzene from the mesopores of a large-pore silica gel KCK-2, which contains practically no micropores. Its specific surface area, as determined from low-temperature adsorption of nitrogen by the BET method, was 320 m²/g. Since this value may be considered physically real, its comparison with the specific surface areas of equivalent model adsorbents with cylindrical pores, computed by the three methods used was of fundamental importance. The initial data for computations are given in Table 5.

Table 5: Initial data of capillary evaporation of benzene at 293 K from mesopores of silicate adsorbents

p/p_s	Method I		Method II		Method III	
	r, Å	t, Å	r, Å	t_e, Å	r, Å	t_e, Å
0.94	363.24	22.96	871.74	31.88	880.73	31.8
0.92	269.48	16.97	487.73	23.46	496.65	23.36
0.90	213.86	14.02	339.28	19.32	348.14	19.20

Table 5 (Continued)

p/p_s	Method I		Method II		Method III	
	r Å	t Å	r Å	t_e Å	r Å	t_e Å
0.88	176.89	12.19	260.15	16.74	268.97	16.61
0.86	150.50	10.90	210.82	14.93	219.61	14.79
0.84	130.69	9.93	177.06	13.57	185.81	13.42
0.82	115.26	9.16	152.44	12.50	161.17	12.34
0.80	102.89	8.54	133.68	11.62	142.38	11.46
0.78	92.75	8.01	118.88	10.89	127.56	10.72
0.76	84.28	7.56	106.89	10.26	115.55	10.08
0.74	77.10	7.17	96.97	9.71	105.62	9.53
0.72	70.92	6.83	88.62	9.23	97.25	9.05
0.70	65.55	6.52	81.49	8.80	90.10	8.62
0.68	60.84	6.25	75.32	8.42	83.91	8.23
0.66	57.87	6.07	71.47	8.17	80.06	7.97
0.64	52.94	5.77	65.16	7.75	73.73	7.54
0.62	49.60	5.55	60.92	7.46	69.48	7.25
0.60	46.58	5.36	57.12	7.19	65.67	6.98
0.58	43.83	5.18	53.69	6.93	62.23	6.72
0.56	41.32	5.01	50.58	6.70	59.10	6.48
0.54	39.02	4.85	47.74	6.48	56.25	6.26
0.52	36.90	4.70	45.13	6.27	53.64	6.05
0.50	34.93	4.56	42.73	6.07	51.22	5.85
0.48	33.11	4.42	40.50	5.88	48.99	5.66
0.46	31.41	4.29	38.44	5.71	46.92	5.48
0.44	29.81	4.17	36.51	5.53	44.98	5.31
0.42	28.32	4.05	34.71	5.37	43.17	5.14
0.40	26.91	3.94	33.01	5.21	41.47	4.98
0.38	25.59	3.83	31.41	5.06	39.86	4.83
0.36	24.33	3.72	29.90	4.92	38.35	4.68
0.34	23.13	3.62	28.47	4.77	36.91	4.53
0.32	21.99	3.52	27.10	4.63	35.53	4.39
0.30	20.91	3.42	25.80	4.50	34.22	4.26
0.28	19.86	3.32	24.55	4.37	32.97	4.12
0.26	18.86	3.23	23.34	4.24	31.76	3.99
0.24	17.88	3.13	22.18	4.11	30.60	3.86
0.22	16.94	3.04	21.06	3.98	29.47	3.73
0.20	16.03	2.94	19.96	3.85	28.37	3.60
0.18	—	—	18.88	3.72	27.29	3.47
0.16	—	—	17.83	3.59	26.23	3.34
0.14	—	—	16.78	3.46	25.17	3.21
0.12	—	—	15.72	3.32	24.12	3.07
0.10	—	—	—	—	23.05	2.93
0.08	—	—	—	—	21.94	2.78
0.06	—	—	—	—	20.76	2.61

The graph of Fig. 3 shows distribution curves calculated by the three methods under consideration. The curves are characterized by a relatively wide distribution with clearly defined maxima. They are associated with the following radius values: 55.5 Å for method I, based on the KELVIN equation, 68.7 Å for method II of DERJAGUIN, BROEKHOFF and DE BOER, and 76.7 Å for method III with additional allowance for the change in surface tension because of the meniscus curvature. In transition from method I to method III the maxima decrease, and the bell-shape curves expand appreciably.

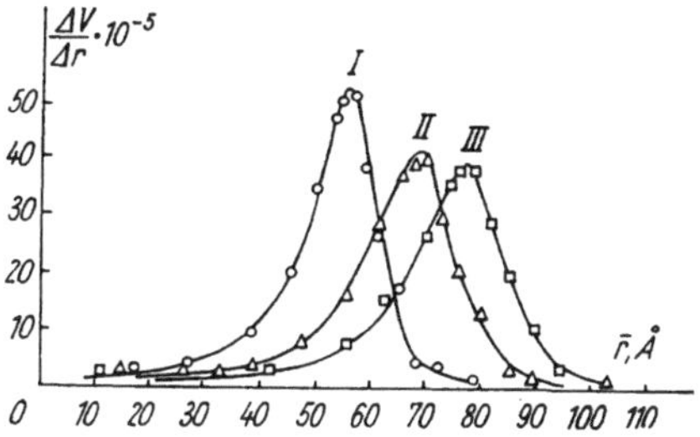

Fig. 3: Mesopore volume distribution according to size for large-pore silica gel KCK-2.

Among the cumulative characteristics of silica gel given in Table 6, of greatest interest are the mesopore specific surface areas obtained by the three methods in question. The specific surface area of silica gel determined by the BET method and equal to 320 m^2/g differs considerably from the specific surface area calculated by method I and practically coincides with the values found by methods II and III. But the respective curves for pore volume distributions according to the sizes differ substantially, mainly in their positions with respect to the x-axis.

Table 6: Silica gel KCK-2, Benzene, 293 K
$S_{BET} = 320$ m^2/g (N_2, 78 K)

Method	Calculation up to:					
	Point of beginning of hysteresis $p/p_s \approx 0.20$			Pore radius $r \approx 16$ Å		
	r, Å	V^*, cm^2/g	S^*, m^2/g	p/p_s	V^*, cm^3/g	S^*, m^2/g
I	15.9	0.945	385	0.20	0.945	385
II	19.8	0.943	310	0.14	0.954	320
III	28.3	0.921	264	0.02	0.960	300

Two typical specimens of active carbons with developed mesoporosity were chosen as examples of carbonaceous adsorbents. During an investigation conducted in cooperation with E. G. ZHUKOVSKAYA a thorough study was made into the adsorption and capillary condensation of nitrogen vapours at 78 K and of benzene at 293 K [18]. The initial computational data are given in Table 2 for nitrogen and in Table 7 for benzene. The branches of capillary evaporation of these isotherms served for a more detailed analysis by the three methods at hand. To the points of the beginning of hysteresis for capillary condensation of nitrogen there corresponded equilibrium relative pressures of about 0.45 for nitrogen and about 0.2 for benzene, respectively.

Table 7: Initial data of capillary evaporation of benzene at 293 K from mesopores of carbonaceous adsorbents

p/p_s	Method I		Method II		Method III	
	r, Å	t, Å	r, Å	t_e, Å	r, Å	t_e, Å
0.96	555.27	39.49	742.33	54.13	751.09	53.98
0.94	371.11	30.82	467.59	41.94	476.26	41.77
0.92	278.57	26.06	343.17	35.27	351.77	35.08
0.90	222.77	22.93	271.75	30.90	280.29	30.69
0.88	185.37	20.67	225.20	27.75	233.70	27.53
0.86	158.53	18.93	192.35	25.33	200.82	25.10
0.84	138.30	17.54	167.87	23.40	176.30	23.16
0.82	122.49	16.39	148.87	21.81	157.28	21.56
0.80	109.78	15.43	133.68	20.47	142.06	20.21
0.78	99.33	14.59	121.23	19.32	129.59	19.05
0.76	90.58	13.86	110.83	18.31	119.16	18.04
0.74	83.14	13.22	101.99	17.42	110.30	17.14
0.72	76.73	12.64	94.38	16.63	102.68	16.34
0.70	71.15	12.12	87.76	15.91	96.03	15.62
0.68	66.24	11.64	81.93	15.26	90.19	14.97
0.66	63.13	11.33	78.24	14.84	86.49	14.54
0.64	57.99	10.81	72.12	14.12	80.35	13.81
0.62	54.48	10.44	67.95	13.61	76.17	13.30
0.60	51.31	10.09	64.17	13.14	72.37	12.83
0.58	48.42	9.77	60.72	12.70	68.91	12.38
0.56	45.77	9.46	57.56	12.28	65.74	11.96
0.54	43.34	9.17	54.65	11.89	62.82	11.57
0.52	41.10	8.90	51.96	11.52	60.12	11.20
0.50	39.02	8.64	49.46	11.17	57.61	10.84
0.48	37.08	8.40	47.13	10.84	55.27	10.51
0.46	35.28	8.16	44.95	10.52	53.08	10.19
0.44	33.58	7.94	42.90	10.21	51.03	9.88
0.42	31.99	7.72	40.97	9.92	49.09	9.58
0.40	30.49	7.51	39.15	9.63	47.26	9.29
0.38	29.06	7.30	37.42	9.36	45.53	9.02
0.36	27.72	7.11	35.78	9.09	43.88	8.75
0.34	26.43	6.91	34.21	8.83	42.30	8.49
0.32	25.20	6.73	32.71	8.58	40.79	8.24
0.30	24.03	6.54	31.27	8.33	39.35	7.99
0.28	22.90	6.36	29.88	8.09	37.95	7.75
0.26	21.81	6.18	28.54	7.85	36.60	7.50
0.24	20.76	6.01	27.23	7.62	35.30	7.27
0.22	19.74	5.83	25.96	7.38	34.02	7.03
0.20	18.74	5.65	24.72	7.15	32.78	6.80
0.18	17.75	5.47	23.50	6.91	31.55	6.56
0.16	16.77	5.29	22.28	6.67	30.33	6.32
0.14	15.82	5.11	21.07	6.43	29.12	6.08

Table 7 (continued)

p/p_s	Method I		Method II		Method III	
	r, Å	t, Å	r, Å	t_e, Å	r, Å	t_e, Å
0.12	—	—	19.85	6.17	27.90	5.82
0.10	—	—	18.60	5.91	26.65	5.56
0.08	—	—	17.30	5.63	25.35	5.28
0.06	--	—	15.91	5.32	23.96	4.97

Of greatest interest was comparison of the quantitative characteristics of mesoporosity obtained from the experimental data on adsorption and capillary evaporation of two different vapours. Besides, it is important to compare the specific surface areas of the mesopores calculated from experiments on capillary evaporation with the corresponding values determined by independent methods. Since active carbons are microporous adsorbents and contain considerable micropore volumes, which basically determine their adsorption properties, the application of such widespread methods as those of BRUNAUER, EMMETT, and TELLER or of the t-method of DE BOER and LIPPENS is physically unjustified [20, 21].

DUBININ and KADLEC proposed the t/F-method [22, 23] as applied to micro- and mesoporous adsorbents, active carbons in particular. Its essence is as follows. Prior to capillary condensation the adsorption of a vapour on active carbon is expressed as the sum of the values of adsorption in micropores and adsorption on the mesopore surface. According to the theory of volume filling of micropores it is possible to calculate adsorption in micropores, and using the difference from the experimental value, adsorption on the mesopore surface. Applying the t-method to adsorption in mesopores, we find the specific surface area of the mesopores not related to the assumption of the geometrical shape of the pores.

In the graphs of Fig. 4 we give, by way of example, the curves for mesopore volume distribution according to the average radii calculated from the three methods considered for AC-11. The graphs for the other carbons are similar.

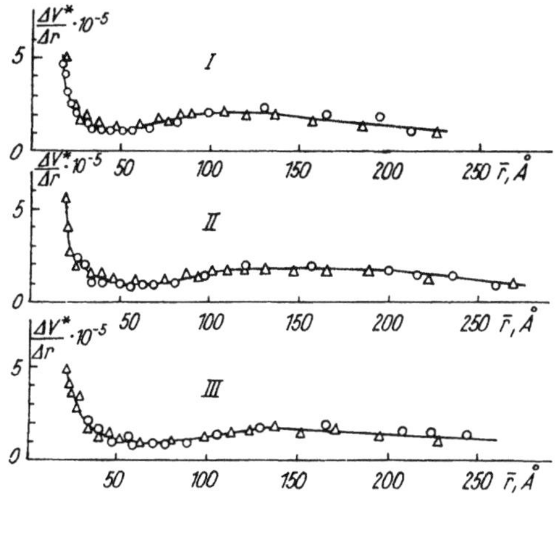

Fig. 4: Mesopore volume distribution according to size for active carbon AC-11 calculated from the capillary evaporation of benzene at 293 K (circles) and nitrogen at 78 K (triangles).

The active carbons investigated are characterized by a wide mesopore volume distribution. Vaguely defined maxima appear on all the curves of medium size range. In the transition from method I to method III the maxima of the curves shift toward the larger radii. In the small-radii region the distribution curves show a rather steep rise up to 5—8 divisions on the ordinate scale. But the level achieved is 5—10 times lower than for the maxima of the above-considered distribution curves for silica gel or the aluminosilicate catalyst. These initial sections of the curves for active carbons should be studied more thoroughly.

Table 8: Active carbons, Nitrogen, 78 K. Calculations up to $r \approx 17$ Å

Adsorbent	Method	V^*, cm³/g	S^*, m²/g	S, m²/g (t/F-method)
AC-11	I	0.401	95	133
	II	0.450	133	
	III	0.456	129	
AC-14	I	0.558	149	190
	II	0.621	190	
	III	0.633	187	

The points calculated from capillary evaporation of benzene are denoted by circles and those for nitrogen, by triangles. The agreement between the results of calculation from two different vapours may be considered satisfactory.

Let us pass over to the cumulative characteristics of mesoporosity, i.e. of the volumes of mesopores and their specific surface areas. Table 8 lists the results of calculations from experiments with nitrogen for two typical specimens of active carbons AC-11 and AC-14. The specific surface areas calculated by methods II and III are practically identically close to the experimental ones calculated by the method. The results for a specimen of AC-13, not listed in the Table, namely 88—89 m²/g for methods II and III and 90 m²/g for the t/F-method, also agree fairly well.

Table 9: Active carbons, Benzene, 293 K. Calculations up to $r \approx 18$ Å

Adsorbent	Method	V^*, cm³/g	S^*, m²/g	S, m²/g (t/F-method)
AC-11	I	0.445	94	140
	II	0.486	113	
	III	0.491	103 (up to $r = 23$ Å)	
AC-14	I	0.676	162	197
	II	0.728	175	
	III	0.721	148 (up to $r = 23$ Å)	

Table 9 shows the results of calculations for the selected carbons from capillary evaporation of benzene from mesopores at 293 K. The conclusions that can be made from the data of this table are similar to those just discussed for the case of nitrogen. Note that the specific surface areas of mesopores determined by the t/F-method from experiments on nitrogen and benzene adsorption practically coincide.

It should be mentioned that the mesopore volumes V^* calculated from experiments with benzene are 10—15% higher than for experiments with nitrogen for all the methods considered. Analysis of the causes for these divergencies is one of the further tasks.

The principal objectives of future investigations

The further theoretical development of the problem must include primarily the use of a unified, if empirical, equation of the t-curve based on the experimental adsorption isotherm, which would describe satisfactorily the adsorption values over the entire range of equilibrium relative pressures used for calculations, from those of the order of hundreds to those as close as possible to unity, for instance 0.98—0.99. That this is necessary follows from Eqs. (7) and (12), which contain an integral with a range from t_e to r in the second term. The integrand contains the function $F(t)$, which depends on the parameters of the equation of the t-curve, i.e. of the adsorption isotherm. This requires applicability of the t-curve equation up to high t values.

Owing to the imperfection of the t-curve equation applied, the r values, as calculated by methods II and III, differ considerably from those obtained by method I at high equilibrium pressures, particularly for the adsorption system benzene-silica gel. Note that the difference in the r values calculated by methods II and III at high p/p_s is insignificant.

Let us try to establish the cause for this divergence, using for this purpose method II and the adsorption system benzene-silica gel. For this system, use is made of an equation of the type (24) (see Appendix), which contains, in place of the parameters 4.05, 27.98, and 0.034 for nitrogen, the respective parameters 9.142, 10.848, and 0.0156 for benzene (see Table 2). Equation (24) can be represented as

$$r - t_e = \frac{2\sigma v}{RT \ln (p_s/p)} + f(t_e, r) , \tag{29}$$

where $f(t_e, r)$ is a correction term to the Kelvin equation. According to the data of Table 5 the difference $r - t_e$ at $p/p_e = 0.94$ for the first term of the equation is 340 Å, while calculation of $f(t_e, r)$ by Eq. (24) with the corresponding values of the parameters yields 477 Å, i.e. the correction exceeds the main value. At relative pressures of over 0.9 the determining role in the expression for $f(t_e, r)$ is played by the term

$$\delta = \frac{0.0156\,(r - t_e)}{\lg (p_s/p)} . \tag{30}$$

Thus, for $r - t_e = 840$ Å we have, from Table 5, $\delta = 488$ Å in place of 477 Å according to precise calculations. The obtained correction for the system benzene-silica gle is this large because of the relatively large parameter $D = 0.0156$ in the Harkins-Jura equation. For the system benzene-carbonaceous adsorbents, $D = 0.0035$, i.e. it is almost 5 times as small, and the indicated divergence is considerably reduced. It can be assumed that at high p/p_s the pore radii calculated by methods II and III are overestimated. Therefore calculations in this field require adjustment.

Calculations of capillary evaporation by method III in the low-pressure range are also subject to reasonable adjustment. As a whole, the proposed version of capillary evaporation analysis with due regard for the dependence of the surface tension on the meniscus curvature is but a first approximation, which requires further theoretical consideration.

Conclusion

From the above analysis of the experimental data it follows that the KELVIN equation modified by the method of DERJAGUIN, BROEKHOFF and DE BOER and with additional inclusion of the change in surface tension with meniscus curvature leads to results which agree better with the data of independent methods, such as those of BRUNAUER, EMMETT, and TELLER for mesoporous and the t/F-method of DUBININ and KADLEC for microporous adsorbents. But despite the similarity of the specific surface areas of the equivalent model adsorbents, as calculated by methods II and III, the curves for pore volume distribution according to the sizes are found to be substantially shifted. In this lies the radical difference between the results of application of methods II and III.

In the future, along with theoretical development of the problem it is necessary to correlate the practical data associated with the properties and structure of the mesopores of adsorbents, catalysts, supports, etc. with the results of estimation of the mesopore parameters by the three methods considered. We made them equally accessible by preliminary computerized calculations of the tables of initial data. It is advisable to use nitrogen and benzene as sorbed substances in order to evaluate their practical merits and demerits.

The co-authors in this work are V. A. GORDEYEVA and L. I. KATAYEVA. The latter performed all computer calculations.

Appendix

Method I

$$\lg (p_s/p) = 16.11/t^2 - 0.1682 \exp[-0.1137t] \quad \text{for } p/p_s = 0.95 - 0.45\,, \tag{18}$$

$$\lg (p_s/p) = 13.99/t^2 - 0.034 \quad \text{for } p/p_s = 0.45 - 0.08\,, \tag{19}$$

$$r - t = \frac{4.05}{\lg (p_s/p)}\,. \tag{20}$$

Method II

for $p/p_s = 0.95 - 0.45$

$$\lg (p_s/p) - 16.11/t_e + 0.1682 \exp[-0.1137t_e] - \frac{2.035}{r - t_e} = 0\,, \tag{21}$$

$$r - t_e = \frac{4.05}{\lg (p_s/p)} + \frac{32.22\,[r/t_e - 1 - \ln (r/t_e)]}{(r - t_e) \lg (p_s/p)} - $$
$$- \frac{2.964 \exp [-0.1137 t_e]\,(r - t_e - 8.795)}{(r - t_e) \lg (p_s/p)} - \frac{26.068 \exp [-0.1137 r]}{(r - t_e) \lg (p_s/p)} , \quad (22)$$

for $p/p_s = 0.45 - 0.08$

$$\lg (p_s/p) - 13.99/t_e^2 + 0.034 - 2.025/(r - t_e) = 0 , \quad (23)$$

$$r - t_e = \frac{4.05}{\lg (p_s/p)} + \frac{27.88\,[r/t_e - 1 - \ln (r/t_e)] - 0.034\,(r - t_e)^2}{(r - t_e) \lg (p_s/p)} . \quad (24)$$

Method III

for $p/p_s = 0.95 - 0.45$

$$\lg (p_s/p) = 16.11/t_e^2 + 0.1682 \exp [-0.1137 t_e] - \frac{2.025}{r - t_e - 3.31} = 0 \quad (25)$$

$$r - t_e = \frac{4.05 (r - t_e)}{[\lg (p_s/p)]\,(r - t_e - 6.62)} + $$
$$+ 1/(r - t_e) \lg (p_s/p) \{32.22\,[r/t_e - 1 - \ln (r/t_e)] - $$
$$- 2.964 \exp (0.1137 t_e)\,[r - t_e - 8.795] - 26.06 \exp (-0.1137 r)\} , \quad (26)$$

for $p/p_s = 0.45 - 0.08$

$$\lg (p_s/p) - 13.99/t_e^2 + 0.034 - 2.025/(r - t_e - 3.31) = 0 , \quad (27)$$

$$r - t_e = \frac{4.05 (r - t_e)}{[\lg (p_s/p)]\,(r - t_e - 6.62)} + \frac{27.98\,[r/t_e - 1 - \ln (r/t_e)] - 0.34 (r - t_e)^2}{(r - t_e)\,(\log (p_s/p)} . \quad (28)$$

References

[1] M. M. Dubinin, Uspekhi Khimii, **21**, 513 (1952)
[2] M. M. Dubinin, Izv. Akad. Nauk SSSR, Ser. Khim., 996 (1974)
[3] M. M. Dubinin, Zh. Fiz. Khim., **30**, 1652 (1956)
[4] M. M. Dubinin and E. G. Zhukovskaya, Zh. Fiz. Khim., **30**, 1840 (1956)
[5] B. V. Derjaguin, Acta physicochim. USSR, **12**, 181 (1940)
[6] B. V. Derjaguin, J. Coll. Interface Sci., **24**, 357 (1967)
[7] J. C. Broekhoff and J. H. de Boer, J. Catal., **9**, 8, 15 (1967); **10**, 153, 368, 377, 391 (1968)
[8] J. C. Broekhoff, "*Adsorption and Capillarity*", *Thesis*, Delft, pp. 1—237, 1969
[9] M. M. Dubinin, L. I. Katayeva and V. I. Ulin, Izv. Akad. Nauk SSSR, Ser. Khim., 510 (1977)
[10] M. M. Dubinin and L. I. Katayeva, Izv. Akad. Nauk SSSR, Ser. Khim., 516 (1977)
[11] R. C. Tolman, J. Chem. Phys., **17**, 333 (1949)
[12] W. S. Ahn, M. S. Jhon, H. Pak and S. Chang, J. Coll. Interface Sci., **38**, 605 (1972)
[13] A. Wheeler, Presentation at Catalysis Symposia, Gibson Island A.A.A.S. Conferences, June 1945 and June 1946

[14] "*Physical and Chemical Aspects of Adsorbents and Catalysts*", Ed. by B. G. LINSEN, Academic Press, London and New York, 1970, p. 1—653
[15] D. DOLLIMORE and C. H. HEAL, J. Appl. Chem. Biotechnol., **14**, 109 (1964)
[16] T. D. OULTON, J. Phys. Coll. Chem., **52**, 1296 (1948)
[17] A. A. ISIRIKYAN, Thesis, Moscow, 1957, p. 1—270
[18] M. M. DUBININ and E. G. ZHUKOVSKAYA, Izv. Akad. Nauk SSSR, Ser. Khim., 1705 (1959)
[19] J. G. T. VAN AKEN, Thesis, Delft, 1969, p. 1—192
[20] S. BRUNAUER, P. H. EMMETT, E. TELLER and J. Amer. Chem. Soc., **60**, 309 (1938)
[21] B. C. LIPPENS and J. H. DE BOER, J. Catal., **4**, 319 (1965)
[22] M. M. DUBININ and O. KADLEC, Carbon, **13**, 263 (1975)
[23] M. M. DUBININ, T. I. ISOTOVA, O. KADLEC, and O. L. KRAYNOVA, Izv. Akad. Nauk SSSR, Ser. Khim., 1232 (1975)

The dynamic adsorption layer and the analogs of the MARANGONI-GIBBS effect

S. S. DUKHIN

(Institute of Colloid Chemistry and Water Chemistry of the Ukrainian Academy of Sciences, Kiev/USSR)

Abstract

The theory of adsorption layer on the moving surface of floating bubble is being developed. The theory is based on the solution of convective diffusion equation which describes diffusion boundary layer of bubble formed by adsorption-desorption processes and surface motion, and of hydrodynamic equations complicated by the account of dynamic adsorption layer effect on surface motion.

The theories of surfactants distributions along a moving surface and weak and strong surface retardation caused by surfactants have also been developed. It has been shown that surfactants adsorption increase towards surface motion direction brings about adsorption drop along interfacial film being shaped between a part of the bubble surface and an approaching particle. Since surface tension drop directed along interfacial film and connected with dynamic adsorption layer prevents interfacial film thinning it must be considered as an analogy of theMARANGONY-GIBBS effect. The quantitative theory of interfacial film stabilization by dynamic adsorption layer has been developed.

Motion at a liquid interface is usually accompanied by extension and contraction of the interface. Thus the translation of a bubble or droplet causes one half of the interface to undergo contraction, while the other half undergoes extension. Fresh surface is filled with adsorbing substance which is desorbed in areas experiencing contraction. These continuous processes are sustained by diffusion currents from the bulk phases into an expanding interface and out of a contracting interface into the bulk phases. Surface motion thus influences the formation of adsorption layers, but there is also a reverse effect, in that the adsorption layers influence the surface motion. Where an interface contracts, the density of the adsorption layer tends to rise, provoking a fall in the surface tension σ and the generation of surface forces which oppose and eventually retard the surface motion. The theory of the dynamic adsorption layer should thus be based upon the simultaneous solution of the diffusion equation complicated by the surface motion effect upon the processes of adsorption-desorption, plus the solution of the hydrodynamic equations complicated by the effect of the dynamic adsorption layer upon the interfacial motion ot the liquid phases.

The theory of the dynamic adsorption layer and of the diffusion field surrounding spherical particles is of particular interest because it is possible to obtain analytic results, and the theory is in any case needed for the solution of practical problems in technology.

The effect of soluble surface active substances on liquid-liquid and gas-liquid interfaces is well known. We have developed the mathematical description of the formation of the dynamic adsorption layer and its associated diffusion field in the bulk liquid in a generalized way without some of the limitations and simplifications [1—4].

Simplifying assumptions concerning weak changes in adsorption, the representation of its dependence on spherical polar angle θ by a cos θ term, and the introduction of a coefficient of retardation to describe the effect of surfactant adsorption on droplet motion have been confirmed in the following extreme cases for all of which the PECLET number $Pe \approx aU/D \gg 1$ and REYNOLDS number $Re \approx aU/\nu \ll 1$ (a = bubble radius, U = bubble velocity, D = diffusion coefficient of the surfactant, ν = kinetic viscosity). a) High surface retardation and moderate surface activity. b) If despite compatively quick establishment of adsorption equilibrium, the dynamic adsorption layer is formed by adsorption kinetics (possible with very low surface activity only). c) The case of weak retardation and low surface activity where the relative change in adsorption is insignificant, but its dependence on polar angle cannot be limited to a cos θ term alone.

With high surface activity or very low adsorption velocity, the surfactant distribution on the moving surface is very uneven. This causes an uneven surface retardation, which together with the uneven distribution of adsorption causes considerable mathematical difficulties.

The effect of the fluid motion on surface and volume concentration distributions, and the counter effect of the dynamic adsorption on the velocity field for $Pe \gg 1$ and either $Re \ll 1$ or $Re \gg 1$ were estimated in the extreme cases of weak retardation of the main part of the surface and intense retardation over the whole surface. In these cases the adsorption and surface velocity fields depend upon polar angle in a complicated way.

If the retardation on the main part of the surface is weak, it is possible to estimate the effect of convective diffusion and adsorption kinetics on the dynamics of the adsorption layer by simultaneous solution of a pair of integral and differential equations [1]. If adsorption kinetics and volume diffusion are neglected, then the results have been verified for $Re \ll 1$, arbitrary Peclet number, and arbitrary surface activity.

These general conclusions are in good accord with published experimental data on the minimum concentration of surfactant necessary to provide an effective retardation of a rising bubble. GRIFFITH's results on the effect of highly surface active materials on droplet motion are consistant with our conclusions [5].

These results are needed to classify convection-diffusion potential formulae [2, 3], the study of the secondary double layer [6], and DORN's effect [7, 8]. They are also in themselves very interesting, since one needs to know the surface concentration distribution on a moving bubble or droplet and details of the surfactant exchange with the bulk phases in order to understand flotation, surfactant extraction by flotation, emulsion breaking, as well as to understand the mechanism of surfactant influence on the spontaneous break-up of falling droplets and droplet fragmentation by ultrasound and turbulence.

Principles of the theory of the dynamic adsorption and diffusion boundary layers

The diffusion field is described mathematically by a solution of the convection-diffusion equation

$$\vec{V} \cdot \operatorname{grad} C = D\,\Delta C\,, \tag{1}$$

in which C and V are the concentration and velocity fields, respectively.

In equation (1) the components of the velocity field are generally complex numbers. Partial differential equations with variable coefficients are difficult to solve, and the difficulties increase when the variable coefficients are complex. The problem is simplified by approximating the velocity components within the diffuse layer using boundary layer theory; for $\delta_D \ll a$, the thickness of the diffuse layer, is much less than the typical distance over which the velocity components change appreciably.

The problem can be solved effectively by conversion of the convection-diffusion equation to the well studied heat conduction equation by introducing the stream function ψ as a new variable. In terms of the stream function the velocity components in spherical coordinates r, θ are

$$V_\theta = -\frac{1}{r \sin\theta}\frac{\partial\psi}{\partial r}; \qquad V_r = \frac{1}{r^2 \sin\theta}\frac{\partial\psi}{\partial\theta}. \tag{1a}$$

If these expressions are substituted into the left hand side of equation (1),

$$V_r \frac{\partial c}{\partial r} + \frac{V_\theta}{r}\frac{\partial c}{\partial\theta} = D\left(\frac{\partial^2 c}{\partial r^2} + \frac{2}{r}\frac{\partial c}{\partial r}\right), \tag{2}$$

then this part of the equation is transformed into $V_\theta\,(\partial c/\partial\theta)$.

On the right hand side of equation (1) the second derivative of ψ appears and one obtains an equation similar to that describing steady heat conduction in the variables θ, ψ. There is also an analog to the coefficient of thermal conductivity whose magnitude is determined by the velocity field.

Let us analyse the diffusion stream on a droplet surface moving in another liquid with $Re \ll 1$. In this case the velocity field can be described by the formulae of HADAMARD and RYBCZYNSKI [9]. The droplet surface is moving, and the surface velocity field is expressible as

$$V(a, \theta) = V_0 \sin\theta, \tag{3}$$

in which $V_0 = \frac{\eta}{2}\frac{U}{\eta + \eta'}$ with U = translational droplet speed and η, η' the coefficients of viscosity of the droplet and of the external medium, respectively. The formulae of HADAMARD and RYBCZYNSKI are also valid for the "moving bubble" problem, and in this case $\eta' \ll \eta$.

Using the HADAMARD-RYBCZYNSKI velocity field, it is easy to show that the tangential component of the velocity in the diffuse boundary layer differs negligibly from the surface velocity field. This is the reason why in the reduction of equation (1) to variables θ, ψ, the coefficient on the right hand side is independent of $\psi = y \sin^2\theta$, where $y = r - a$:

$$\frac{\partial c}{\partial\theta} = Kna^2 \sin^3\theta \frac{\partial^2 c}{\partial\psi^2}, \tag{4}$$

in which $K = P_e^{-1}$; $n = U/V_0$.

With the introduction of a new variable

$$t = \int \sin^3\theta \, d\theta = \frac{I}{3}\cos^3\theta - \cos\theta + a_1$$

we obtain the heat conduction equation with constant coefficient Kna^2

$$\frac{\partial c}{\partial t} = Kna^2 \frac{\partial^2 c}{\partial \psi^2}. \tag{5}$$

In order to complete the analogy with the heat flow equation, it is necessary to choose the integration constant a_1 in such a way that t is everywhere positive and $\theta = 0$ coincides with $t = 0$. This happens at $a_1 = 2/3$, so that explicitly

$$t = p_1(\theta) = \frac{2}{3} + \frac{\cos^3 \theta}{3} - \cos \theta = \frac{2 + \cos \theta}{(1 + \cos \theta)^2} \frac{\sin^4 \theta}{3}. \tag{6}$$

The dynamic adsorption layer and the boundary diffusion field depend strongly on the position along the spherical surface because of the continuous adsorption of a surfactant on one side of a moving droplet and ts desorption on the other side. The boundary conditions thus need to take into account convective transport along the surface as well as surface-bulk phase interchange. At $r = a$ this boundary condition becames

$$\frac{\partial \Gamma}{\partial t} = - \operatorname{div}_s (\Gamma V_t - D_s \operatorname{grad} \Gamma) + j_n \tag{7}$$

in which D_s = surface diffusion coefficient, and j_n is the normal component of the surfactant flux in the diffusion boundary layer. Our model thus regards surface transport as due to a combination of convective transport plus surface diffusion.

The value of the surfactant flux j_n is determined by the slowest of the two processes: adsorption-desorption kinetics or diffusion through the boundary layer. If the kinetics are rapid, then local equilibrium is maintained so that $\Gamma(\theta)$ and $C(a, \theta)$ are related locally in the same way as the corresponding equilibrium values Γ_0 and C_0. Except near saturation the condition for local equilibrium can be regarded as linear,

$$\Delta(\theta)/c(a, \theta) = \Gamma_0/c_0 = \alpha , \tag{8}$$

in which α is a measure of surface activity characteristic of the dissolved substance.

If the adsorption-desorption kinetics are rapid and local equilibrium is maintained, then j_n is diffusion controlled,

$$j_n = -D \frac{\partial c}{\partial r} (a, \theta) , \tag{9}$$

and both sides of equation (7) are functions of the boundary concentration $C(a, \theta)$. Substitution of (8) and (9) into (7) yields a boundary condition for the solution of equation (5).

Theory of the dynamic adsorption layer and surface retardation of drops or bubbles

If the PECLET number is small, $Pe \ll 1$, then two simplifying approximations are possible.

As it will be shown in the sequence, if $Pe \ll 1$, then the surface concentrations differ negligibly from their equilibrium values:

$$(\Gamma(\theta) - \Gamma_0)/\Gamma_0 = [c(a, \theta) - c_0]/c_0 \ll 1 . \tag{10}$$

Furthermore, if $Pe \ll 1$, then the convection-diffusion equation (1) reduces to the LAPLACE equation, and the hydrodynamic and diffusion fields may be truncated to the first two spherical harmonics and their derivatives (terms in $\sin\theta$ and $\cos\theta$ only):

$$c(r, \theta) = c_0 + \Delta c \frac{a^2}{r^2} \cos\theta , \qquad \Gamma(\theta) = \Delta_0 + \Delta\Gamma \cos\theta \tag{11}$$

in which because of (8)

$$\Delta\Gamma = \alpha\, \Delta c . \tag{12}$$

With these simplifications, it has been shown that the effect of the adsorption layer on the droplet hydrodynamics is quantitatively described by the introduction of a coefficient of retardation. The velocity fields inside and outside the drop are respectively

$$V'_\theta(r, \theta) = V_0(2r^2/a^2 - 1) \sin\theta , \qquad V'_r(r, \theta) = V_0(1 - r^2/a^2) \cos\theta , \tag{13}$$

$$V_\theta(r, \theta) = U_L \left(1 - \frac{2\eta + 3\eta' + 3\gamma}{4(\eta + \eta' + \gamma)} \frac{a}{r} - \frac{\eta' + \gamma}{4(\eta + \eta' + \gamma)} \frac{a^3}{r^3}\right) \sin\theta , \tag{14}$$

$$V_r = -U_L \left(1 - \frac{2\eta + 3\eta' + 3\gamma}{2(\eta + \eta' + \gamma)} \frac{a}{r} + \frac{\eta' + \gamma}{2(\eta + \eta' + \gamma)} \frac{a^3}{r^3}\right) \cos\theta , \tag{15}$$

in which U_L is the translational speed of the droplet

$$U_L = \frac{2}{3} \frac{\Delta\varrho g a^2}{\eta} \frac{\eta + \eta' + \gamma}{2\eta + 3\eta' + 3\gamma} \tag{16}$$

and

$$V_0 = \frac{1}{3} \frac{\Delta\varrho g a^2}{2\eta + 3\eta' + 3\gamma} . \tag{17}$$

$\Delta\varrho$ is the density difference between the two bulk phases, g is the acceleration of gravity, and γ is the retardation coefficient

$$\gamma = \frac{\partial\sigma}{\partial c} \frac{\Delta c}{3 V_0} . \tag{18}$$

Explicit expressions for Δc, $\Delta\Gamma$, γ and V_0 are derivable from the boundary condition (9), which under the present simplifications becomes

$$D \frac{\partial c}{\partial r}(a, \theta) = \frac{1}{a \sin\theta} \frac{\partial}{\partial\theta} \left[\sin\theta \left(\Gamma(\theta)\, V_0 \sin\theta - D_s \frac{\partial\Gamma}{\partial\theta}\right)\right] . \tag{19}$$

In view of (10), (11), and (12) it is elementary to show that

$$\Delta c = \frac{2\Gamma_0 V_0}{D + 2D_s\alpha/a} \tag{20}$$

whence from (18)

$$\gamma = \frac{2}{3} \frac{\Gamma_0 R T \alpha}{D + 2D_s\alpha/a} , \tag{21}$$

where R is the gas constant and T the absolute temperature.

We need to show that if $Pe \ll 1$, then (10) is satisfied for arbitrary surface activity

$$\frac{\Delta c}{c_0} = \frac{2\Gamma_0}{c_0 a} \frac{V_0 a}{D} \frac{1}{1 + 2D_s/D \cdot \Gamma_0/c_0 a} < \frac{V_0}{u} \frac{D}{D_s} Pe \ll 1 . \tag{22}$$

The final result for the surface concentration distribution for $Pe \ll 1$ and the establishment of practically instantaneous adsorption equilibrium is derived from (12), (20), (17), and (21)

$$\Gamma(\theta) = \Gamma_0 - \Gamma_0 \frac{2\Delta \varrho g a^2 \alpha}{3[D + 2D_s\alpha/a]\,[2\eta + 3\eta' + 2RT\Gamma_0\alpha/(D + 2D_s\alpha/a)]} \cos\theta . \tag{23}$$

For low surface activity

$$\alpha \ll a \tag{24}$$

then surface diffusion makes no contribution to either the surface concentration or to the surface retardation, viz. (20), (21), provided that D and D_s have the same order of magnitude.

For high surface activity,

$$\alpha \gg a, \tag{25}$$

then surface diffusion is the dominant factor which determines adsorption distribution and surface retardation. The convective surface transport is negligible.

Analogous assumptions (3) concerning the velocity field on the retarded surface and of a constant thickness for the boundary layer make possible a derivation of the retardation coefficient for $Pe \gg 1$

$$\gamma = 2RT\Gamma_0^2\delta_D/3D_a c_0 . \tag{26}$$

If in addition to $Pe \gg 1$, also

$$\gamma \ll \tfrac{3}{2}\eta , \tag{27}$$

then the condition for low surface retardation can be stated in the form (10)

$$(\Gamma_0/c_0)^2 c_0 \ll (27/4)\, \eta D a/RT\delta_D . \tag{28}$$

If $\gamma \to \infty$ equation (16) reduces to the terminal velocity of a falling spherical particle

$$U_{st} = \frac{2\Delta\varrho g a^2}{9\eta} . \tag{29}$$

The theory of the dynamic adsorption and diffusion boundary layers of a bubble with $Pe \gg 1$, $Pe \ll 1$ and weak surface retardation

When the surface redardation is weak and equation (28) is satisfied, then the convection-diffusion equation (2) under the transformation $t = \frac{2}{3} + \frac{1}{3}\cos^3\theta - \cos\theta$; $\psi = y \sin^2\theta$; $(y = r - a)$; is transformed into the heat flow equation (5) with constant coefficient. The field equation is thus a simple one, but the complexity of equation (7) requires the investigation of extreme cases which will simplify the boundary condition. The ratio of the second to the first terms of the right in equation (19) may be written as $(D_s U/D V_\theta \sin\theta\, Pe)$ $(p \ln \Gamma/\partial\theta)$, so that with $Pe \gg 1$ (except near the poles $\theta = 0$, $\theta = \pi$), equation (19) may be simplified by dropping the surface diffusion term

$$D \frac{\partial c}{\partial r}(a, \theta) = \frac{V_0}{a \sin\theta} \frac{\partial}{\partial\theta} [\sin^2\theta \cdot \Gamma(\theta)] . \tag{30}$$

For lower values of α and Pe, the adsorption changes become relatively small, and the equilibrium concentration along the $c(a, \theta)$ droplet surface (10) is derived from (30)

$$\frac{\partial c}{\partial y}(a, \theta) = \Lambda \cos\theta , \tag{31}$$

where

$$\Lambda = \frac{2 V_0 \Delta}{Da} . \tag{32}$$

Transformed to the variables ψ, t, the boundary condition (31) becomes

$$\left.\frac{\partial c}{\partial \psi}\right|_{\psi=0} = \Lambda \frac{\cos\theta}{\sin^2\theta} = \Lambda \frac{f(t)}{1 - f^2(t)} \tag{33}$$

in which

$$f(t) = \cos\theta . \tag{34}$$

To avoid the determination of the roots of the cubic equation (6), we shall represent a root ($\cos\theta$) by $f(\theta)$, the exact form of which is not required. In addition, it will be assumed that in regions far from the bubble

$$c|_{\psi\to 0} = c_0 \tag{35}$$

as using a source function

$$\frac{1}{2} \frac{\exp[-\psi^2/4\pi n a^2(t - t')]}{[\pi K n a^2(t - t')]^{1/2}} \tag{35}$$

and an initially homogeneous concentration field, and taking into account the fact that the boundary condition (33) is equivalent to a nonsteady source with capacity $Kna^2 \dfrac{\partial c(t)}{\partial \psi}$, the solution of equation (5) is

$$c(t, \psi) = (Kna^2/4\pi)^{1/2} \int_0^t \left.\frac{\partial c(t)}{\partial \psi}\right|_{\psi=0} \frac{\exp[-\psi^2/4Kna^2(t - t')]}{(t - t')^{1/2}} dt' + c_0 =$$

$$= \Lambda (Kna^2/\pi)^{1/2} \int_0^t \frac{f(t')}{1 - f^2(t')} \frac{\exp[-\psi^2/4Kna^2(t - t')]}{(t - t')^{1/2}} dt' + c_0 . \tag{36}$$

Because the functional form of $f(t)$ is unknown, we change the variable of integration from t' to θ' making use of equation (6)

$$t' = \frac{2}{3} + \frac{\cos^3\theta'}{3} - \cos\theta' . \tag{37}$$

Then by definition from (6, 34) $f(t') = \cos\theta'$ and

$$\delta c(\theta, y) = c_0 - c(\theta, y) =$$

$$= \frac{2\Gamma_0 V_0 n^{1/2}}{D(\pi Pe)^{1/2}} \int_0^\theta \frac{\exp[-y^2 \sin^4\theta\, Pe/4a^2 n \chi_1(\theta, \theta')]\, d\theta'}{\chi_1(\theta, \theta')^{1/2}}, \tag{38}$$

where

$$\chi_1(\theta, \theta') = \cos\theta' - \cos\theta - (\cos^3\theta' - \cos^3\theta)/3\,. \tag{39}$$

From equation (38) it is apparent that the concentration has been reduced exponentially at a distance from the surface $\delta_D = a/Pe^{1/2}$. From (38) we also have the local adsorption equilibrium

$$\frac{\Gamma_0 - \Gamma(\theta)}{\Gamma_0} = \frac{c_0 - c(a, \theta)}{c_0} = \frac{2}{\sqrt{\pi}} \frac{\Gamma_0}{c_0} \frac{Pe^{1/2}}{an^{1/2}} \cdot I(\theta) \tag{40}$$

where

$$I(\theta) = \int_0^\theta \frac{\cos\theta' \sin\theta'\, d\theta'}{\chi_1^{1/2}(\theta, \theta')}. \tag{41}$$

From equations (40, 41) it is seen that for low surface activity

$$\Gamma_0/c_0 \ll \delta_D\,, \tag{42}$$

then equation (10) specifying relatively insignificant adsorption change is satisfied.
On the other hand, if instead of (42),

$$\Gamma_0/c_0 \gg \delta_D\,, \tag{43}$$

then the adsorption over most of the bubble surface is considerably less than Γ_0. In other words, if (43) is valid, then

$$c(a, \theta) \ll c_0 \tag{44}$$

and to a first approximation the simplified boundaıy condition is

$$c(a, \theta) = 0\,. \tag{45}$$

A solution to (5) with the boundary conditions (45) and (33) is

$$c(y, \theta) = \frac{2c_0}{\sqrt{\pi}} \int_0^N e^{-z^2}\, dz\,, \tag{46}$$

where

$$N = \left(\frac{V_0}{aD}\right)^{1/2} \frac{\sqrt{3}}{2} \frac{1 + \cos\theta}{\sqrt{2 + \cos\theta}}\, y\,. \tag{47}$$

The last formula yields an equation for the angular dependence of the thickness of the diffuse boundary layer

$$\delta_D(\theta) = \sqrt{\frac{\pi}{3}}\, \delta_D n^{1/2} \frac{\sqrt{2 + \cos\theta}}{1 + \cos\theta}\,. \tag{48}$$

Using (46) to derive $\frac{\partial c}{\partial r}(a, \theta)$ and introducing it into (30), we have the adsorption distribution equation

$$\frac{1}{a \sin\theta} \frac{\partial}{\partial\theta} [\Gamma(\theta) V_0 \sin^2\theta] = \frac{D}{\delta_D n^{1/2}} \sqrt{\frac{3}{\pi}}\, c_0 \frac{1 + \cos\theta}{\sqrt{2 + \cos\theta}}, \tag{49}$$

which is easily integrated

$$\Gamma(\theta) = \frac{2\delta_D n^{1/2}}{\sqrt{3\pi}} c_0 \frac{1 - \cos\theta}{\sin^2\theta} \sqrt{2 + \cos\theta}. \tag{50}$$

The constant of integration has been adjusted to make $\Gamma(\theta)$ finite at $\theta \to 0$. The singularity at $\theta \to \pi$ is eliminated when surface diffusion is taken into account. Equation (50) shows that if (43) is valid, then (44) is satisfied everywhere except near $\theta = \pi$.

The results for these extreme cases have a simple physical interpretation. The tangential current of adsorbed material along a droplet surface is proportional to Γ_0, and the rate of bulk to surface exchange is proportional to $t_0 C_0$.

It is furthermore clear that when the value of the ratio Γ_0/C_0 is sufficiently small, then tangential transport is so small and the bulk to surface exchange so large that the adsorption is practically always at its equilibrium value Γ_0. If Γ_0/C_0 is large enough, then bulk to surface exchange is insufficient, and the steady state adsorption distribution differs widely from the equilibrium one.

Now let us examine the dynamic adsorption layer without $\frac{\partial c}{\partial r}(a, \theta)$ restricting the value of the surface activity. Equation (5) and the left hand side of (30) have to be related to $c(a, \theta)$ or equivalently to $\Delta(\theta)$. By operational methods the solution of equation (5) regarding the boundary condition of infinity (35) is $c(a, \theta)$:

$$c(t, \psi) = \frac{2}{\sqrt{\psi}} \int\limits_{\psi/2\sqrt{Kna^2t}}^{\infty} c\left(t - \frac{\psi^2}{4Kna^2\zeta^2}, 0\right) e^{-\zeta^2} d\zeta - c_0 \operatorname{erf} c \frac{\psi}{2\sqrt{Kna^2t}}. \tag{51}$$

From (51) we determine the derivative $\left.\frac{\partial c}{\partial \psi}\right|_{\psi=0}$ and avoid an ambiguity in the limit $\psi \to \infty$ by transforming to the variable $t' = t - \psi^2/4Kna^2\zeta^2$:

$$\frac{\sqrt{\pi}}{2} \left.\frac{\partial c}{\partial \psi}\right|_{\psi=0} = \frac{c_0 - c(\theta, 0)}{2\sqrt{Kna^2t}} + \frac{1}{2\sqrt{Kna^2}} \int\limits_0^t \frac{\partial c(t', 0)}{dt'} \frac{dt'}{\sqrt{t - t'}}. \tag{52}$$

Into (30) we substitute $\frac{\partial c}{\partial y} = \frac{\partial c}{\partial \psi} \frac{\partial \psi}{\partial y}$ and recall that $\frac{\partial \psi}{\partial y} = a V_0 \sin^2\theta$, $\frac{\partial t}{\partial \theta} = \sin^3\theta$. Then after some manipulation (30) becomes

$$\frac{\sqrt{\pi}}{2} \frac{d}{dt} [\Gamma(\theta) \sin^2\theta] = \frac{\delta_D n^{1/2}}{2} \left[\frac{c_0 - c(\theta, 0)}{\sqrt{t}} + \int\limits_0^t \frac{dc(t', 0)}{dt'} \frac{dt'}{\sqrt{t - t'}}\right] \tag{53}$$

Equation (53) is integrated on both sides from 0 to t

$$\frac{\sqrt{\pi}}{2}\alpha c(\theta, 0)\sin^2\theta = \frac{\delta_D n^{1/2}}{2}\left[(c_0 - c(0,0))\, 2\sqrt{t} + \int_0^t dt' \int_0^{t'} \frac{dc}{dt''}\frac{dt''}{\sqrt{t'-t''}}\right]. \quad (54)$$

Inverting the order of integration in the double integral and following by an integration by parts one obtains

$$\int_0^t dt' \int_0^{t'} \frac{dc(t'', 0)}{dt''}\frac{dt''}{\sqrt{t'-t''}} = -2c(0,0)\sqrt{t} + \int_0^t \frac{c(t', 0)}{\sqrt{t-t'}}dt'. \quad (55)$$

Upon substitution of (55) into (54) we have finally

$$\alpha^{-1}\Gamma(\theta)\sin^2\theta = c(a,\theta)\sin^2\theta = m\left[c_0\sqrt{t} - \frac{1}{2}\int_0^t \frac{c(t',0)}{\sqrt{t-t'}}dt'\right], \quad (56)$$

where

$$m = 2\delta_D n^{1/2}/\alpha\sqrt{\pi}.$$

Before reducing the integral equation (56) to canonical form, let us consider the special cases $m \ll 1$ and $m \gg 1$. With $m \ll 1$, the integral in equation (56) may be neglected, and the result agrees with (50), because the condition $m \ll 1$ is identical with (43).

If $c(t, 0)$ in the integral term of (56) is substituted by $c(a, \theta) = \alpha\Gamma(\theta)$ according to (50) one can see that in the right hand side of this equation the relation of the second term to the first one is of m order everywhere exept near $\theta = \pi$. The logarithm divergence there can be eliminated by taking into account the surface diffusion. Therefore the integral of (56) may be neglected.

If $m \gg 1$ the solution must be of the form $c(a, \theta) = c_0 + \delta c(a, \theta)$ with $\delta c(a, \theta) \ll c_0$ everywhere except at $\theta = \pi$ (the justification for this will be given later). Inserting this form for $c(a, \theta)$ into (56) we have for $c\delta(a, \theta)$

$$c_0\sin^2\theta = -\frac{m}{2}\int_0^t \frac{\delta c(a,\theta)}{\sqrt{t-t'}}dt'. \quad (57)$$

In the left hand side of (57) we write

$$\varphi(t) = \sin^2\theta \quad (58)$$

whence there results an ABEL integral equation

$$\delta c(a, t) = \frac{2c_0\sin\pi/2}{m\pi}\left[\frac{\varphi(0)}{t^{-1/2}} + \int_0^t \frac{\varphi'(\tau)\,d\tau}{\sqrt{t-\tau}}\right]. \quad (59)$$

From (58) we have $\varphi(0) = 0$. Because $d\varphi/d\tau = 2\sin\theta'\cos\theta'(d\theta'/d\tau)$ then after insertion of this quantity into (59) we recover (40), for the condition $m \gg 1$ being identical with (42).

We conclude that the dynamic adsorption equation (56) leads in the limiting cases to the same results obtained by other methods. Expressed in terms of the variables $x = \cos\theta$; $y(x) = \sin^2\theta$ and $c(a, \theta)$, equation (56) is transformed into the VOLTERRA integral equation

$$y(x) = mc_0 x \sqrt{\frac{2}{3} + \frac{x^3}{3} - x} + \frac{m}{2} \int_1^x \frac{y(x')\,dx'}{[x - x' - (x^3 - x'^3)/3]^{1/2}} . \tag{60}$$

The theory of the dynamic adsorption layer of a bubble with $Re \gg 1$ and weak surface motion retardation

At the present time there exists no quantitative theory of droplet motion in liquids with a large REYNOLDS number. For bubbles, however, the case is simpler, gas viscosities being so much smaller than liquid viscosities. It has been shown that to a first approximation the flow around a gas bubble at large REYNOLDS number is identical with the flow around a body immersed in an ideal liquid, provided that no surfactants are present on the bubble surface.

For $Re < 800$, the bubble retains its spherical shape and the velocity field is that for an ideal liquid

$$V_r^{(0)} = -U(1 - a^3/r^3)\cos\theta , \qquad V_\theta^{(0)} = U(1 + a^3/2r^3)\sin\theta . \tag{61}$$

U is the fluid velocity and a is the bubble radius.

In the second approximation it is necessary to set the tangential stress on the bubble surface equal to zero. The resulting boundary condition is not satisfied by (61):

$$\eta\left(\frac{1}{r}\frac{\partial V_r}{\partial\theta} + \frac{\partial V_0}{\partial r} - \frac{V_0}{r}\right)_{r=a} = 0 . \tag{62}$$

The true solution can be written in perturbation form $\vec{V}(r, \theta) = \vec{V}^{(0)}(r, \theta) + \vec{V}^{(1)}(r, \theta)$ in which

$$V^{(1)}(r, \theta) \ll V^{(0)}(r, \theta) \tag{63}$$

everywhere within the hydrodynamic boundary layer $\delta \sim a/Pe^{1/2}$. Within this boundary layer $y = (r - a) \ll a$, equation (61) is greatly simplified

$$V_\theta^{(0)} = V_0 \sin\theta , \qquad V_r^{(0)} = -3U\frac{y}{a}\cos\theta , \tag{64}$$

where

$$V_0 = \tfrac{3}{2}U . \tag{65}$$

Assuming (63—65), it may be shown (10) that $V_\theta^{(1)}$ satisfies

$$\frac{3U}{a}V_\theta^{(1)}\cos\theta + \frac{3}{2}\frac{U}{a}\sin\theta\frac{\partial V_\theta^{(1)}}{\partial\theta} - \frac{3U}{a}y\cos\theta\frac{\partial V_\theta^{(1)}}{\partial y} = \nu\frac{\partial^2 V_\theta^{(1)}}{\partial y^2} . \tag{66}$$

Transforming to variables $t = p_1(\theta)$, $\psi = y \sin^2 \theta$, $f = V_\theta^{(1)} \sin \theta$, (66) becomes the heat conduction equation

$$\frac{\partial f}{\partial t} = K_1 \frac{\partial^2 f}{\partial \psi^2} \tag{67}$$

in which

$$K_1 = 2a\nu/3U . \tag{68}$$

The boundary condition for (67) is obtained by substituting $V_\theta = V_\theta^{(0)} + V_\theta^{(1)}$ into (62) and transforming to the new variables.

The result is

$$\left.\frac{\partial f}{\partial \psi}\right|_{\psi=0} = \frac{3U}{a}. \tag{69}$$

As $\psi \to \infty$ we must also have $f \to 0$. Using a source function

$$\frac{1}{2} \frac{\exp\left[-\psi^2/4K_1(t-z)\right]}{\left[\pi K_1(t-z)\right]^{1/2}}$$

and recognizing that (69) is equivalent to a non-steady heat source of strength $K_1 \frac{\partial c(t)}{\partial \psi}$, then for an initially homogeneous field the solution to (67) is

$$f(t, \psi) = \left(\frac{K_1}{\pi}\right)^{1/2} \frac{3U}{a} \int_0^t \frac{\exp\left[-\psi^2/4K_1(t-z)\right]}{(t-z)^{1/2}} dz \tag{70}$$

whence

$$V_\theta^{(1)}(\theta, y) = -\frac{2U}{\sin\theta}\left(\frac{3}{2\pi Pe}\right)^{1/2} \int_0^{P_1(\theta)} \frac{\exp\left[-y^2 \sin^4\theta/4K_1(p_1(\theta)-z)\right]}{(p_1(\theta)-z)^{1/2}} dz \tag{71}$$

and on the bubble surface

$$V_\theta^{(1)}(0, \theta) = -\left(\frac{8}{\pi Pe}\right)^{1/2} \frac{U \sin\theta(2+\cos\theta)^{1/2}}{1+\cos\theta}. \tag{72}$$

To derive a convection diffusion equation, it is important to simplify the expression for $V_\theta(y, \theta)$ inside the diffuse boundary layer:

$$V_\theta(y, \theta) = V_0 \sin\theta + \frac{\partial V_\theta}{\partial y}(0, \theta)\, y . \tag{73}$$

Using (62) and taking into account (63) and $Re \gg 1$ there results

$$\frac{\partial V_\theta}{\partial y}(\theta, 0) = \frac{V_\theta(0, \theta)}{a} = \frac{V_\theta^{(0)}(0, \theta) - V_\theta^{(1)}(0, \theta)}{a} \approx \frac{V_0}{a} \sin\theta . \tag{74}$$

According to (73, 74), $V_\theta(\theta, y)$ changes only slightly in the diffuse boundary layer and has the same angular dependence as in the case $Re < 1$; $Pe \gg 1$. It is therefore possible to reduce the convective diffusion equation to the form (4) in this case as well; but care is needed, because the coefficient on the right hand side of (4) has a different interpretation. Furthermore, it is easier if $Re \gg 1$ to justify the neglect of surface diffusion with respect to convection in the boundary condition (19) than if $Re \ll 1$ and $Pe \gg 1$. Thus if V_0 is unrelated to U, the convective diffusion equation and the boundary condition (50) with $Pe \gg 1$, $Re \ll 1$ and also with $Re \ll 1$ are completely identical. This implies that the adsorption fields (40) and (50) are equally valid if $Re \gg 1$. On the other hand, if $Re \gg 1$, then the velocity U which appears in the PECLET number can no longer be expressed by equation (16).

According to (65), $n = U/V_0 = 2/3$. It is worth mentioning that at $Re \ll 1$ and weak retardation U_L is expressed by equation (16). In the case of weakly retarded bubble $\eta' = \gamma = 0$. This simplifies equations (16), (17) and, finally, $n = 2$.

The range of validity of (40) and (50) expressing the adsorption distribution will change from the case $Re \gg 1$ to the case $Re \ll 1$, the diffusion layer thickness δ is smaller in the former case than in the latter.

The effect of the dynamic adsorption layer on coagulation processes

Both the surface concentration of adsorbed material and the surface tension are functions of position on the dynamic surface. Every fluid element on the surface experiences a force directed along the gradient of increasing surface tension. The tangential liquid motions caused by these surface tension gradients can either hinder or accellerate the thinning of films [11—13], and of these effects, the MARANGONI-GIBBS effect is the best known. It plays an important role in the stabilization of foams and emulsions.

As an example, suppose that a soap film is stretched and thinned. The increase in surface area causes a drop in surface concentration of adsorbed surfactant and a consequent rise in the surface tension, an effect which tends to resist the process of film stretching.

The kinetics of coalescence of liquid emulsion droplets (phase 1) suspended in a second immiscible liquid (phase 2) is influenced by the presence at the interface of a third component, soluble in both phases and able to lower the interfacial tension. It is found experimentally, that if the third component is removed from the interface by diffusion into phase 2, the coalescence rate increases, while if it diffuses into phase 1, the rate decreases. This observation may be explained by noting that diffusion into the outer phase causes an increase in surfactant concentration in the gap between two approaching droplets, while diffusion into the inner phase has the opposite result. The gap region in the former case, enriched in surface active material, produces a lower local interfacial tension, the surface spreads away from the gap region, which then thins and increases the rate of droplet encounter and coalescence. If surface active material diffuses into the droplet interior, an analogous argument shows that the gap between two approaching droplets increases in thickness due to surface motion, and coalescence is correspondingly inhibited.

These effects have been thoroughly investigated by MACKAY and MASON [14], who studied the kinetics of layer thinning between a flat surface and an approaching droplet. An interferometric method was used and the results compared with theoretical predictions based upon laminar flow theory for the bulk fluid in the gap and stability theory for the fluid interface. The theory was simplified by idealizing the gap to that between two parallel flat plates.

In first approximation the experimental results support the theory in so far as the viscous resistance of the thinning gap is concerned. The effect of a third soluble component on the kinetics of thinning was also at least qualitatively in accord with expectations for a diffusion model, but quantitatively it was not possible to overcome the mathematical difficulties associated with surface motion caused by surface tension. The fact that the latter is present was confirmed by microscopic observation of dust particles at the droplet interface.

Reference [15] presents results of a similar nature. The effect of surfactant diffusion on droplet coalescence confirms the results of MacKay and Mason. In addition, the influence of surface inactive materials such as inorganic salts was investigated. As it is expected, the sign of the effect is reversed when a surface inactive substance is substituted for a surface active one. Thiessen [15] points out the relevance of studies of this type to the extraction of salts from aqueous solutions using organic solvents. The isolation of metals by this technique is a currently popular one.

With regard to foam stability, the attention of investigators has been drawn [16 — 18] to the process of bubbles approaching a gas-liquid interface, the thinning and destruction of the interfacial film, its kinetics, and the influence of surface active substances upon the process. Predictions for the thinning rate based upon viscous flow theory have been confirmed experimentally. The theory takes into account surface tension gradiants generated by surface active materials in the system and which act to impede the thinning velocity. Klassen's results [19] regarding the thinning kinetics between two converging bubbles are in good accord with the idea of surface tension gradients coupled with viscous forces.

In flotation work the mechanism of thinning of the aqueous layer between a mineral particle and a bubble is unclear. The film thins as if under the influence of inertial forces, and it is destroyed as if by particle impact. As the inertial forces for tiny particles are small compared to viscous traction forces, Sutherland and Work [20] have hypothesized that attraction forces exist between particle and bubble causing their mutual adhesion. These forces are sufficiently long range as to exceed the thickness of the boundary layers. This theory has been employed to explain an experiment performed by Ewans and Ewers [21].

A small section of the surface of an inclined solid was coated with a thick layer of inorganic, soluble surfactant. The surface was flushed with water and the thickness of the draining water film proved to be uniform over the inclined surface except over that section covered with surfactant, where the draining film was much thinner. Because in the transition region between thick and thin films the air/water interface must exhibit a local curvature, the investigators concluded that the resultant capillary pressure was compensated by long range forces originating in the surfactant patch, said forces being of longer range than the film thickness equal to tens of microns. While Sutherland and Work did not explain the nature of these long range forces, they did point out the significance of their findings for the theory and practice of flotation. The long range effect which causes the thinning, is associated with the desorption of the surfactant from the solid surface and its diffusive transport into the air/water interface. The local lowering of the surface tension generates a Marangoni effect with the outward flux of film directed away from the surfactant patch towards the thicker regions of water film, thus enhancing film thinning.

The influence of the ionic adsorption layer upon coagulation processes [22—25]

Any portion of an dynamic ionic adsorption layer is an electric double layer diverted from electroneutrality. The adsorbed layer aquires the charge of the fast diffusing ion, while the diffuse layer takes the charge of the slow diffusing ion. It is possible to survey adsorption layer interactions qualitatively and to describe their kinetics without rigorous mathematical analysis. The initial adsorption of surface active ions is followed by the adsorption of the counter ions which reside in the diffuse double layer. Macroscopically equivalent numbers of oppositely charged ions are involved so as to preserve overall electric neutrality, each ion being transported by diffusion. Counter-ions are rapidly diffusing compared with the surface active species, but the build-up of electric fields at the interface modifies the transport in such a way as to accellerate the slowly diffusing species and decellerate the rapidly diffusing ones so that finally the fluxes of each are equal. Thus the condition of approximate equality of the currents of adsorbing ions and counter-ions is provided by the appearance of the excessive surface charge the sign of which coincides with that of fast-diffusing ions.

It remains to examine the effect of the ionic character of the surfactant upon surface forces. First diffusion — electric analog of the MARANGONI-GIBBS effect could be mentioned. Since the film stretching provides adsorption, both film surfaces are charged simultaneously and are pushed away from each other which makes film thinning difficult. This repulsion of non-equilibrium double layers takes place at distances which are many times greater in comparison with the thickness of the double layer.

In the experiment of EWANS and EWERS described in the previous section, if ionic surfactants are used, the mechanism of film thinning is more complicated. The solid substrate from which the ionic surfactant is desorbed aquires a charge opposite to that generated by the adsorption of the surfactant at the air/water interface. The resulting electrostatic attraction contributes to making the film thinner [24], said effect being superimposed upon the MARANGONI hydrodynamic thinning described previously. The combined result is understandable only after thorough mathematical analysis of the role played by each mechanism in the overall effect. Only in the case of high concentrations and high surface activity the electrostatic contribution to the thinning kinetics could be neglected. THIESSEN [15], in discussing the kinetic influence of inorganic electrolytes, needed to invoke the arguments of reference [25] to explain her results.

The MARANGONI-GIBBS effect in flotation

Tiny particles immersed in a fluid move with the stream-lines of the fluid, meaning that in the vicinity of a bubble they tend to stream past it. Larger particles which possess appreciable intrinsic inertia will deviate from the stream lines of the fluid and tend to impinge upon the bubble. There thus exists a critical particle size a_{cr} for which if $a < a_{cr}$ the inertial capture of the particle by the bubble is impossible [26]. The flotation of tiny particles is nevertheless possible if the distance between the bubble and particle surfaces becomes smaller than a critical value h_{cr}. The film of thickness h_{cr} will be thinned, become unstable, and collapse if a long range attractive force exists between the particle and the bubble, drawing them together, and h_{cr} is determined by the effective range of these long range forces.

In the theory of flotation of small particles developed by DERYAGUIN, DUKHIN and RULEV [27, 28] it was not necessary to take into account the influence of the dynamic adsorption layer. The adsorption change along the bubble surface brings about the change of adsorption in the interfacial film as well. The adsorption in the center of the interfacial film is smaller then in its periphery so the central surface tension is greater than the peripherical one. But the increase of the surface tension towards the centre of the film should be followed by the surface flow in the same direction involving all film liquid. And the flow of liquid in the gap between the particle and the bubble prevents the film thinning and the flotation. It would be natural to call this effect the interfacial film stabilization by the dynamic adsorption layer in the elementary flotation act. There is an analog of the MARANGONI-GIBBS effect in the elementary flotation act of the bubble surface. This mechanism describes how a dynamic adsorption layer can contribute to the stabilization of an interface film via the MARANGONI-GIBBS effect, and this contribution will be zero if the surface concentrations on the bubble remain uniform.

The quantitative analysis of this effect requires the calculation of adsorption drop between the centre of the film $\Gamma^f(0, t)$ and its periphery $\Gamma^f(a, t)$. The calculation has been performed for the disk-like particle, because the film could be regarded as flat-parallel, and the diffusion process inside the film has an axis symmetry. Thus, the

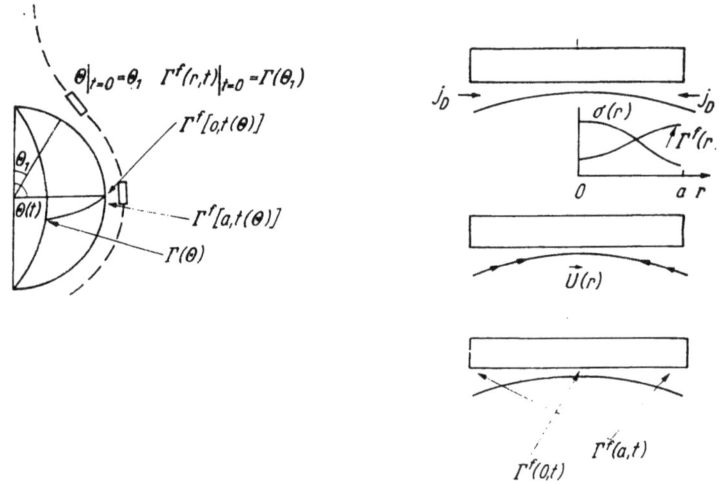

Fig. 1: It explains the mechanism of interphase film stabilization by dynamic adsorption layer.

(a) trajectory of flat particle along bubble surface. θ_1 is the angle which characterizes the particle entering bubble diffusion layer at the moment $t = 0$, $\Gamma(\theta)$ is adsorption distribution along bubble surface, $\Gamma^f[0, t(\theta)]$, $\Gamma^f[a, t(\theta)]$ – time dependences of adsorption in the centre and in the periphery of the interphase film, resp.

(b) adsorption distribution Γ^f, surface tension σ and velocity $\vec{u}(r)$ in the interphase film, j_D is the diffusion stream.

equation of the non-steady diffusion is

$$\frac{\partial \Gamma^f}{\partial t} = D_s\left(\frac{\partial^2 \Gamma^f}{\partial r^2} + \frac{1}{r}\frac{\partial \Gamma^f}{\partial r}\right), \tag{75}$$

in which D_s is the coefficient of the surface diffusion of the surfactant molecules. The initial and boundary conditions are, respectively

$$\frac{\partial \Gamma^f}{\partial r}(0, t) = 0,\ \Gamma^f(a, t) = \Gamma(t)\,, \qquad \Gamma^f(r, 0) = \Gamma(\theta_1)\,. \tag{76}$$

The solution of equation (7) can be presented with the help of DUAMEL integral

$$\delta\Gamma^f(r, t) = \sum_{n=1}^{\infty} \frac{2J_0(\mu_n r/a)}{\mu_n J_1(\mu_n)}\,\chi_n(t)\,, \tag{77}$$

where

$$\chi_n(t) = \frac{\mu_n^2}{\tau_D}\int_0^t (\Gamma(\tau) - \Gamma(\theta_1))\,e^{-\mu_n^2(\tau-t)/\tau_D}\,d\tau + \Gamma(t) - \Gamma(\theta_1)\,. \tag{78}$$

J_0 and J_1 are the BESSEL functions of the first type, μ_n are the roots of J_0, $\tau_D = a^2/D_s$ is the time relevant to that of the diffusion in the film. The stabilizing effect of the non-equilibrium adsorption layer can be neutralized by the pressing force due to the normal component of the liquid velocity. The liquid moves towards the bubble surface near the upper hemisphere. In doing that liquid flows around the particle, as though pressing it to the bubble. We can get the formula (29) for the force which presses the disk-like particle with thickness b to the bubble surface at $Re \gg 1$

$$F_1 = 48\eta\,\frac{abU}{R}\cos\theta\,. \tag{79}$$

The flow pressing particle to the bubble surface causes also excessive pressure in the interfacial film $\delta p(r)$ which should compensate the pressing force

$$F_2 = \int_0^a 2\pi r\,\delta p(r)\,dr\,. \tag{80}$$

There is an extreme value of the film thickness h_{lim}. At $h > h_{\text{lim}}$ the film gets thinner due to the pressing force. At $h = h_{\text{lim}}$ film thickness is stabilized because the gap out-flow, caused by pressure drop, is balanced by the in-flow due to the surface-tension drop. We have calculated extreme film thickness for the case of the plane-parallel gap between the disk-like particles and the almost flat ones of the greater surface section bubble. This extreme film thickness is determined by the film thickness stability condition. It corresponds to the equation of zero liquid flow through any cylinder section of the film

$$2\pi r\int_0^{h_{\text{lim}}} u(r, z)\,dz = 0 \tag{81}$$

in which the radial tangential velocity distribution $u(r,z)$ is determined by the co-action of the pressure gradients and the surface tension. The influence of $\partial p/\partial r$ on $u(r, z)$ is described in the NAVIER-STOKES equation (82) where the influence of $\partial\sigma/\partial r$ is considered while it is solved within the boundary condition (83) because $\partial\sigma/\partial r$ must be balanced by

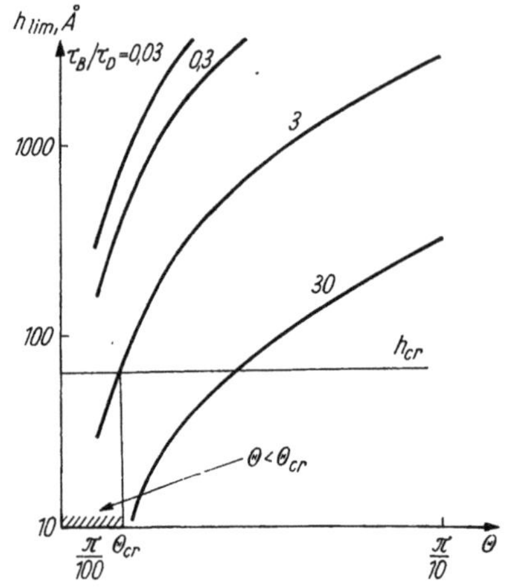

Fig. 2: An extreme thickness of interphase film $h_{\lim}$ for the different bubble surface sections characterized by the angle θ.

The region where the impossibility of deposition has not been proved, $\theta < \theta_{cr}$.

the liquid viscous tension on the surface-air interface. The Navier-Stokes equations simplification by approximation of the thin films hydrodynamics is favourable for the identification of the $u(r, z)$ function:

$$\eta \frac{\partial^2 u}{\partial z^2} = \frac{\partial p(r)}{\partial r}, \tag{82}$$

$$u_r|_{z=h} = 0; \qquad \left.\frac{\partial u_r}{\partial z}\right|_{z=0} = \eta^{-1} \frac{\partial \sigma}{\partial r}. \tag{83}$$

Here $z = 0$ and $z = h$ correspond to the surfaces of the interfacial film which, accordingly, bounder the air and the solid particle. Two unknown constants which appear with the solution of the equation (82), are determined by the boundary condition (83). Substitution of the expression for $u(r, z)$ in equation (81) and the intergration gives us the connection between pressure and adsorption distribution within the interfacial film:

$$\delta p(r) = \frac{3}{2h_{\lim}} \frac{\partial \sigma}{\partial r} d\Gamma(r). \tag{84}$$

This equation can lead to the formula for h_{lim}, because the adsorption distribution is expressed by (77), and the pressure distribution is connected through (80) with the pressure force:

$$-F_1 = F_2 = \frac{3}{2h_{\lim}} \int_0^a 2\pi\, r\, \delta\Gamma(r)\, dr. \tag{85}$$

The lefthand side in this equation can be expressed with the help of equation (80) and the well-known formula for the bubble velocity

$$U_b = \frac{1}{g} g a^2/\nu, \tag{86}$$

in which g is gravity acceleration, ν is the kinematic viscosity of water. The righthand side can be expressed by the substitution of the absorption distribution (77). Integration is performed with the formula

$$\int_0^a J_0\left(\frac{\mu_n \Gamma}{a}\right) r\, dr = \frac{a}{\mu_n} J_1(\mu) . \tag{87}$$

The obtained relations make possible to express h_{lim}

$$h_{\text{lim}}(\theta) = \frac{\gamma}{\eta} \frac{3\pi R}{16} \frac{a}{b} \sum_{n=1}^{\infty} \frac{\chi_n(t)}{\mu_n} . \tag{88}$$

Here γ is expressed by (26).

We shall use the dependence $\theta(t)$ characterizing both the interfacial film and the particle transport along the bubble surface

$$\operatorname{tg} \frac{\theta}{2} = \operatorname{tg} \frac{\theta_2}{2} e^{t/\tau_b} , \tag{89}$$

where $\tau_b = R/V_0$ is the characteristic time the bubble surface motion.

This dependence can be obtained by integrating of the equation of the motion of the bubble surface section

$$R \frac{d\theta}{dt} = V_0 \sin \theta . \tag{90}$$

The angular dependence of the adsorption on the upper-half of the bubble surface can be approximated by the formula

$$\Gamma(\theta) = \Gamma_0 - \Delta\Gamma \cos \theta \tag{91}$$

in which

$$\Delta\Gamma = c_0 \left(\frac{\Gamma_0^2}{c_0}\right) \frac{Pe^{1/2}}{R} . \tag{92}$$

The substitution of the expression (91) for $\Gamma(\theta)$ into equation (78) and integration gives

$$\chi_n(t) = 2\Delta\Gamma \operatorname{tg}^2 \frac{\theta_1}{2} (e^{-2t/\tau_b} - e^{-\mu_n^2 t/\tau_D}) \frac{1}{1 + \mu_n^2 \tau_b / 2\tau_D} . \tag{93}$$

On the basis of algebraic transformations using (92), (89), (26) and (93), the formula (88) could be reduced to the following one

$$h_{\text{lim}} = \frac{\gamma}{\eta} \frac{9\pi R}{48} \frac{a}{b} \operatorname{tg}^2 \frac{\theta}{2} \sum_{n=1}^{\infty} \frac{1 - \left[\dfrac{\operatorname{tg}(\theta_1/2)}{\operatorname{tg}(\theta/2)}\right]^{2(1+\mu_n^2\tau_b/2\tau_D)}}{\mu_n^2 (1 + \mu_n^2 \tau_b / 2\tau_D)} . \tag{94}$$

This formula is considerably simplified if $\theta > 2\theta_1$, which permits us to drop the compound term dependable on θ. We have calculated the dependence of $h_{\text{lim}}(\theta)$ at $\theta < 2\theta_1$ and the number of values $\tau_b/2\tau_D$ with

$$\frac{\gamma}{\eta} = 0.3 ; \qquad \frac{a}{b} = 3 ; \qquad R = 0.1 \text{ cm} .$$

The curves shown on Figure 2, are calculated for the values $\tau_b/\tau_D = 3;\ 0.3;\ 0.03$ which correspond to the radii 1 μm, 3.3 μm, 10 μm, presuming that $D_s = 10^{-5}\dfrac{\text{cm}^2}{\text{sec}^2}$; $\tau_b = 3 \cdot 10^{-3}$ sec; $V_0 = \frac{3}{2}\,U = 30$ cm/sec at $R = 0.1$ cm. The particle could approach the bubble at a distance less then h_{cr} (which is necessary for flotation) only being close to the upper pole of the bubble.

The angles region, determined by the condition

$$h_{lim}(\theta_{\text{cr}}) = h_{\text{cr}} \tag{95}$$

is so narrow (Fig. 2), that the flotation efficiency is very inconsiderable. The formula for θ_{cr} could be easily obtained from (94) if we put $h_{\text{lim}} = h_{\text{cr}}$

$$\theta_{\text{cr}} = \left[\frac{\eta b}{\gamma a}\,\frac{64 h_{cr}}{3\pi R}\Bigg/\sum_{n=1}^{\infty}\frac{1}{\mu_n^2(1 + \mu_n^2\tau_b/2\tau_D)}\right]^{1/2}. \tag{96}$$

The time of existence of the interfacial film (the time of the bubble motion with the adjacent particle surface section) approximately equals to the order of τ_b. The time of equalization of concentration in the interfacial film brought about by the diffusion, equals to the order of τ_D. If $\tau_D/\tau_b \gg 1$ there is no time for the diffusion to level up in order to effect concentration drop. So, at great values of this ratio its influence upon the stabilization effect is inconsiderable (see (96)).

From (90) it could be seen that at $\theta \to 0$ the rate of the surface motion and accordingly, the particle with the interfacial film is also reduced. Therefore, the closer the particle to the upper pole of the bubble, the more considerable is the equalization of concentration drop in the interfacial film. Consequently, h_{lim} is reduced with θ, what could be seen from (94), and the particle deposition is possible near the upper pole.

During the transition from the plate-like particle to the spherical one a quick weakening of the interfacial film stabilization effect should be expected. With the same excessive pressure in the gap, caused by the pressing force, the film thinning would be more intensive, because the thickness of the interfacial film (in case of sphere) is rapidly increased, being directed from its centre. This makes the out-flow of liquid easier. Let us presume that the interfacial film of radius a is transferred along the bubble surface together with the spherical particle of radius a, as well as in case of the disk-like particle. Then, boundary conditions (76) are satisfied. Equation (75) should be defined more precisely, since the farther is the centre of the film the thicker is the film itself. But, if $D_s \sim D$ and the approximate condition

$$\Delta/c \gtrapprox a\,, \tag{97}$$

is satisfied the interchange between the centre of the interfacial film and its periphery is performed by the surface diffusion mainly. In this case equation (75) is applicable accordingly, its solution (77) (by the substitution of D_s for D). If conditions (97) and, and (42) are simultaneously satisfied; i.e. if

$$a < \delta\,, \tag{98}$$

equation (91) can also be applied.

The radial flow in the gap between the plane and the sphere can be approximately considered as plane-parallel near the centre of the film. This approximation becomes less precise while approaching the film periphery. Here an error in the adsorption distribution definition is increased, because identification of $\Gamma^f[t(\theta)]$ and $\Gamma(\theta)$ at $r = a$ is rather conventional. So let us consider the radial flow in the film as plane-parallel, i.e. let us use equation (82) and boundary condition (83), having substituted h by $h + \Gamma^2/a$. Then (87) will preserve its meaning, and instead of (85) we get

$$-F_1 = F_2 = \frac{3}{2}\frac{\partial\sigma}{\partial r}\int_0^a 2\pi r \frac{\delta\Gamma(r)}{h + r^2/a}\,dr\,. \tag{99}$$

In order to perform integration instead of (87) we shall apply

$$\int_0^a \frac{J_0\left(\frac{\mu_n r}{a}\right)}{h + r^2/a} = aK_0\left(\mu_n\sqrt{\frac{h}{a}}\right), \tag{100}$$

where K_0 is the modified BESSEL function of the second type. The expression for the pressing force is also changed:

$$F_1 = 6\pi\eta a\frac{3a}{R}U_b\,. \tag{101}$$

According to all changes we get the analog of (96) for the case of the spherical particle

$$\theta_{\text{cr}}^{sph} = \left[\frac{4}{3}\frac{\eta}{\gamma}\frac{a}{R}\Bigg/\sum_{n=1}^{\infty}\frac{K_0\left(\mu_n\sqrt{\frac{h_{\text{cr}}}{a}}\right)}{\mu_n J_1(\mu_n)\,(1 + \mu_n^2\tau_b/\tau_D)}\right]^{1/2}. \tag{102}$$

Comparing (96) and (102) one can see that the zone of deposition for the spherical particle is approximately $(a/h_{\text{cr}})^{1/2}$ times wider.

Estimation of the effectiveness of the particles capture considering the interfacial film stabilization by the dynamic adsorption layer

The so-called effectiveness of the particles capture by the bubble is the measure of intensity of the particles deposition on the bubble surface:

$$E = b_{\text{cr}}^2/R_p^2 \tag{103}$$

in which b_{cr} is the distance of the so-called critical trajectory of a particle from axis of symmetry far from the bubble, R_p is the radius of bubble during potential flow. At $b > b_{\text{cr}}$ the trajectories of the particles go beyond the limits. At $b < b_{cr}$ they are limited by the bubble surface.

The equation for the streamlines with potential flow near the bubble is

$$b^2 = \sin^2 \theta r^2(1 - R_p^3/r^3) \tag{104}$$

This formula is simplified for r close to R_p

$$b^2 = \sin^2 \theta(H + 1)\, 3R_p a\ , \tag{105}$$

in which $H = h/a$, h is the distance between the centre of the particle and the surface of the bubble. At small distances the particle trajectory deviates from the streamlines because of hydrodynamic interaction. It requires more precise definition [27]

$$\frac{1}{H+1} F_p(H) \frac{dH}{d\theta} + 2 \frac{\cos\theta}{\sin\theta} = 0\ . \tag{106}$$

In the absence of the already mentioned interfacial film stabilizing effect the extreme trajectory is determined by the fact that the particle approaches the bubble at the distance of H_{cr} at $\theta = \pi/2$. And the integration results in:

$$\int_{H_{cr}}^{H} \frac{F_p(H)\, dH}{H+1} = \ln \frac{1}{\sin^2\theta}. \tag{107}$$

Having broken up the zone of integration into two intervals $[H_{cr}, 1]$ and $[1, H]$, and presuming that $F_p(H) = 1$ at $H > 1$ we apply the method which was used by Deryaguin [30]. At $H > 1$ we shall get:

$$\int_{H}^{1} \frac{F_p(H)\, dH}{H+1} = \ln \frac{2}{\sin^2\theta(H+1)}. \tag{108}$$

Since hydrodynamic interaction in the region of $H > 1$ may be neglected, equation (108) in this region should coincide with that of the critical streamline, i.e. with the equation (104) at $b = b_{cr}$.

Having put together (107), (108), (104) we get:

$$E_p = 6 \frac{a}{R_p} e^{-I_p}\ , \tag{109}$$

in which

$$I_p = \int_{H_{cr}}^{1} \frac{F_p(H)\, dH}{1+H}. \tag{110}$$

The influence of the interfacial film stabilizing effect makes the extreme trajectory limitation by $\theta_{cr} < \frac{\pi}{2}$. Considering that boundary condition we get:

$$\int_{H_{cr}}^{H} \frac{F_p(H)\, dH}{1+H} = \ln \frac{\sin^2\theta_{cr} \cdot 2}{\sin^2\theta} \tag{111}$$

and, accordingly, instead of (109)

$$E_p^* = 6 \frac{a}{R_p} e^{-I_p} \sin^2 \theta_{cr} . \tag{112}$$

Thus, the decrease of the capture effectiveness due to the film stabilization effect is expressed like this

$$\frac{E_p^*}{E_p} = \sin^2 \theta_{cr} \sim \theta_{cr}^2 \tag{113}$$

in which θ_{cr} is determined by (102).

References

[1] S. S. DUKHIN, Diffusionno-electricheskaya teoria neravnovesnjikh electropoverkhnostnjikh sil i electicheskikh javlenij, doctorskaya dissertatsia, Inst. Fiz. Khim. AN SSSR, Moscow, 1966

[2] B. V. DERYAGUIN, S. S. DUKHIN, V. A. LISITCHENKO, Zh. Fiz. Khim. **33**, 2280 (1959)

[3] B. V. DERYAGUIN, S. S. DUKHIN, V. A. LISITCHENKO, Zh. Fiz. Khim. **34**, 324 (1960)

[4] S. S. DUKHIN, L. V. BUJKOV, Zh. Fiz. Khim. **38**, 3011 (1964)

[5] R. M. GRIFFITH, Chem. Eng. Sci., **17**, 1054, 1962

[6] S. S. DUKHIN, Zh. Fiz. Khim. **34**, 1052 (1960)

[7] S. S. DUKHIN, "*Sb. Issledovanija v oblasti poverkhnostnikh sil*", Nauka, M., 1964, 47—49

[8] S. S. DUKHIN, B. V. DERJAGUIN, "*Electrophores*", Nauka, M., 1976

[9] Rybczynski, Bull. de Cracovie (A) (1911), p. 40
Hadamard Comp. Rend. **152**, 1735 (1911)

[10] S. S. DUKHIN, B. V. DERJAGUIN, Zh. Fiz. Khim. **35**, 1246, 1961

[11] K. S. ALAN, S. G. MASON, J. Colloid. Sci. **17**, 383 (1962)

[12] G. E. CHARLES, S. G. MASON, J. Colloid. Sci. **15**, 236 (1960)

[13] H. GROOTHIUS, F. S. ZUIDERIWEG, Chem. Eng. Sci. **12**, 288 (1960)

[14] D. C. MAC KAY, S. G. MASON, J. Colloid. Sci. **18**, 674 (1963)

[15] D. THIESSEN, Z. Phys. Chem., Lzg., **223**, 218 (1963)

[16] P. A. REHBINDER, E. K. VENSTREM, Kolloidn. Zh., **53**, 145, 1930

[17] E. K. VENSTREM, P. A. REHBINDER, Zh. Fiz. Khim. **2**, 754, 1931

[18] K. S. ALLAN, C. E. CHARLES, S. G. MASON, J. Colloid. Sci. **16**, 150 (1964)

[19] V. I. KLASSEN, Voprosy teorii aeratsii i flotatsii. Goskhimizdat, M., 1949

[20] K. C. SUTHERLAND, E. V. WORK, "*Printsipy flotatsii*", Metallurgizdat, Moskva, 1968

[21] L. F. EWANS, W. E. EWERS, "*Recent Developments in Mineral Dressing*" (London, Institution of Mining and Metallurgy, 1953, 457—63; L. F. Ewans, Ind. Enq. Chem., **46**, 2420 (2954)

[22] S. S. DUKHIN, Kolloidn. Zh. **23**, 409 (1961)

[23] S. S. DUKHIN, "*Sb. Issledovanija v oblasti poverkhnostnikh sil*", M., 1961, 38

[24] S. S. DUKHIN, Dokl. AN SSSR, **130**, 1298 (1960)

[25] S. S. DUKHIN, "*Sb. Issledovanja v oblasti poverkhnostnikh sil*", M., 1961, Izd-vo AN SSSR

[26] B. V. DERYAGUIN, S. S. DUKHIN, Izv. AN SSSR, seria metallurgija, toplivo, No. 1, 82 (1959)

[27] B. V. Deryaguin, S. S. Dukhin, N. N. Rulev, Kolloidn. Zh. **38**, 251, (1976)
[28] B. V. Deryaguin, S. S. Dukhin, N. N. Rulev, V. P. Semenov, Kolloidn. Zh. **38**, 258 (1976)
[29] J. Happel, G. Brener, *"Gidrodinamika pri malikh chislakh Reynoldsa"*, MIR, M., 1976, 174
[30] B. V. Deryaguin, L. P. Smirnov, *"Sb. Issledovanija v oblasti poverkhnostnikh sil"*, Nauka, M., 1967, str. 188

A microscopic theory of dispersion interactions in capillary systems

F. M. Kuni, A. I. Rusanov

(Departments of Physics and Chemistry,
Leningrad State University, Leningrad/USSR)

Abstract

A systematic review is given of modern results on dispersion interactions (ordinary and retarding) in capillary systems. The concept of molecular forces and the molecular nature of capillary systems is used as a starting point for the microscopic theory. A consistent molecular statistical description of capillary systems is based on the Gibbs grand canonical ensemble. A functional method is used which derives the necessary general relationships of statistical mechanics in a compact form. The solution of the basic problem of the influence of the medium on the interaction of molecular bodies is achieved as a strict result of the investigation of collective phenomena in systems of many molecules. This result is formulated as the interaction principle in the language of fundamental physical notions reflecting the role of the medium as an intermediator of interaction. A wide range of key problems in the theory of capillary systems is considered from the viewpoint of the interaction principle: molecular correlations in capillary systems, the molecular structure of flat, slightly and strongly curved surface layers, the interaction of macroscopic particles. The notions used in the interaction principle appear in these problems as compressibilities and adsorptions. They are parameters which describe those collective phenomena caused by the influence of the medium. The construction of the pair effective molecular potential is especially considered from data on X-ray scattering. During the whole article, a comparison is made with the alternative macroscopic approach which regards matter as not consisting of molecules, but as a continuum described by a macroscopic characteristic, the dielectric permittivity. This comparison involves not only the disjoining pressure of a film for which the macroscopic theory was originally formulated, but also the majority of other results on dispersion interactions in capillary systems also obtained recently within the framework of the macroscopic approach. The agreement of results is a reliable confirmation of the potential in which the molecular nature of capillary systems is expressed.

Introduction

A remarkable feature of capillary systems is their great dependence on molecular interactions. Their nature, in contrast to ideal systems, is completely determined by molecular forces. The significance of Gibbs' theory of capillarity consists in that it has expressed this peculiarity of capillary systems in the form of general thermodynamic relationships.

The thermodynamic description applied by Gibbs did not depend explicitly on molecular forces. Further progress in the understanding of capillary phenomena was not possible without more concrete ideas on molecular interactions in condensed systems. Gibbs himself had in view such a way. The most important step in this direction was the creation by GIBBS of statistical mechanics, which discovered new possibilities for explaining material properties on the microscopic level.

Though the state of capillary systems is determined by the interaction of molecules over the whole range of distances between them, the long range dispersion interactions are manifested more directly under the usual experimental conditions. The high sensitivity to dispersion contributions (meaning an integrated dependence on intermolecular forces for all distances) determines the special importance of dispersion interactions. For this reason, the study of dispersion interactions takes a central place in the physics and chemistry of liquids, solutions, heterogeneous and colloidal systems.

The presence of the medium usually surrounding the interacting molecules, microscopic particles, or macroscopic bodies, considerably complicates the formulation of the theory of dispersion interactions in condensed systems since it attaches to the theory a field character typical for a problem with an infinitely large number of interacting degrees of freedom; and demands, even in zero approximation, taking into account the dominant collective phenomena.

Theoretical studies of dispersion interactions in a condensed state have a long history. The modern stage for which the field emphasis has become determining began with the creation of the macroscopic theory of dispersion interactions proposed by LIFSHITZ [1] and then developed by DZYALOSHINSKI, LIFSHITZ and PITAEVSKY [2]. Considering the interaction of bodies to be realized by means of a fluctuating magnetic field, the authors of this theory based the description on a macroscopic property of a substance, the dielectric permittivity. Given its values over the continuum of frequencies, the dielectric permittivity yields much information about the properties of substances.

Because it is unrelated to the low density approach, the macroscopic theory has made it possible to transcend the limits of traditional gas theory, such as virial expansion in the density, and to obtain reasonable results on dispersion interaction in condensed systems.

Of course a knowledge of the frequency dependence of the dielectric permittivity necessary for a macroscopic theory is still far from being complete. Moreover, even aside from finding the dielectric permittivity frequency dependence as an independent matter not related directly to the problem of interaction, there is in the macroscopic theory another principal difficulty pertaining to the field character of the problem under consideration. Although unrestricted by the low density approach, the macroscopic theory, nevertheless, considers the medium to be quite homogeneous up to the boundary surfaces of interacting bodies. In the case of the interaction of two semi-infinite bodies analysed in the macroscopic theory (this was the only example for which the theory was formulated), the medium was a flat layer between the bodies, and this layer was assumed to be completely homogeneous up to the boundary surfaces of the bodies, themselves also taken as homogeneous.

In reality, however, the influence of the medium on the interaction between bodies is much more complex.

First, the medium changes its structure in the presence of foreign bodies and becomes inhomogeneous. It is essential that this inhomogeneity has a microscopic scale. Local physical properties of the medium change over distances comparable with molecular dimensions. Having become inhomogeneous, the medium conveys the interaction between bodies in a different manner.

A concrete physical manifestation of the medium inhomogeneity is the formation of the solvate and adsorption layers around the bodies. The adsorption layers are of great importance for capillary systems. The change in interaction caused by the adsorption layers is the origin of some of the interesting qualitative results for capillary systems.

Secondly, the influence of the medium is also manifested by arrangement of the surface of interacting bodies and by forming new surface layers of the bodies. This rearrangement, in turn, influences the interaction of the bodies. The influence of the surface layers of the bodies on their interaction may be very essential in capillary systems.

Thus, we see that the macroscopic theory has solved only that part of the real problem that is connected with the role of the medium as a homogeneous mediator of interaction. The influence on dispersion interaction of structural changes in the medium and the interacting bodies is not treated by the macroscopic theory.

The solution of the problem of dispersion interaction in condensed systems was given in our work [3—26]. The theory of dispersion interaction developed in these articles may be called microscopic. It starts directly from the first principles of statistical mechanics for systems of many interacting molecules. The problem of dispersion interaction is formulated and solved as a problem of collective phenomena in the Gibbs statistical ensemble for systems with an infinitely large number of interacting molecules.

Strictly speaking, the interaction energy is composed of energy sums for possible double, triple and so on interactions of single molecules, which result from mutual perturbations of the internal quantum-mechanical structure of single molecules. However, such a strict representation of the interaction energy is only suitable in the most general formulation. It is not acceptable for the formulation of a physical theory, and not only because of its complexity. One of the main purposes of any theory consists in recovering the information put into it from a comparison of the theory with experiment. However, it is clear that fine details included in the energy of a system as a complex function of an enormous number of particle coordinates cannot be reestablished from experimental data however detailed they may be.

It is quite sufficient, for the prediction of true local one- and two-particle properties of a real system, to represent the interaction energy as the sum of pair effective intermolecular potentials. The description of the interaction energy of a system in terms of the pair effective potential is adequate for the description of the system in terms of the frequency dependent dielectric permittivity. In both cases the information is given in terms of a function dependent on one independent variable whose role is that of the intermolecular distance, and another variable playing the role of the frequency. A closer connection between the pair effective potential and the frequency dependent dielectric permittivity may be established by the comparison of the results of the microscopic theory with the macroscopic one.

The introduction of the pair effective potential corresponding to the molecular nature of a substance and taking into account the mean influence of multi-particle forces was considered in references [27—29]. The pair effective potential of molecular interaction constructed in these articles on the basis of quantum mechanical calculations differs from the corresponding gas potential by a renormalization factor. This factor depends on the distance between the interacting molecules and on the local structure of the solvate clouds surrounding the molecules. For great distances between the molecules, the renormalization factor approaches a certain limiting value not dependent on the distance. Hence, the asymptotic dependence of the pair effective potential and the gas potential becomes the same for large distances. However, even for large distances, the limiting value for the renormalization factor depends, as before, on the state of a system.

The pair effective potential in non-uniform systems may also depend on the location of the interacting molecules. This remark also applies to the dielectric permittivity. The dependence of the latter on the spatial coordinates is approximated by a step-function in the macroscopic theory: the dielectric permittivity changes step-wise due to the transition from one phase to the other through the boundary surface. Analogous approximations should be used for the pair effective potential.

As in the macroscopic theory by LIFSHITZ, DZYALOSHINSKY, and PITAEVSKY, our microscopic theory is asymptotic. It refers to distances which are large compared with the mean intermolecular distance. Dispersion interactions are manifested for such arge distances.

The mathematical basis for the microscopic theory is the method of asymptotic summation of diagrams developed in references [7—9, 17]. The idea of this method consists in finding out the essential dispersion interaction diagrams of statistical mechanics and complete summation. The idea is realizable due to the possibility of gathering the sums of the asymptotically essential diagrams into expressions corresponding to observable physical quantities. The fact that the asymptotically essential diagrams of all orders (i.e., the interactions of any large number of molecules) are taken into account and that the sums are strictly reduced to observable quantities, guarantees the complete applicability of the method to condensed systems.

Thus, the range of asymptotic distances at which dispersion forces act offers a unique possibility of a rigorous solution of the problem within the framework of a theory for systems with an infinitely large number of interacting degrees of freedom.

Naturally, the complete solution of the dispersion interaction problem achieved in the microscopic theory has provided wide applications of the theory. The problem of the dimensions of the interacting bodies possible for consideration was already not a major one. Actually, the bodies might be taken in the whole range beginning from single molecules up to infinitely large macroscopic bodies. As a result, a large number of practical problems was solved within the framework of the microscopic theory. Among them was a problem on the interaction of two semi-infinite bodies [6, 16, 18—19] for which, as it was mentioned above, the macroscopic theory was formulated. However, it has now become possible to make progress by taking into account the influence on the interaction of the non-uniformity of both the medium layer and of the interacting bodies themselves.

The universality of the microscopic theory was most clearly expressed in those papers [22—24, 26] in which this theory was represented in the whole as the interaction principle.

The rigorous results for the case of molecular correlations in liquid obtained by using the microscopic theory [7—9] stimulated the generalization of the macroscopic theory for this case. Such a generalization was carried out by KEMOKLIDZE and

PITAEVSKY [30], the results obtained confirming the conclusions of the microscopic theory.

Very recently, the macroscopic theory was used for determining the local structure of surface layers [31, 32]. An agreement was also achieved with the results [3—6, 24—26] obtained earlier with the aid of the microscopic theory.

Naturally, only the main asymptotic terms were compared in the above cases since the corrective terms have been found only in the microscopic theory.

Summing up the results of the modern stage of dispersion interaction theory, it should be said that deep understanding and good quantitative description of dispersion interactions at large distances have been achieved by the parallel efforts of the microscopic and macroscopic approaches. At the same time, it is very encouraging that many of the rigorous results obtained on dispersion interactions have already been checked in experiments or may be checked in the near future due to the modern accuracy of measurements.

The main task of the further development of the theory will be to obtain a more exact knowledge of the pair effective potential of substances and their frequency-dependent dielectric permittivity. These facts will complement each other and will lead to a complete understanding of the complicated picture of dispersion interactions.

The purpose of this review article is to describe that part of the theory of dispersion interactions where rigorous and complete results have been obtained so far. The description will be given in the language of the microscopic approach, as pointed out in the title, and will be given "from zero", i.e., beginning from microscopic theory rooted in GIBBS' classical statistical mechanics.

More detailed comparisons with the results of the alternative macroscopic approach will be given in the course of the article.

The interaction principle [22—24, 26] has been chosen here as a general viewpoin for the whole subject including many concrete applications.

Having used consistently the whole mathematical apparatus and the asymptotic summation method for the derivation of the interaction principle (Section 1—2), we acquire with its aid the possibility of direct solution of various problems (Sections 3—8). This leads to a general formulation and makes possible the presentation of much material in little space.

The use of functional methods for establishing the general formalism of statistical mechanics also contributes to compactness of the method. This is not surprising, since functional methods correspond to non-uniform systems in which the internal properties are not numerical parameters, but functions dependent on the point of observation. Just such systems are capillary systems.

In the concluding section (Section 9), a general evaluation of the microscopic theory is given, and the meaning of the parameters used in the theory is established.

Paying attention mainly to the fundamental formulation of problems and their analytical solution, we completely omit in this article numerical results, graphs, and tables. Among the references to the literature on microscopic theory, those papers are also referred to in which the corresponding results were obtained before the establishment of the interaction principle.

1. Generating functional method in statistical mechanics

The microscopic theory of dispersion forces is a theory of systems with many interacting molecules. The description of such systems is given by statistical mechanics.

For the derivation of the necessary relationships of statistical mechanics, it is convenient to follow a functional viewpoint and use the generating properties of the grand partition function

$$\Xi = \sum_{N=0}^{\infty} \frac{1}{N!} \int \dots \int d1 \dots dN\, \zeta(1) \dots \zeta(N) \exp\left[-\beta U(1, \dots, N)\right], \tag{1.1}$$

being considered as a functional fo the function

$$\zeta(i) = z_{s_i} \exp\left[-\beta u(i)\right], \tag{1.2}$$

which plays the role of the generalized activity of a system. Here the argument i denotes a set of variables referring to the degrees of freedom of the ith molecule, i.e. its type s_i, and its spatial coordinates, x_i, including the radius-vector of its location, $\vec{r}_i$, and generally speaking, its orientation angles, Q_i.

The symbol $\int di$ denotes summing up with respect to kinds and integrating with respect to spatial coordinates: $\int di \equiv \sum_{s_i} \int dx_i$. Next, $u(i)$ is an external field, z_{s_i} is the activity connected with the chemical potential, μ_{s_i}, by the relationship

$$z_{s_i} \equiv \exp\left(\beta\mu_{s_i}\right)/\Lambda_{s_i}^3(T), \tag{1.3}$$

where $\Lambda_{s_i}(T)$ is a known function of temperature, T, resulting from an integration with respect to the momenta canonically conjugated to the spatial coordinates. Finally, $\beta \equiv I/k_B T$ is the inverse temperature, $U(I, \dots, N)$ is the interaction energy for a system with N molecules.

The distribution functions

$$\varrho(1, \dots, n) = \Xi^{-1} \sum_{N=n}^{\infty} \frac{1}{(N-n)!} \int \dots \int d(n+1) \dots dN \times \\ \times \zeta(1) \dots \zeta(N) \exp\left[-\beta U(1, \dots, N)\right], \tag{1.4}$$

describing the molecular distribution probability may be represented in the generating functional method as

$$\varrho(1, \dots, n) = \zeta(1) \dots \zeta(n)\, \Xi^{-1}\, \delta^n \Xi / \delta\zeta(1) \dots \delta\zeta(n). \tag{1.5}$$

It is convenient to introduce the notion of a complex, $\{n\}$, implying a group of n molecules of assigned types and spatial coordinates. Then, $\varrho(1, \dots, n) \equiv \varrho\{n\}$.

Formulae (1.1)—(1.5) will serve as a basis for further consideration.

From (1.2) and (1.3) we have the operator relationship

$$\beta^{-1}\, \partial/\partial\mu = \int dx_i\, \zeta(i)\, \delta/\delta\zeta(i), \tag{1.6}$$

where the derivative with respect to $\zeta(i)$ is taken at constant temperature and the derivative with respect to μ_{s_i} is taken at constant temperature and external field. Applying this relationship to the functional representation (1.5) yields

$$\int dx_1 [\varrho\{n+1\} - \varrho\{n\}\, \varrho(1)] = \beta^{-1}\, \partial\varrho\{n\}/\partial\mu_{s_1} - n_{s_1}\{n\}\, \varrho\{n\}, \tag{1.7}$$

where $n_{s_i}\{n\}$ is the number of molecules of the kind s_1 in the complex $\{n\}$. Equation (1.7) expresses the compressibility theorem in the most general form.

A realistic description of intermolecular forces is given by the pair effective (taking an average of non-additive contributions into account) interaction potential, $\Phi(i, j)$. In terms of this potential, we have

$$U(1, \dots, N) = \tfrac{1}{2} \sum_{i \neq j = 1}^{N} \varphi(i, j) . \qquad (1.8)$$

By substituting (1.8) into (1.1), the exponent in (1.1) is represented as the sum of all possible products of the MAYER functions $f(i, j) = \exp[-\beta\varphi(i, j)] - 1$. This leads to the known expression for Ξ in the form of a sum of all the diagrams the points of which correspond to generalized activities, ζ, and the lines linking the points correspond to the MAYER functions, f. Such lines are called the MAYER bonds.

Every point in the diagrams for Ξ is a field point with the understanding that it comprises summation with respect to the kinds and integration with respect to the spacial coordinates of molecules.

Evidently, the application of the operator $\zeta(i)\, \delta/\delta\zeta(i)$ to a field point makes it fixed. At such a point the kind and coordinates of a molecule, on which depends the generalized activity ζ corresponding to this point, assume fixed values of s_i and x_i.

Since in the functional method the distribution functions are found from Eq. (1.5) by successive application of the operator on Ξ (all the arguments $1, 2, \dots, n$ are supposed to be different), this furnishes a simple way for the derivation of their diagram expansions. In this way, an expression for $\varrho(1, \dots, n)$ is obtained as the sum of diagrams which consist of the fixed points $\zeta(1), \dots, \zeta(n)$, of the field points ζ, of the MAYER bonds f, and in which every field point is linked (immediately or through other field points) with at least one fixed point. It is evident that the restriction pointed out concerning the absence of field points not linked with fixed points is due to the multiplier Ξ^{-1} in (1.5). The diagram expansion for the distribution functions constructed by the generating functional method will be used below as the mathematical formalism for calculations. It is important that the final results of calculations, after summing up the essential diagrams of all orders, will be represented in terms of physically observable quantities. The existence of these quantities is not connected with the convergence of expansions. For this reason the subsequent application of the diagram expansions as a working mechanism for calculations will not be restricted by the range of the actual convergence of the expansions. This will also be valid for condensed systems for which activities playing the role of the diagram expansion parameters are not small.

Describing correlative and structural properties, the distribution functions make possible finding any local properties of a system. Corresponding expressions follow directly from the definition of the distribution functions, Eq. (1.4).

An important local property of capillary systems is the pressure tensor. In the general case when the interactions are not central and molecules have orientational degrees of freedom, the expression for the pressure tensor at a certain point of space, r, has the form

$$\hat{P}(\vec{r}) = k_B T \hat{1} \sum_p \varrho_p(\vec{r})$$
$$- \frac{1}{2} \int d1\, d2\, \frac{\partial \Phi(1, 2)}{\partial \vec{r}_2} \otimes (\vec{r}_2 - \vec{r}_1) \int_0^1 d\eta\, \delta[\vec{r} - \eta(\vec{r}_2 - \vec{r}_1) - \vec{r}_1]\, \varrho(1, 2) . \qquad (1.9)$$

Here $\hat{1}$ is the unit tensor, $\varrho_p(\vec{r})$ is the one-particle distribution function for the component p at the point $\vec{r}$ averaged (integrated) over all orientations. The derivative with respect to $\vec{r}_2$ means the gradient along the coordinate $\boldsymbol{r}_2$; the symbol $\otimes$ denotes the direct product of vectors (resulting in a tensor), and δ denotes the usual δ-function.

The pressure tensor satisfies the mechanical equilibrium condition

$$Div\ \hat{P} = 0\ , \tag{1.10}$$

which follows from Eqs. (1.4) and (1.9). Physically, the equilibrium condition expresses the fundamental law of momentum conversation in equilibrium systems.

The equilibrium conditions in the form (1.10), where the term with the external field gradient is absent, implies that in Eq. (1.9) for the pressure tensor the contributions are taken into account from all parts of space where the substance is capable of the force field action.

In the case of central interaction, when $\Phi(i, j)$ depends on spatial coordinates only through the distance between molecules, only the two-particle distribution function should be averaged with respect to angular coordinates in (1.9). Then Eq. (1.9) changes, after the replacement of integration over $\vec{r}_2$ by integration over $\vec{R} \equiv \vec{r}_2 - \vec{r}_1$, into the usual expression of Irving and Kirkwood

$$\hat{P}(\vec{r}) = k_B T \hat{1} \sum_p \varrho_p(\vec{r}) - \frac{1}{2} \sum_{p,q} \int d\vec{R}\, \frac{\vec{R} \otimes \vec{R}}{R} \frac{d\Phi_{pq}(R)}{dR} \int_0^1 d\eta\, \varrho_{pq}(\vec{r} - \eta\vec{R}, \vec{r} - \eta\vec{R} + \vec{R})\,, \tag{1.11}$$

where subscripts p and q refer to components, the distribution functions have been averaged with respect to molecular orientation. The pressure tensor defined by Irving-Kirkwood's formula is symmetrical.

2. Interaction principle

The established relationships of statistical mechanics help to answer the central question about dispersion forces in condensed systems. This question may be put in the most general form as follows.

It is necessary to find the interaction potential for two molecular bodies of any nature, shape, and dimensions in a liquid or gas medium of arbitrary composition.

Since the topic in question is dispersion forces, it is implied that the bodies are separated by a distance considerably ecxeeding the size of single molecules.

For the future, it will be necessary to know in an explicit form only the dispersion component of the pair effective intermolecular potential, i.e., that component which begins to dominate at large distances. Recognizing that the form of the distance dependence for this component is the same as in gases, we may write for it the expression

$$\Phi(i, j) = -A_{s_i s_j} r_{ij}^{-\lambda}\,. \tag{2.1}$$

Here $r_{ij} \equiv |\vec{r}_i - \vec{r}_j|$ is the intermolecular distance, $A_{s_is_j}$ is the force constant, the parameter λ takes values 6 or 7 corresponding to ordinary or retarding dispersion forces, respectively. For the sake of brevity, the same designation of the force constant is used in both cases. It should be kept in mind, however, that the constant has different numerical values and even different dimensions for these two cases.

Evidently, the interacting bodies (we will call them bodies I and 2) may be understood, in general, as the molecular complexes $\{n_1\}$ and $\{n_2\}$. The system consisting of the bodies 1 and 2 may be understood as the molecular complex $\{n_1 + n_2\}$.

The desired potential of interaction of the bodies 1 and 2, W_{12}, is the average force potential. As it is known, the average force potential is directly related to the distribution functions of statistical mechanics. In the case of two-body interaction, this relationship may be written as

$$W_{12} = -\beta^{-1} \ln [\varrho\{n_1 + n_2\}/\varrho\{n_1\}\, \varrho\{n_2\}], \tag{2.2}$$

which follows from the definition of the distribution functions, Eq. (1.4), and their meaning as the probabilities of molecular distribution.

In the range of large distances between the bodies, which is of interest to us, it is possible to use the method of asymptotic estimation [7—9] of the diagrams for the expansion obtained in Section I for the distribution function $\varrho\{n_1 + n_2\}$.

Evidently, small factors in the form of MAYER bonds, being stretched when increasing the distance between the bodies, will certainly be present in all the diagrams except for the ones which reduce to the product $\varrho\{n_1\}\, \varrho\{n_2\}$. The main contribution will be given by the diagrams with the minimum number of bonds being stretched. These are the diagrams which may be divided into two parts by removing one MAYER bond, each part containing one of the complexes $\{n_1\}$ and $\{n_2\}$. Only one bond may be stretched in these diagrams as the distance between the bodies increases without limit, and just these diagrams turn out to be asymptotically essential.

The result of summation of the asymptotically essential diagrams may be descriptively represented as

$\varrho\{n_1+n_2\} = \varrho\{n_1\}\varrho\{n_2\}$ + 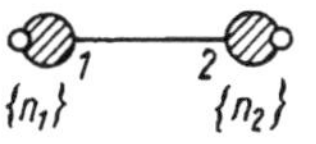

(2.3)

The blocks shown here denote the sums of diagrams in the expansions for $\varrho\{n_1\}$ and $\varrho\{n_2\}$, respectively, the line designates the MAYER bond stretched $f(1, 2) = \beta A_{s_1s_2} r_{12}^{-\lambda}$, the ends 1 and 2 of this line may be linked to any of the points of the corresponding block. Subsequent linking of the line to various points of the blocks may be expressed in the functional method by multiplying the ends of the line by the operators $\zeta(1)\, \delta/\delta\zeta(1)$ and $\zeta(2)\, \delta/\delta\zeta(2)$ with subsequent integration over all the values of the arguments 1 and 2.

The action of these operators on the functions $\varrho\{n_1\}$ and $\varrho\{n_2\}$, respectively, is easily found by the functional representation [Eq. (1.5)] for these functions:

$$\zeta(1) \frac{\delta}{\delta\zeta(1)} \varrho\{n_1\} = \zeta(1) \frac{\delta}{\delta\zeta(1)} \left\{ \left[\prod_l^{n_1} \zeta(l) \right] \frac{1}{\Xi} \frac{\delta^{n_1} \Xi}{\prod\limits_l^{n_1} \delta\zeta(l)} \right\}$$

$$= \sum_l^{n_1} \delta(1, l)\, \varrho\{n_1\} + \varrho\{n_1 + 1\} - \varrho\{n_1\}\, \varrho(1)\,. \tag{2.4}$$

Here l refers to the points of the block $\{n_1\}$, $\delta(1, l)$ is the usual δ-function for continuous variables and KRONEKER's symbol for discrete ones.

Using the designation

$$\Delta\varrho(1/\{n_1\}) = \sum_{l}^{n_1} \delta(1, l) + \frac{\varrho\{n_1 + 1\}}{\varrho\{n_1\}} - \varrho(1) \tag{2.5}$$

the result obtained is written in the form

$$\zeta(1) \frac{\delta}{\delta\zeta(1)} \delta\{n_1\} = \varrho\{n_1\} \, \Delta\varrho(1/\{n_1\}) \tag{2.6}$$

and similarly for the function $\varrho\{n_2\}$.

Obviously, $\varrho\{n_1 + 1\}/\varrho\{n_1\}$ is the conditional one-particle distribution function in the field of the complex $\{n_1\}$. Then, the function $\Delta\varrho(1/\{n_1\}$ defined by Eq. (2.5) is the excess one-particle distribution function in the presence of body 1, excess as compared with the case when the body 1 is absent.

The substitution of Eq. (2.3) into Eq. (2.2), taking into account Eq. (2.6), and the similar equation for the function $\varrho\{n_2\}$ leads, after expanding the logarithm in a series and rejecting all but the main dispersion interaction terms, to the following important result [22]:

$$W_{12} = - \sum_{s} \int_{V_1'} d\vec{r}_1 \, \Delta\varrho_s(\vec{r}_1/\{n_1\}) \sum_{t} \int_{V_2'} d\vec{r}_2 \, \Delta\varrho_t(\vec{r}_2/\{n_2\}) \, A_{st} r_{12}^{-\lambda} \, . \tag{2.7}$$

The integration over angular coordinates has been carried out in Eq. (2.7). Respectively, $\Delta\varrho_s(\vec{r}_1/\{n_1\})$ and $\Delta\varrho_t(\vec{r}_2/\{n_2\})$ are the excess local densities of the components s and t at the points $\vec{r}_1$ and $\vec{r}_2$ due to the presence of the interacting bodies. The factor $-A_{st} r_{12}^{-\lambda}$ describes the interaction of individual excess molecules at given points. The integral contribution of interaction is taken over all the regions V_1' and V_2' in which the excess local densities differ effectively from zero.

In Fig. 1 the boundary surfaces of the interacting bodies are shown with solid lines. Outer and inner dotted lines show the effective boundaries of the adjacent surface layers from the sides of the body and the medium. The regions inside the dotted lines correspond to the bulk state of the bodies. These regions are practically not realized in bodies of microscopic dimensions. The regions V_1' and V_2' contributing to interaction lie inside the outer dotted lines. The integration in Eq. (2.7) is carried out over these regions.

Because of its generality and simplicity, the result expressed in Eq. (2.7) may be called the interaction principle. It is formulated verbally as follows.

Any two molecular bodies in an arbitrary liquid or gaseous medium interact as they would interact in a vacuum if the local density of every component in the bodies and in the surrounding surface layers were the same as the excess quantity for the density with respect to its value for the homogeneous medium in the absence of the bodies.

Fig. 1

This principle completely solves the central problem of interaction, the determination of the effect on interaction of the medium which actually surrounds the bodies. Indeed, Eq. (2.7) takes into consideration both the effect of subsequent force transmission by the molecules of the medium which changes its structure and forms the solvate and adsorbed layers round the bodies and the force transmission effect connected with rearranging the interacting bodies themselves, their own surface layers forming inside the bodies.

A formal analogy with the Archimedes hydrostatic law achieved in the interaction principle is explained by the fact that only the medium molecules near the interacting bodies become intermediators of the body interaction in the case of dispersion forces. This analogy vanishes in the case of electric forces. For example, in the case of dipole forces the line 1 —— 2 in Eq. (2.3) should be replaced by the sum of diagrams:

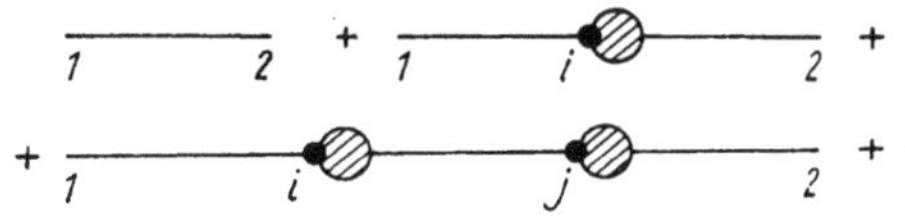

(2.8)

where the blocks, in their turn, represent the diagram sums in the expansions for $\varrho(i)$, $\varrho(j)$, ..., and the lines denote, as before, the stretched Mayer bonds. The diagrams presented in (2.8) describe the interaction including the effect of those medium molecules which are at a distance from the bodies exceeding the molecular size. Taking into account such interactions demands the introduction of additional coefficients connected with the dielectric permittivity of the medium [22].

3. Molecular correlations

Since the physical notions used in the formulation of the principle are fundamental and comprehensive and are applicable irrespective of the size and structure of the interacting bodies, it is possible to attack, from a single point of view, a wide variety of key problems, different in their scale, which have not been solved by other, more narrow methods.

Our further development will be devoted to numerous applications of the interaction principle. We begin by considering in this section the case when both the interacting molecular groups are of a microscopic scale.

Evidently, in this case the dimensions of the body I and the surrounding surface layer are small as compared with the distance between the bodies 1 and 2. Then, integrating over $\vec{r}_1$ in (2.7), one may put $\vec{r}_{12} = |\vec{R}_1 - \vec{r}_2|$, where the vector $\vec{R}_1$ characterizes the position of the body 1, and place $|\vec{R}_1 - \vec{r}_2|$ before the integral sign with respect to $\vec{r}_2$. Thus, the peculiarity of a microscopic scale is displayed in that the total excess molecule number appears in the interaction principle. The subsequent use of the compressibility theorem established in Section 1, Eq. (1.7), leads, taking (2.5) into account, to the formula

$$\int_{V_1'} d\vec{r}_1\, \Delta\varrho_s(\vec{r}_1/\{n_1\})\, \vec{r}_{12}^{\,-\lambda} = |\vec{R}_1 - \vec{r}_2|^{-\lambda} \beta^{-1} \frac{\partial \ln \varrho\{n_1\}}{\partial \mu_s} . \tag{3.1}$$

Similar operations may be made for the body 2 whose dimensions, in the case under consideration, are also microscopic.

The substitution of (3.1) and an analogous formula for the body 2 into the interaction principle, Eq. (2.7), yields

$$W_{12} = -\beta^{-2} R_{12}^{-\lambda} \sum_{s,t} A_{st} \frac{\partial \ln \varrho\{n_1\}}{\partial \mu_s} \frac{\partial \ln \varrho\{n_2\}}{\partial \mu_t} . \tag{3.2}$$

Here $R_{12} \equiv |\vec{R}_1 - \vec{R}_2|$ and the vector $\vec{R}_2$ characterize the position of the molecular group $\{n_2\}$; R_{12} is assumed to be large as compared with the dimensions of the interacting molecular groups.

Equation (3.2) in combination with Eq. (2.2) yields an expression for the distribution function $\varrho\{n_1 + n_2\}$ describing the correlations of the interacting molecular complexes which is of greatest interest in the case under consideration. This expression is [7—9, 23]

$$\varrho\{n_1 + n_2\} = \varrho\{n_1\}\, \varrho\{n_2\} + \beta^{-1} R_{12}^{-\lambda} \sum_{s,t} A_{st} \frac{\partial \varrho\{n_1\}}{\partial \mu_s} \frac{\partial \varrho\{n_2\}}{\partial \mu_t} . \tag{3.3}$$

As it is seen from (3.3), the total probability of the simultaneous presence of the two molecular groups is composed of the product of probabilities for single groups and a term reflecting their mutual correlation.

If single molecules are taken as correlating molecular complexes, Eq. (3.3) becomes

$$\varrho_{pq}(\vec{r}_1, \vec{r}_2) = \varrho_p(\vec{r}_1)\, \varrho_q(\vec{r}_2) + \beta^{-1} r_{12}^{-\lambda} \sum_{s,t} A_{st} \frac{\partial \varrho_p(\vec{r}_1)}{\partial \mu_s} \frac{\partial \varrho_q(\vec{r}_2)}{\partial \mu_t} , \tag{3.4}$$

where averaging is made with respect to molecular orientation, the subscripts s and t refer to molecular kinds.

Equations (3.2)—(3.4) are applicable to any capillary system, with any degree of spatial non-uniformity and any number of components. For the simplest case of a one-component, uniform medium, Eq. (3.4) yields

$$G(r_{12}) = \frac{1}{k_B T} A \chi^2 r_{12}^{-\lambda} , \tag{3.5}$$

where $G(r_{12}) = \varrho(\vec{r}_1, \vec{r}_2)/\varrho^2 - 1$ is the correlation function, ϱ is the density (the number of molecules per unit volume), and $\chi = (k_B T/\varrho)\, \partial\varrho/\partial\mu$ is the isothermal compressibility. The quantity χ is defined as dimensionless and is related to the usual compressibility, x, by $\chi = x k_B T \varrho$.

Equation (3.5) determines the strict asymptotic behaviour of the correlation function at distances large compared with molecular dimensions. The problem of such an asymptotic is of major significance for the theory of condensed systems and statistical physics. This problem was discussed earlier [33—35] within the framework of known approximations in the statistical theory of liquids, and the results obtained depend sensitively on the approximation to be chosen.

First established in our papers [7—9], the strict result, Eq. (3.5), was later confined on the ground of the macroscopic theory of molecular forces by Kemoklidze and Pitaevsky [30]. As in the whole macroscopic theory, this work was based on the concept of a liquid as the medium of which the properties are completely determined by a macroscopic quantum, the dielectric permittivity. The microscopic notions of molecular forces and the pair effective potential were not introduced explicitly. However, the final result of this work coincides with Eq. (3.5) if the pair effective potential is chosen correspondingly. For example, for the case of retarding dispersion forces ($\lambda = 7$), we should put

$$A = \frac{23\hbar C}{4\pi} \alpha^2 , \qquad \alpha \equiv \frac{1}{4\pi\varepsilon^{5/4}} \frac{\partial\varepsilon}{\partial\varrho} , \tag{3.6}$$

where $\hbar$ is PLANCK's constant divided by 2π, c is the vacuum velocity of light, and ε is the static dielectric permittivity of a liquid. A more complicated expression, including the frequency integral for the dielectric permittivity, is obtained for the case of ordinary dispersion forces without retardation ($\lambda = 6$). We will not write this expression here.

The coincident of results mentioned is a reliable verification of the pair effective potential conception used in the microscopic approach. Besides, Eqs. (3.6) make possible the actual determination of the force constant, A, for this potential and show that the microscopic and macroscopic approaches are complementary.

In rarefied systems, as it is seen from (3.6), the quantity α becomes the molecular polarizability, and the constant A becomes the usual interaction constant for two isolated molecules. Thus, the departure of the quantity α defined according to (3.6) from the molecular polarizability is a measure of non-additivity of dispersion interactions in condensed systems.

The possibilities of the microscopic theory and the interaction principle are not exhausted with the main asymptotic contributions to molecular correlations. Finding the corrective asymptotic terms, which are completely omitted in the macroscopic approach, demands taking into consideration the finite size of the surface layers surrounding the interacting molecular groups. In the work beginning with Eq. (3.1), this size was considered to be negligible as compared with the distance between the molecular groups.

It is easy to find in Eqs. (3.1)—(3.5) the corrective asymptotic terms which becomes essential as the distance between the interacting molecular groups becomes smaller.

We exhibit only the final result for the correlation function, $G(r_{12})$, of a homogeneous liquid, confining ourselved for the sake of simplicity, to the case of dispersion forces without retardation [7]:

$$G(r_{12}) = \frac{1}{k_B T} A\chi^2 r_{12}^{-6} + \frac{10}{k_B T} A\varrho\chi\gamma r_{12}^{-8} . \tag{3.7}$$

The parameter γ is connected with the correlation function second moment by the relationship

$$\gamma = 4\pi \int_0^\infty dr r^2 G(r) . \tag{3.8}$$

Equation (3.7) makes Eq. (3.5) with $\lambda = 6$ more accurate.

As estimations show, near the critical point the role of the corrective terms in (3.7) turns out to be essential even at distances considerably exceeding the average intermolecular distance. The microscopic approach thus makes it possible to ascertain the applicability range of the theory itself: it is limited by the vicinity of the critical point.

In the above statistical mechanical description, single molecules played the role of elements from which the systems considered were constructed. It is easy to see, however, that none of the specific properties of molecules were actually used. Thus, not only molecules, but also disperse particles may be imagined as primary elements in all the results obtained. Respectively, the complexes $\{n\}$, $\{n_1\}$, $\{n_2\}$, and so on, may be understood as groups consisting of the corresponding number of disperse particles with given kinds, location and orientation in space. This more general point of view automatically permits the consideration of a wide range of disperse systems of great importance in modern science and technology (see, e.g., the monograph [36]). In particular, Eqs. (3.5)—(3.7) will refer to uniform monodisperse systems characterized by the number of particles per unit volume, ϱ, and the isothermal compressibility, χ.

4. Structure factor

In the previous section attention was payed to simple liquids not only because of their own importance, but also because they are the most available source of knowledge of the pair effective potential. The latter and its force constant are necessary for the description of more complicated capillary systems.

Equations (3.6) make it possible to find the force constant for the case of retarding dispersion interactions. An analogous relationship for the force constant, as it was mentioned, could be obtained by comparing the microscopic and macroscopic approaches for the case of dispersion interactions without retardation ($\lambda = 6$). In the latter case, finding the force constant for the pair effective potential would demand additionally a knowledge of the frequency dependence of the dielectric permittivity.

A more direct possibility for the determination of the pair effective potential of dispersion interactions for ordinary or retarding forces, is given by experiments on the scattering of X-rays by simple liquids. In such experiments, the structure factor, $I(k)$, is measured which depends on the wave number, k, and is related to the correlation function, $G(r)$, introduced in the previous section by the Fourier transformation

$$k[I(k) - 1] = 4\pi\varrho \int_0^\infty dr rG(r) \sin kr \,. \tag{4.1}$$

At present, the accuracy of the experiment is high enough, and a detailed determination of $I(k)$ may be expected for small k satisfying the condition $kr_0 \ll I$ where r_0 characterizes the molecular size. The experimental knowledge of the structure factor for small k is sufficient for the determination of the force constant for the pair effective potential. Indeed, this determination is achieved by the comparison of experimental data for the structure factor with the theoretical data which follows, as it will now be shown, from asymptotic Eq. (3.5) for its Fourier transformation at $r \gg r_0$.

Finding the asymptotics of the structure factor, $I(k)$, defined by Eq. (4.1) for small k is a simple mathematical task when $G(r)$ diminishes as $r \to \infty$ as an even power of I/r. Just such a situation takes place in the case of ordinary dispersion forces, i.e. with $\lambda = 6$. Asymptotic Eq. (3.5) leads then to the following asymptotics of the structure factor for small k:

$$I(k) = a_0 + a_2 k^2 + a_3 |k|^3 + \dots \tag{4.2}$$

The coefficients of the first two terms of the expansion are related to the zeroth and second moments of the correlation function $G(r)$ by the formulae

$$a_0 = 1 + 4\pi\varrho \int_0^\infty dr r^2 G(r) , \tag{4.3}$$

$$a_2 = - \tfrac{2}{3} \pi\varrho \int_0^\infty dr r^4 G(r) . \tag{4.4}$$

By using the compressibility theorem, Eq. (1.7), the coefficient a_0 is easily reduced to χ: $a_0 = \chi$. The coefficient a_2 is expressed, due to Eq. (3.8), through the parameter γ from the corrective term of Eq. (3.7) by $a_2 = -\varrho\gamma/6$.

Directly related to the pair effective potential is the third expansion term in (4.2), the coefficient of which depends explicitly on the force constant:

$$a_3 = \frac{\pi^2}{12 k_B T} A \varrho \chi^2 . \tag{4.5}$$

This term differs from the two previous ones by a non- analytical dependence on the wave number at the point $k = 0$ where its third derivative is discontinuous. This peculiarity of the third expansion term makes it easily recognizable in the background of the analytical dependence on the wave number, which results in high accuracy of the determination of the force constant for the pair effective potential from experimental data on $I(k)$ at small k.

According to Eq. (3.5), the correlation function $G(r)$ for the case of dispersion forces with retardation, i.e. for $\lambda = 7$, diminishes with $r \to \infty$ as an odd power of I/r. As it was shown by Fisher [37], Eq. (3.5) yields in this case the following asymptotic expression for the structure factor, $I(k)$, for small k:

$$I(k) = I_0(k) + I_1(k) . \tag{4.6}$$

The function $I_0(k)$ is even, regular at the point $k = 0$, and has the expansion

$$I_0(k) = a_0 + a_2 k^2 + a_4 k^4 + \dots , \tag{4.7}$$

where two first coefficients are given by the same formulae as for the case of dispersion forces without retardation.

The function $I_1(k)$ possesses a logarithmic singularity in its fourth derivative for the point $k = 0$ and has the form

$$I_1(k) = - \frac{\pi}{20 k_B T} A \varrho \chi^2 k^4 \ln |k| + 0(|k|^5) . \tag{4.8}$$

This function reflects the peculiarity of the asymptotic formula (3.5) for the case considered of retarding interactions and gives an opportunity to find the force constant for $\lambda = 7$ from experimental data on scattering. The singularity mentioned of the fourth derivative makes it possible to extract from the structure factor the contribution described by the function $I_1(k)$ explicitly dependent on the force constant. As anticipated in the explanation following Eq. (2.1), the force constant in (4.8) differs in its dimension from the force constant in (4.5).

The method described in this and in the previous section do not exhaust all the possible ways of finding the force constant. As it was noted at the beginning of this paper, direct theoretical calculations based on the concepts of modern quantum chemistry [27, 28] are also an important source of knowledge of the pair effective potential.

5. Molecular structure of systems with flat interfaces

We now turn to a wide circle of problems related to the investigation of the molecular structure of capillary systems.

This investigation has two principal difficulties. The first is due to the fact that physical conditions in capillary systems change at distances of the order of molecular dimensions. For this reason, the customary local uniformity approximation in statistical mechanics becomes invalid. The second difficulty consists in large molecular correlations in capillary systems due to the high density, which makes impossible the use of well developed methods in the theory of gases.

The difficulties pointed out are a particular manifestation of the general difficulty formulated at the beginning of the article as the main problem of interaction theory. In order to understand the direct connection between problems on the molecular structure of capillary systems and the interaction principle, we represent the total probability of the joint location of two molecular groups as the product of the probability of one of them, say, of the second one, on the conditional probability of the first group in the presence of the second one

$$\varrho\{n_1 + n_2\} = \varrho\{n_2\}\, \varrho(\{n_1\}/\{n_2\}) \,. \tag{5.1}$$

If we now identify the molecular group $\{n_2\}$ with the body producing the surface force field in a capillary system and the group $\{n_2\}$ with the molecular complex of which the location probability is of interest, then the conditional distribution function $\varrho(\{n_1\}/\{n_2\})$ will just describe the molecular structure of the capillary system. For establishing the connection with the interaction principle, it is left to express the conditional probability by means of (5.1) and (2.2) through the potential W_{12} determined by the interaction principle:

$$\varrho(\{n_1\}/\{n_2\}) = \varrho\{n_1\} \exp(-\beta W_{12}) \,. \tag{5.2}$$

From the standpoint of the interaction principle, the peculiarity of the determination of the molecular structure of capillary systems consists only in the fact that the second of the interacting molecular groups plays the role of a macroscopic body producing the surface force field.

Since the molecular group $\{n_1\}$ of which the location probability is to be found has microscopic dimensions, the approximation Eq. (3.1), is applicable to it. However, body 2 in the interaction principle now has macroscopic dimensions which are no longer small as compared with the distance between the interacting bodies 1 and 2. Hence, another expression is needed for body 2 in the interaction principle.

Since, in the range of action of dispersion forces interesting for us, the distance between the bodies 1 and 2 may be considered, as before, to be large as compared with the effective thickness of the surface layers adjacent to the boundary of body 2, the necessary expression for body 2 in the interaction principle, Eq. (2.7), will be

$$\int_{V_2'} d\vec{r}_2\, \Delta\varrho_t(\vec{r}_2 \,/\, \{n_2\})\, r_{12}^{-\lambda} = [\varrho_t^{(\beta)} - \varrho_t^{(\alpha)}] \int_{V_2} d\vec{r}_2\, r_{12}^{-\lambda} + \Gamma_t^{(\alpha\beta)} \int_{S_2} dS_2 r_{12}^{-\lambda}\,. \tag{5.3}$$

Here the subscripts α and β refer to bulk properties inside the medium and the macroscopic body 2, respectively, ϱ_t is the molecule number density of the component t, S_2 is the boundary surface between body 2 and the medium (drawn in Fig. 1 with the solid line), dS_2 is an area element of the surface S_2, V_2 is the volume inside the surface S_2, $\Gamma_t^{(\alpha\beta)}$ is the adsorption of the component t on the boundary between the body and the medium (the effect of their mutual perturbation is thus taken into account). It is supposed, for the sake of simplicity, that the quantity $\Gamma_t^{(\alpha\beta)}$ is constant along the surface S_2.

Equation (5.3) makes it possible to replace effectively the interaction with the macroscopic body and the surface layers adjacent to its surface by the interaction with the homogeneous bulk and surface masses. The presence in Eq. (5.3) of the total adsorption $\Gamma_t^{(\alpha\beta)}$ including both the adsorptions from the medium and the body makes the right-hand side of Eq. (5.3) independent, in the approximation adopted, on the exact position of the boundary surface.

Thus, we see that the interaction principle, Eq. (2.7), together with its probability treated according to (5.2) and with Eqs. (3.1) and (5.3), gives a universal and very operative method for solving problems involving the molecular structure of capillary systems, the molecules in phase α and the molecules of the surface layer under solution actually reducing to the calculation of the bulk and surface integrals in Eq. (5.3). The final results depend on the interfacial shape in a capillary system and on its curvature. Here we will study the results on the molecular structure of systems with flat interfaces. The results of investigations of capillary systems with slightly and strongly curved interfaces will be presented in two subsequent sections.

The calculation of the distribution functions of a flat surface layer with the above method leads to the formula [3, 5, 10—12, 17, 20, 23]:

$$\varrho\{n\} = \varrho^{(\alpha)}\{n\} + \frac{2\pi}{(\lambda-2)(\lambda-3)}\, h^{-\lambda+3} \sum_{s,t} A_{st} \frac{\partial\varrho^{(\alpha)}\{n\}}{\partial\mu_s} [\varrho_t^{(\beta)} - \varrho_t^{(\alpha)}]$$
$$+ \frac{2\pi}{\lambda-2}\, h^{-\lambda+2} \sum_{s,t} A_{st} \frac{\partial\varrho^{(\alpha)}\{n\}}{\partial\mu_s}\, \Gamma_t^{(\alpha\beta)}\,, \tag{5.4}$$

where $\{n\}$ is the surface layer molecular group which was designated as $\{n_1\}$ in the interaction principle, $\varrho\{n\}$ is the location probability for this group which was understood in Eqs. (5.1) and (5.2) as the conditional distribution function and designed as $\varrho(\{n_1\}/\{n_2\})$. Now the surface layer itself plays the role of the medium, the intermediator of interaction between the molecular complex $\{n\}$ and the body creating the surface force field. Respectively, the subscript α now refers to the bulk phase from which the surface layer is formed. Evidently, $\varrho^{(\alpha)}\{n\}$ is understood as the probability of location of the molecular group $\{n\}$ in the uniform bulk phase α; earlier, the distribution function $\varrho\{n_1\}$ in Eq. (5.2) corresponded to this probability. Finally, h designates the distance from the center of the molecular group $\{n\}$ up to the boundary surface, the distance being supposed to be large as compared with the dimensions of the group. All other notation has its original meaning.

The complex $\{n\}$ in Eq. (5.4) may contain an arbitrary number of molecules of different kinds. If this number is more than unity, $\varrho\{n\}$ in Eq. (5.4) should be symmetrized by reflecting it at the point inside the complex, the distance from which to the boundary surface was adopted as h. For example, in the case when the complex $\{n\}$ consists of two molecules of kind p and q at the points $\vec{r}_1$ and $\vec{r}_2$, respectively, it is convenient to take as h the distance from the boundary surface to the point $\vec{r} = (\vec{r}_2 + \vec{r}_2)/2$. Then the two-particle distribution function symmetrized by reflection at the point r will be $\varrho_{pq}(\vec{r}_1, \vec{r}_2) + \varrho_{pq}(\vec{r}_2, \vec{r}_1)$. It should be stressed that this symmetrization does not mean a loss of information since any physical local characteristics satisfy the necessary symmetry condition.

Equation (5.4) is asymptotic for large h. It contains both the main asymptotic term proportional to $h^{-\lambda+3}$ and the corrrective one proportional to $h^{-\lambda+2}$. The quantities $0(h^{-\lambda+1})$ of higher order in h are neglected.

As numerical comparison of the main and corrective asymptotic terms shows [15], Eq. (5.4) begins to work already at distances of the order of the average intermolecular distance in the phase α. Just these distances correspond to the range of dispersion forces.

Evidently, Eq. (5.4) takes into account the possibility of transition of molecules from one phase into another. If however, the macroscopic body inducing the surface force field is insoluble consideration do not penetrate into the bulk of the body, then set $\varrho_t^{(\alpha)} = 0$ for all the components present only in the body and $\varrho_t^{(\beta)} = 0$ for all the components present only in the surface layer. In particular, when the body is a rarefied gaseous medium, $\varrho_t^{(\beta)} = 0$ for all components. In this case, the body role in creating the surface force field is effectively manifested by the absence of layer molecules in the body.

Equation (5.4) takes into account the effect of mutual perturbation of the body and the surface layer. This effect is described by the term containing the adsorptions, $\Gamma_t^{(\alpha\beta)}$, the body itself may be in an arbitrary aggregation state.

Equation (5.4) contains much information and makes possible the determination of all the local properties of the surface layer.

Choosing one molecule of the kind p as the complex $\{n\}$ in (5.4), we obtain a formula for the one-particle distribution function, ϱ_p, the local density of component p in the surface layer. By summing it up over all the components, we obtain a formula for the total local density, ϱ, of the surface layer. By dividing ϱ_p by ϱ, we obtain the formula for the local mole fraction, x_p, of the component p in the surface layer [20]:

$$x_p = x_p^{(\alpha)} + \frac{2\pi}{(\lambda - 2)(\lambda - 3)} h^{-\lambda+3} \sum_{s,t} A_{st} \frac{\partial x_p^{(\alpha)}}{\partial \mu_s} [\varrho_t^{(\beta)} - \varrho_t^{(\alpha)}] + \\ + \frac{2\pi}{\lambda - 2} h^{-\lambda+2} \sum_{s,t} A_{st} \frac{\partial x_p^{(\alpha)}}{\partial \mu_s} \Gamma_t^{(\alpha\beta)}. \tag{5.5}$$

Knowing the one- and two-particle distribution functions, it is possible to calculate the local pressure tensor, $\hat{P}$, characterizing the mechanical state of the surface layer. The starting points for calculation are the general expressions of statistical mechanics for the pressure tensor, Eqs. (1.9) or (1.11), given in Section 1. In these expressions contributions are taken into account from all the molecule pairs interacting through the point of observation of the pressure tensor. Until the distances between interacting molecules are small compared with the distance, h, from the observation point to the boundary surface, Eq. (5.4) may be used. In other cases, when, in particular, direct interactions with the macroscopic body creating the surface force field are taken into account, it is necessary to use Eq. (3.4) obtained in Section 3. Naturally, the use of Eq. (3.4) is also necessary for calculating other local properties (e.g., the local density of energy) which, as with the pressure tensor, do not actually possess a strictly local structure.

For capillary systems with a flat interface, the pressure tensor is reduced, evidently, to two components: the normal one, P_N, and the tangential one, P_T. The final results of calculations for these components at large distances h from the boundary surface [4, 5, 13, 21] are given below:

$$P_N = P^{(\alpha)}\,, \tag{5.6}$$

$$P_T = P^{(\alpha)} + \frac{2\pi}{(\lambda-2)^2}\,h^{-\lambda+3}\sum_{s,t} A_{st}\varrho_s^{(\alpha)}[\varrho_t^{(\beta)} - \varrho_t^{(\alpha)}] + \frac{2\pi(\lambda-3)}{(\lambda-2)^2}\,h^{-\lambda+2}\sum_{s,t} A_{st}\varrho_s^{(\alpha)}\Gamma_t^{(\alpha\beta)}\,. \tag{5.7}$$

Here $P^{(\alpha)}$ is the pressure in phase α. The thermodynamic relationship $\partial P^{(\alpha)}/\partial\mu_s = \varrho_s^{(\alpha)}$ is taken into account in Eq. (5.7).

As should be expected from the mechanical equilibrium condition, Eq. (1.10), the normal component of the pressure tensor in a flat surface layer turns out to be independent of h.

6. Molecular structure of systems with slightly curved interfaces

We now consider the case when the curvature radius, denoted as a^*, of the boundary surface takes finite, but such large absolute values that $|a^*| \gg h$, the distance h maintaining its former order of magnitude.

Both concave and convex surfaces are under consideration. On this account, the surface curvature radius, as well as the principal curvature radii, may take both positive and negative signs.

According to the suppositions made, we now will take into account terms of the order of $h^{-\lambda+3}h/a^*$. As before, the quantities $0(h^{-\lambda+1})$ will be neglected. In addition we will neglect quantities $0(h^{-\lambda+3}h^2/a^{*2})$ and $0(h^{-\lambda+2}h/a^*)$ for the present. In the limit $a^* \to \infty$ the results of this section will converge into the results for the case of flat interfaces.

Instead of Eq. (5.4), we now will have [25, 26]

$$\varrho\{n\} = \varrho^{(\alpha)}\{n\} + \frac{2\pi}{(\lambda-2)(\lambda-3)}\,h^{-\lambda+3}\left[1 + \frac{(\lambda-3)}{(\lambda-4)}\frac{h}{a^*}\right] \times \sum_{s,t} A_{st}\frac{\partial\varrho^{(\alpha)}\{n\}}{\partial\mu_s}[\varrho_t^{(\beta)} - \varrho_t^{(\alpha)}] + \frac{2\pi}{\lambda-2}\,h^{-\lambda+2}\sum_{s,t} A_{st}\frac{\partial\varrho^{(\alpha)}\{n\}}{\partial\mu_s}\Gamma_t^{(\alpha\beta)}\,. \tag{6.1}$$

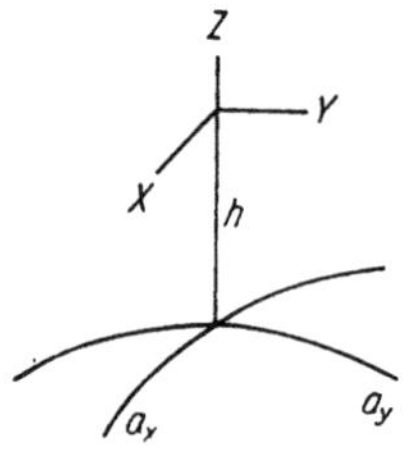

Fig. 2

As in the case of flat interfaces, it is easy to obtain from Eq. (6.1) the corresponding formulae for the total local density and the local composition, as well as the formula for the local pressure tensor. Evidently, the pressure tensor will be diagonal in the curvilinear coordinate system with the origin at the observation point, with the Z-axis perpendicular to the boundary surface, and with the X, Y-axes along the principal directions on this surface, at the point of its intersection by the Z-axis.

Such a curvilinear system X, Y, Z is presented in Fig. 2. The lines are the principal directions of the boundary suıface at the point where it is intersected by the Z-axis. These principal directions are marked with the principle surface curvature radii, corresponding to them a_x and a_y. The distance h is also shown from the intersection point of the boundary surface with the Z-axis to the origin of the curvilinear coordinate system X, Y, Z. This origin is combined with the observation point.

We demonstrate the final result of calculation of the pressure tensor diagonal components in the above curvilinear coordinate system [25, 26]:

$$P_{zz} = P^{(\alpha)} + \frac{4\pi}{(\lambda - 2)^2 (\lambda - 4)} h^{-\lambda+3} \frac{h}{a^*} \sum_{s,t} A_{st} \varrho_s^{(\alpha)} [\varrho_t^{(\beta)} - \varrho_t^{(\alpha)}] , \tag{6.2}$$

$$P_{xx} = P^{(\alpha)} + \frac{2\pi}{(\lambda - 2)^2} h^{-\lambda+3} \left[1 + \frac{(\lambda - 5)}{2(\lambda - 4)} \frac{h}{a_x} + \frac{(\lambda - 3)}{2(\lambda - 4)} \frac{h}{a_y} \right] \sum_{s,t} A_{st} \varrho_s^{(\alpha)} [\varrho_t^{(\beta)} - \varrho_t^{(\alpha)}] + \frac{2\pi(\lambda - 3)}{(\lambda - 2)^2} h^{-\lambda+2} \sum_{s,t} A_{st} \varrho_s^{(\alpha)} \Gamma_t^{(\alpha\beta)} , \tag{6.3}$$

$$P_{yy} = P^{(\alpha)} + \frac{2\pi}{(\lambda - 2)^2} h^{-\lambda+3} \left[1 + \frac{(\lambda - 3)}{2(\lambda - 4)} \frac{h}{a_x} + \frac{(\lambda - 5)}{2(\lambda - 4)} \frac{h}{a_y} \right] \sum_{s,t} A_{st} \varrho_s^{(\alpha)} [\varrho_t^{(\beta)} - \varrho_t^{(\alpha)}] + \frac{2\pi(\lambda - 3)}{(\lambda - 2)^2} h^{-\lambda+2} \sum_{s,t} A_{st} \varrho_s^{(\alpha)} \Gamma_t^{(\alpha\beta)} . \tag{6.4}$$

Unlike the distribution functions and the normal component of the pressure tensor, $P_N \equiv P_{zz}$, its tangential components, P_{xx} and P_{yy}, depend not only on the surface curvature radius, a^*, but also on the principal curvature radii, a_x and a_y. Obviously, of these three variables giving the surface geometry, only two are independent since, as known, $2/a^* = 1/a_x + 1/a_y$. Naturally, Eqs. (6.3) and (6.4) are symmetrical to interchanges of a_x with a_y.

The term of order $h^{-\lambda+3}h/a^*$, playing the role of the corrective term with respect to the main term of order $h^{-\lambda+3}$ in the equations for the distribution functions and the pressure tensor tangential components is the main term in the equation for the normal pressure tensor component. Indeed, according to Eq. (5.6) of the previous section, the normal component of the pressure tensor may differ from the pressure in phase only due to the curvature of the boundary surface.

It is much more difficult than in the flat case to verify the correspondence of Eqs. (6.2)—(6.4) for the pressure tensor to the mechanical equilibrium condition. Proceeding to this important problem, we consider, as a preliminary, two particular cases: cylindrical and spherical interfaces.

In the case of a cylindrical interface, we have $a_x = \infty$ and $a_y = a$ where a is the radius of the cylindrical boundary surface, and the X-axis is chosen parallel to the axis of the surface.

Let us denote the tangential components, P_{xx} and P_{yy} as P_{T1} and P_{T2}, respectively. It may be shown that the mechanical equilibrium condition, Eq. (1.10), for systems of cylindrical symmetry has the form

$$\frac{d}{dr}(rP_N) = P_{T2}\,, \tag{6.5}$$

where r is the distance from the observation point for the pressure tensor to the axis of the cylindrical surface.

Substituting (6.2) and (6.4) into Eq. (6.5), we find that the constant terms and the terms of order $h^{-\lambda+3}$ in the left and right-hand sides of Eq. (6.5) coincide. As for the terms of order $h^{-\lambda+3}h/a$ and $h^{-\lambda+2}$, they contain a contribution from absent in Eq. (6.2) the P_N terms of order $h^{-\lambda+3}h^2/a^2$ and $h^{-\lambda+2}h/a$, respectively. Evidently, we may use Eq. (6.5) for finding these unknown terms. As a result instead of (6.2), we will have in the case of a cylindrical boundary surface [25, 26]

$$P_N = P^{(\alpha)} + \frac{2\pi}{(\lambda-2)^2(\lambda-4)}\,h^{-\lambda+3}\left(\frac{h}{a} + \frac{3}{2}\frac{h^2}{a^2}\right)\sum_{s,t} A_{st}\varrho_s^{(\alpha)}[\varrho_t^{(\beta)} - \varrho_t^{(\alpha)}]$$
$$+ \frac{2\pi}{(\lambda-2)^2}\,h^{-\lambda+2}\,\frac{h}{a}\sum_{s,t} A_{st}\varrho_s^{(\alpha)}\Gamma_t^{(\alpha\beta)}\,. \tag{6.6}$$

The case of a spherical interface is considered in a similar way. We have $a_x = a_y = a$ where a is the radius of a spherical interface. The tangential components, P_{xx} and P_{yy}, now coincide: $P_{xx} = P_{yy} = P_T$, and the mechanical equilibrium condition, Eq. (1.10), has the form

$$\frac{1}{2r}\frac{d}{dr}(r^2P_N) = P_T\,, \tag{6.7}$$

where r is the distance from the pressure tensor observation point to the center of the spherical interface. Equation (6.7) may be used both for the verification of the terms already found in (6.2)—(6.4) and for the determination of unknown corrective terms to P_N. Taking these corrective terms into account, we will have for the case of a spherical boundary surface under consideration [25, 26].

$$P_N = P^{(\alpha)} + \frac{4\pi}{(\lambda-2)^2(\lambda-4)} h^{-\lambda+3}\left(\frac{h}{a} + 2\frac{h^2}{a^2}\right) \sum_{s,t} A_{st}\varrho_s^{(\alpha)}[\varrho_t^{(\beta)} - \varrho_t^{(\alpha)}]$$
$$+ \frac{4\pi}{(\lambda-2)^2} h^{-\lambda+2}\frac{h}{a}\sum_{s,t} A_{st}\varrho_s^{(\alpha)}\Gamma_t^{(\alpha\beta)}. \tag{6.8}$$

For the general case of a silghtly curved interface of arbitrary shape, the mechanical equilibrium condition, Eq. (1.10), has the form

$$-\frac{1}{(a_x-h)(a_y-h)}\frac{d}{dh}[(a_x-h)(a_y-h)P_N] = \frac{P_{xx}}{a_x-h} + \frac{P_{yy}}{a_x-h}. \tag{6.9}$$

As in the above particular cases, Eq. (6.9) may be used, on the one hand, for the verification of the terms already found in (6.2)—(6.4) and, on the other hand, for finding higher corrective terms to P_N. Made more accurate in this way, the expression for the pressure tensor normal component becomes [25, 26]

$$P_N = P^{(\alpha)} + \frac{2\pi}{(\lambda-2)^2(\lambda-4)} h^{-\lambda+3}\left(\frac{h}{a_x} + \frac{h}{a_y} + \frac{3}{2}\frac{h^2}{a_x^2}\right.$$
$$\left. + \frac{3}{2}\frac{h}{a_y^2} + \frac{h^2}{a_x a_y}\right)\sum_{s,t} A_{st}\varrho_s^{(\alpha)}[\varrho_t^{(\beta)} - \varrho_t^{(\alpha)}] +$$
$$+ \frac{2\pi}{(\lambda-2)^2} h^{-\lambda+2}\left(\frac{h}{a_x} + \frac{h}{a_y}\right)\sum_{s,t} A_{st}\varrho_s^{(\alpha)}\Gamma_t^{(\alpha\beta)}. \tag{6.10}$$

The above results of our investigations of molecular structure in systems with flat and slightly curved interfaces remained for years the success of the microscopic theory. Only very recently [31] analogous investigations have been made on the ground of the macroscopic theory of dispersion interactions [1, 2]. The local density and pressure tensor were the objects of investigation. Obtained in the main asymptotic term (which was considered only in [31]), the dependences of these quantities on the distance from the boundary surface and on its curvature coincide with the above ones.

7. Molecular structure of systems with strongly curved interfaces

The case of strongly curved interface differs in principle from the cases of flat and slightly curved interfaces, considered above.

The most interesting region for strongly curved interfaces is the central part of a capillary system, the inner part of a drop, of the adsorbent pore or cavity, and so on. Here, the effects of the overlapping of surface regions, as well as the molecular correlations and the effects of mutual perturbation of phases inside the capillary system, should be taken into account. We will investigate the central part of a capillary system assuming the surface curvature radius to be great as compared with molecular dimensions.

Since the form of the relationships to be obtained depends strongly on the interface shape, we consider only two most typical cases: cylindrical and spherical capillary systems.

We start from a cylindrical capillary system. The calculation of the distribution functions for such a system in the vicinity of its symmetry axis yields the formula [26]

$$\varrho\{n\} = \varrho^{(\alpha)}\{n\} + 2\pi c_\lambda \frac{(\lambda-5)!!}{(\lambda-2)!!} a^{-\lambda+3} \sum_{s,t} A_{st} \frac{\partial \varrho^{(\alpha)}\{n\}}{\partial \mu_s} [\varrho_t^{(\beta)} - \varrho_t^{(\alpha)}]$$
$$+ 2\pi c_\lambda \frac{(\lambda-3)!!}{(\lambda-2)!!} a^{-\lambda+2} \sum_{s,t} A_{st} \frac{\partial \varrho^{(\alpha)}\{n\}}{\partial \mu_s} \Gamma_t^{(\alpha\beta)} . \qquad (7.1)$$

Here $c_\lambda = \pi$ for ordinary dispersion forces ($\lambda = 6$) and $c_\lambda = 2$ for dispersion forces with retardation ($\lambda = 7$). The radius of the cylindrical boundary surface, a, now takes only positive values. The superscript α refers to a hypothetical bulk phase with given values of the chemical potentials. Such a phase is actually not realized even in the central part of the capillary system because of the great curvature of the interface.

In Eq. (7.1) both the main term of order $a^{-\lambda+3}$ and the corrective term of order $a^{-\lambda+2}$ do not depend on the location of the molecular complex $\{n\}$ with respect to the axis. The dependence on the distance, r, from the complex to the axis appears only in the quantities $0(a^{-\lambda+3}r^2/a^2)$ and $0(a^{-\lambda+2}r^2/a^2)$ which are neglected in the complex central part under consideration where $r/a \ll 1$. Thus, the radius a actually plays the role of distance from the boundary surface with respect to the central part. The quantities $0(a^{-\lambda+1})$ are also neglected in Eq. (7.1).

The absence of terms of order $a^{-\lambda+3}r/a$ and $a^{-\lambda+2}r/a$ in Eq. (7.1) is the consequence of symmetry. Because of the given symmetry, it is also possible not to symmetrize additionally the distribution function $\varrho\{n\}$ by reflecting at a point inside the complex $\{n\}$, as it was necessary to do in Eqs. (5.4) and (6.1).

With the aid of Eq. (7.1), one may obtain the formulae for other local quantities in the vicinity of the symmetry axis of a cylindrical capillary system. We only cite here the formula for the local composition [26]

$$x_p = x_p^{(\alpha)} + 2\pi c_\lambda \frac{(\lambda-5)!!}{(\lambda-2)!!} a^{-\lambda+3} \sum_{s,t} A_{st} \frac{\partial x_p^{(\alpha)}}{\partial \mu_s} [\varrho_t^{(\beta)} - \varrho_t^{(\alpha)}]$$
$$+ 2\pi c_\lambda \frac{(\lambda-3)!!}{(\lambda-2)!!} a^{-\lambda+2} \sum_{s,t} A_{st} \frac{\partial x_p^{(\alpha)}}{\partial \mu_s} \Gamma_t^{(\alpha\beta)}$$

and the formula for the local pressure tensor $\hat{P}$, for its normal component, P_N [26]:

$$P_N = P^{(\alpha)} + \pi c_\lambda \frac{(\lambda-5)!!}{(\lambda-2)!!} a^{-\lambda+3} \sum_{s,t} A_{st} \left\{ \left[-2 + \frac{\lambda-1}{\lambda-2}(1 - 2^{-\lambda+2})\right] \varrho_s^{(\alpha)} \varrho_t^{(\alpha)} \right.$$
$$\left. + \left[2 - \frac{\lambda-1}{\lambda-2}(1 - 2^{-\lambda+3})\right] \varrho_s^{(\alpha)} \varrho_t^{(\beta)} - \frac{\lambda-1}{\lambda-2} 2^{-\lambda+2} \varrho_s^{(\beta)} \varrho_t^{(\beta)} \right\}$$
$$+ \pi c_\lambda \frac{(\lambda-3)!!}{(\lambda-2)!!} a^{-\lambda+2} \sum_{s,t} A_{st} \left\{ \left[2 - \frac{\lambda-1}{\lambda-2}(1 - 2^{-\lambda+2})\right] \varrho_s^{(\alpha)} \right.$$
$$\left. - \frac{\lambda-1}{\lambda-2} 2^{-\lambda+2} \varrho_s^{(\beta)} \right\} \Gamma_t^{(\alpha\beta)} \qquad (7.3)$$

and for the tangential component, P_{T1}, along the direction parallel to the axis [26]

$$P_{T1} = P^{(\alpha)} + \pi c_\lambda \frac{(\lambda-5)!!}{(\lambda-2)!!} a^{-\lambda+3}$$
$$\times \sum_{s,t} A_{st} \left\{ 2\left[-1 + \frac{1}{\lambda-2}(1-2^{-\lambda+2})\right] \varrho_s^{(\alpha)}\varrho_t^{(\alpha)}\right.$$
$$\left. + 2\left[1 - \frac{1}{\lambda-2}(1-2^{-\lambda+3})\right] \varrho_s^{(\alpha)}\varrho_t^{(\beta)} - \frac{1}{\lambda-2} 2^{-\lambda+3}\varrho_s^{(\beta)}\varrho_t^{(\beta)}\right\}$$
$$+ \pi c_\lambda \frac{(\lambda-3)!!}{(\lambda-2)!!} a^{-\lambda+2} \sum_{s,t} A_{st} \left\{2\left[1 - \frac{1}{\lambda-2}(1-2^{-\lambda+2})\right]\varrho_s^{(\alpha)}\right.$$
$$\left. - \frac{1}{\lambda-2} 2^{-\lambda+3}\varrho_s^{(\beta)}\right\} \Gamma_t^{(\alpha\beta)} . \tag{7.4}$$

As for the second tangential component, P_{T2}, along the direction perpendicular to the axis, it coincides with P_N in the approximation adopted, as it is seen, for example, from the mechanical equilibrium condition, Eq. (6.5).

As in Eq. (7.1), the quantities $0(a^{-\lambda+3}r^2/a^2)$, $0(a^{-\lambda+2}r^2/a^2)$, and $0(a^{-\lambda+1})$ are neglected in Eqs. (7.2)–(7.4).

If, in particular, the body creating the surface force field plays the role of an adsorbent, and the chemical potential values in an adsorbate are given, Eqs. (7.1)–(7.4) describe the dependence of the pore radius, a, on respective local quantities near the axis of a cylindrical pore. This corresponds to usual experimental conditions when the medium with given chemical potential is in touch with the adsorbent surface with a number of pores of different size, and the distribution of properties depending on the pore size is to be found.

The case of a spherical capillary system is considered in a similar way. For the distribution functions of the system in the vicinity of its symmetry center, we have [24, 26]

$$\varrho\{n\} = \varrho^{(\alpha)}\{n\} + \frac{4\pi}{\lambda-3} a^{-\lambda+3} \sum_{s,t} A_{st} \frac{\partial \varrho^{(\alpha)}\{n\}}{\partial \mu_s} [\varrho_t^{(\beta)} - \varrho_t^{(\alpha)}]$$
$$+ 4\pi a^{-\lambda+2} \sum_{s,t} A_{st} \frac{\partial \varrho^{(\alpha)}\{n\}}{\partial \mu_s} \Gamma_t^{(\alpha\beta)} , \tag{7.5}$$

where a is the radius of the spherical boundary surface.

Due to the symmetry of the system, the distribution functions do not depend, within the approximation adopted, on the distance r from the center of the spherical system, and these functions are not to be symmetrized by reflecting in a point inside the complex n. The dependence of the distribution functions on the distance r appears only in the terms of order $a^{-\lambda+3}r^2/a^2$ and $a^{-\lambda+2}r^2/a^2$. However, these terms, as well as the terms of order $a^{-\lambda+1}$, are neglected in Eq. (7.5). With respect to the systems central part under consideration, the radius a may be understood as the distance from the boundary surface.

Equation (7.5) determines any local quantity near the center of a spherical capillary system. We cite only the expression for the local pressure tensor. In the case of spherical symmetry, the pressure tensor in the vicinity of the symmetry center is proportional to the unit tensor [this may be easily shown, e.g., by means of the mechanical equilibrium condition, Eq. (6.7)]. The latter means that the pressure tensor is reduced to one component, the local pressure, P. The expression for the local pressure, P, is [24, 26]

$$P = P^{(\alpha)} - \frac{2\pi\lambda}{3(\lambda-2)(\lambda-3)} a^{-\lambda+3}$$
$$\times \sum_{s,t} A_{st} \left\{ 2\left[-1 + 2^{-\lambda+2} + \frac{3(\lambda-2)}{\lambda} \right] \varrho_s^{(\alpha)} \varrho_t^{(\alpha)} \right.$$
$$\left. + 2\left[1 - 2^{-\lambda+3} - \frac{3(\lambda-2)}{\lambda} \right] \varrho_s^{(\alpha)} \varrho_t^{(\beta)} + 2^{-\lambda+3} \varrho_s^{(\beta)} \varrho_t^{(\beta)} \right\}$$
$$- \frac{4\pi\lambda}{3(\lambda-2)} a^{-\lambda+2} \sum_{s,t} A_{st} \left\{ \left[1 - 2^{-\lambda+2} - \frac{3(\lambda-2)}{\lambda} \right] \varrho_s^{(\alpha)} \right.$$
$$\left. + 2^{-\lambda+2} \varrho_s^{(\beta)} \right\} \Gamma_t^{(\alpha\beta)} . \tag{7.6}$$

Quantities of the same order are neglected in Eq. (7.6), as in Eq. (7.5).

Recently, the calculation of the local density in the central part of a drop was carried out by STORONKIN and one of the authors [32] on the basie of the macroscopic theory. The dependence of the local density on the drop radius, found in the main asymptotic term for ordinary dispersion forces without retardation, coincides with the above one. The comparison of the coefficients in corresponding asymptotic terms is a direct way for the construction of the pair effective potential in the case when one of the interacting molecules is in the central part of the drop and the other is in the medium surrounding the drop.

The qualitative behaviour of the local density and the local pressure tensor in the central part of a small system as presented by the above formulae (density and pressure decrease with decreasing system size) was recently confirmed by the results of a direct numerical experiment for a small drop obtained by the molecular dynamics technique [38].

8. Interaction of macroscopic particles

For a full presentation of the applications of the interaction principle, there remains to consider the situation when both the interacting molecular groups are of macroscopic scale. We will conventionally call these groups macroscopic particles granting that they may actually be very large and even infinitely large bodies.

For macroscopic particles, the main interest in the interaction principle is reduced to the interaction potential, W_{12}, directly determined by the principle. The probability aspects which played an important role in previous applications of the interaction principle, sink into the background.

The macroscopic dimensions of both interacting particles permit the use of Eq. (5.3) in the interaction principle for integration with respect to $\vec{r}_2$ and a similar equation for integration with respect to $\vec{r}_1$. As a result, we have a universal and simple method for finding the interaction potential, W_{12}, of two macroscopic particles in the medium. This finding is actually reduced to the calculation of four elementary integrals determined by the geometry of the particles and their mutual location:

$$
\begin{aligned}
I^{(\lambda)}_{V_1V_2} &= \int_{V_1} d\vec{r}_1 \int_{V_2} d\vec{r}_2\, r_{12}^{-\lambda}\,, & I^{(\lambda)}_{S_1V_2} &= \int_{S_1} dS_1 \int_{V_2} d\vec{r}_2\, r_{12}^{-\lambda}\,, \\
I^{(\lambda)}_{V_1S_2} &= \int_{V_1} d\vec{r}_1 \int_{S_2} dS_2\, r_{12}^{-\lambda}\,, & I^{(\lambda)}_{S_1S_2} &= \int_{S_1} dS_1 \int_{S_2} dS_2\, r_{12}^{-\lambda}\,,
\end{aligned} \tag{8.1}
$$

where, by analogy to Eq. (5.3), S_1 denotes the boundary surface between the particle 1 and medium, dS_1 is the area element for this surface, V_1 is the volume inside the surface S_1. Accordingly, the interaction potential W_{12}, is

$$
\begin{aligned}
W_{12} = &-I^{(\lambda)}_{V_1V_2} \sum_{s,t} A_{st}[\varrho_s^{(\gamma)} - \varrho_s^{(\alpha)}]\,[\varrho_t^{(\beta)} - \varrho_t^{(\alpha)}] - I^{(\lambda)}_{S_1V_2} \sum_{s,t} A_{st}\Gamma_s^{(\alpha\gamma)}[\varrho_t^{(\beta)} - \varrho_t^{(\alpha)}] \\
&- I^{(\lambda)}_{V_1S_2} \sum_{s,t} A_{st}[\varrho_s^{(\gamma)} - \varrho_s^{(\alpha)}]\,\Gamma_t^{(\alpha\beta)} - I^{(\lambda)}_{S_1S_2} \sum_{s,t} A_{st}\Gamma_s^{(\alpha\gamma)}\Gamma_t^{(\alpha\beta)}\,,
\end{aligned} \tag{8.2}
$$

where the superscripts α, β, γ refer to bulk properties inside the medium and the particles; $\varrho_s^{(\alpha)}$, $\varrho_s^{(\gamma)}$, $\varrho_s^{(\beta)}$ are the molecule number densities of the corresponding components in the phases α, γ, β, respectively; $\Gamma_s^{(\alpha\gamma)}$ and $\Gamma_s^{(\alpha\beta)}$ are adsorptions of the corresponding components on the boundary of the medium with the particles under consideration.

We present the final results of calculations using the above method for a number of the most interesting examples.

Let us begin with the case of interaction of two semi-infinite bodies separated by a flat medium layer (a film) of thickness H. This case is interesting in that it is found in experiments on measuring the film disjoining pressure, Π, the force between the bodies per unit area of their surface. Such measurements were carried out first for retarding forces [39] and later for dispersion forces without retardation [40—42]. We demonstrate the final result directly for the experimental quantity Π which is connected with the body interaction potential determined by Eqs. (8.1), (8.2) by the evident relationship:

$$
\Pi = -\frac{d}{dH}\,(W_{12}/S)\,, \tag{8.3}
$$

where S is the surface area of the body.

The result mentioned may be written as the asymptotic relationship [6, 14, 16, 18 19, 23]

$$
\Pi = B_1 H^{-\lambda+3} + B_2 H^{-\lambda+2}\,, \tag{8.4}
$$

where quantities $0(H^{-\lambda+1})$ are neglected, and the coefficients B_1 and B_2 are

$$
B_1 = -\frac{2\pi}{(\lambda-2)\,(\lambda-3)} \sum_{s,t} A_{st}[\varrho_s^{(\gamma)} - \varrho_s^{(\alpha)}]\,[\varrho_t^{(\beta)} - \varrho_t^{(\alpha)}]\,, \tag{8.5}
$$

$$
B_2 = -\frac{2\pi}{\lambda-2} \sum_{s,t} A_{st}\{\Gamma_s^{(\alpha\gamma)}[\varrho_t^{(\beta)} - \varrho_t^{(\alpha)}] + [\varrho_s^{(\gamma)} - \varrho_s^{(\alpha)}]\,\Gamma_t^{(\alpha\beta)}\}\,. \tag{8.6}
$$

For large film thicknesses the dominant first term in Eq. (8.4) conforms with the results of the macroscopic theory [1, 2] obtained without taking into account the non-uniformity of the film. The latter has been made, however, in the second term of our formula, Eq. (8.4), which is important for small thicknesses of the film. The second term contains the adsorptions $\Gamma_s^{(\alpha\gamma)}$ and $\Gamma_t^{(\alpha\beta)}$ on the boundary between the medium and the bodies under consideration and, consequently, describes the physical effects connected with the formation of adsorbed layers.

The term containing the product of adsorptions [the last item in Eq. (8.2)] is omitted in the asymptotic relationship (8.4). This term is of order $H^{-\lambda+1}$, which is beyond the limits of asymptotic accuracy of the interaction principle. The fact that, in contrast to the general formula (8.2), the interaction of the adsorbed layers of a flat film is outside the range of applicability of the method itself, is explained by the peculiarity of the flat case under consideration. This peculiarity is manifested in that the volume and surface integrals in the right-hand side of (5.3) have different orders with respect to the inverse distance from the point $\vec{r}_1$ to the surface S_1 when this distance increases to infinity: the surface integral diminishes by one order more rapidly than the volume one.

Observed in experiments on drawing solids together, the dependence of the disjoining pressure on the film thickness corresponded only to the main asymptotic term in Eq. (8.4). This means that the film investigated in the experiments had sufficiently large thickness as compared with molecular dimensions. It may be expected that finer effects connected with adsorbed layer formation and described by the second term of Eq. (8.4) will be manifested in measurements with smaller thicknesses.

If the disjoining pressure has been measured experimentally, Eqs. (8.4)—(8.6) may be used for finding the force constants in the pair effective intermolecular potentials. This furnishes an additional method beyond that mentioned in Section 4 for the force constant determination. Obviously, there is an additional difficulty of such determination in the presence of various kinds of molecules since the constants B_1 and B_2 determined from an experiment depend simultaneously on several force constants. On this account, there should be measurements for a large number of samples or additional interpolation equations (of the type of known as multiplicative correlations) for the molecular force constants of different kinds.

For the case when the phases γ and β are one-component and insoluble in the liquid film, Eqs. (8.4)—(8.6) permit the computation of the tension, σ, of the whole film (together with the surface layers of the phases γ and β). This is made with the aid of the thermodynamic relationship [43, Eq. (14. 48)]:

$$\frac{d\sigma}{d\Pi} = H - \frac{\Gamma^{(\alpha\gamma)}}{\varrho^{(\gamma)}} - \frac{\Gamma^{(\alpha\beta)}}{\varrho^{(\beta)}}, \tag{8.7}$$

which is the connection between σ and Π. The tension calculated in this way has the form [19]

$$\sigma = \sigma_1 + \sigma_2 + \frac{\lambda - 3}{\lambda - 4} B_1 H^{-\lambda+4} + \\ + \frac{\lambda - 2}{\lambda - 3} B_2 H^{-\lambda+3} - B_1 \left[\frac{\Gamma^{(\alpha\gamma)}}{\varrho^{(\gamma)}} + \frac{\Gamma^{(\alpha\beta)}}{\varrho^{(\beta)}}\right] H^{-\lambda+3}, \tag{8.8}$$

where the quantities $0(H^{-\lambda+2})$ are omitted, the integration constant, $\sigma_1 + \sigma_2$, is equal to the sum of the surface tensions for the thick film which is formed from the thin film under consideration as $H \to \infty$.

Since the surface tension, due to its generating properties, is a characteristic thermodynamic function, Eq. (8.8) together with (8.5)—(8.6) completely describes the thermodynamics of thin films.

One of the most important applications of Eqs. (8.4)—(8.6) and (8.8) is the investigation of the problem of film stability. As it is known from thermodynamics, this problem is reduced to establishing the sign of the disjoining pressure. If the disjoining pressure is positive and, respectively, the film tension increases when thinning the film, the latter is stable. In the opposite case the film is unstable.

The determination of the sign for the disjoining pressure from Eqs. (8.4)—(8.6) was made in [19]. We cite only the final conclusions obtained in this work.

The disjoining pressure of a symmetrical film of any composition is negative in its main asymptotic term both for ordinary and retarding dispersion forces. Only rare cases of mixtures (e.g., the hydrogen-helium system) for which the magnetic polarizability of some components is important may serve as an exception for this rule.

As for unsymmetrical films, the positive sign of the main asymptotic term for the disjoining pressure may also be expected for them. It is necessary for this that $\varrho_s^{(\gamma)} - \varrho_s^{(\alpha)}$ and $\varrho_t^{(\beta)} - \varrho_t^{(\alpha)}$ whould have different signs at least for some of the components. For example, the constant B_1 will be positive if the phases α, β, γ contain one component and $\varrho^{(\gamma)} > \varrho^{(\alpha)} > \varrho^{(\beta)}$.

The problem is thus solved for large thicknesses of a film when the coefficient B_1 in (8.4) determines the sign and magnitude of the disjoining pressure. When thinning the film, the corrective term in (8.4) should be taken into account when the coefficient B_2 is given by Eq. (8.6). The most interesting case is that of a free film when $\varrho_s^{(\gamma)} = 0$ and $\varrho_t^{(\beta)} = 0$ to which we confine ourselves below. Evidently, free films are at the same time symmetrical.

If a free film is composed of one component, then always $B_2 < 0$. For such a film, consequently, the corrective term for the disjoining pressure is also negative. Thus, a free-component film is always unstable.

The appearance of the positive sign of B_2 may only be associated with a change in composition in the surface parts of the film and with the corresponding volume effects. In particular, B_2 will be positive when a surfactant forms a sufficiently dense structure on the film surface.

The presence of the positive sign in the corrective terms in Eqs. (8.4) and (8.8) may lead to the appearance of a positive disjoining pressure and to film stability in the absence of ionic electrostatic repulsion. It is essential that this kind of stability, in accordance with the meaning of the corrective terms, may only be explained by taking the film non-uniformity into account.

Our next example of the use of the method described in this section for calculation of the interaction potential of macroscopic particles will be the case of interaction of a spherical body of a radius a and a semi-infinite body, both being in the medium. The boundary surface of the semi-infinite body is assumed to be flat. The mutual location of the bodies is characterized by the parameter H, the shortest distance between the boundary surfaces of the bodies.

We cite the final result directly for the force, F_{12}, acting between the bodies [23]:

$$F_{12} = -\frac{d}{dH} W_{12} = \frac{(2\pi)^2}{(\lambda-2)(\lambda-3)(\lambda-5)} \left\{ H^{-\lambda+5} - (H+2a)^{-\lambda+5} - \frac{\lambda-5}{\lambda-4}(H+a)\left[H^{-\lambda+4} - (H+2a)^{-\lambda+4}\right] \right\}$$
$$\times \sum_{s,t} A_{st}[\varrho_s^{(\gamma)} - \varrho_s^{(\alpha)}][\varrho_t^{(\beta)} - \varrho_t^{(\alpha)}]$$
$$- \frac{(2\pi)^2 a}{(\lambda-2)(\lambda-3)} [H^{-\lambda+3} - (H+2a)^{-\lambda+3}] \sum_{s,t} A_{st} \Gamma_s^{(\alpha\gamma)} [\varrho_t^{(\beta)} - \varrho_t^{(\alpha)}]$$
$$+ \frac{(2\pi)^2}{(\lambda-2)(\lambda-4)} \left\{ H^{-\lambda+4} - (H+2a)^{-\lambda+4} - \frac{\lambda-4}{\lambda-3}(H+a)\left[H^{-\lambda+3} - (H+2a)^{-\lambda+3}\right] \right\} \sum_{s,t} A_{st}[\varrho_s^{(\gamma)} - \varrho_s^{(\alpha)}] \Gamma_t^{(\alpha\beta)}$$
$$- \frac{(2\pi)^2 a}{\lambda-2} [H^{-\lambda+2} - (H+2a)^{-\lambda+2}] \sum_{s,t} A_{st} \Gamma_s^{(\alpha\gamma)} \Gamma_t^{(\alpha\beta)} . \qquad (8.9)$$

Finally, as one more example, we consider the interaction of two spherical bodies of the same radius, a, in the medium. The mutual location of the bodies will be given, as before, by the shortest distance, H, between their boundary surfaces.

This time, we give the result in the form of the interaction potential and confine ourselves, for the sake of simplicity, to the case $\lambda = 6$ corresponding to ordinary dispersion forces without retardation [23]

$$W_{12} = \frac{\pi^2}{6} \left\{ \ln \frac{(H+2a)^2}{H^2 + 4Ha} - \frac{4a^2}{H^2 + 4Ha} \left[1 - \frac{2a^2}{(H+2a)^2}\right] \right\}$$
$$\times \sum_{s,t} A_{st}[\varrho_s^{(\gamma)} - \varrho_s^{(\alpha)}][\varrho_t^{(\beta)} - \varrho_t^{(\alpha)}] - \frac{2\pi^2}{3} \frac{a^4}{(H+2a)^3} \left[\frac{1}{H^2} - \frac{1}{(H+4a)^2}\right]$$
$$\times \sum_{s,t} A_{st} \left\{ \Gamma_s^{(\alpha\gamma)} [\varrho_t^{(\beta)} - \varrho_t^{(\alpha)}] + [\varrho_s^{(\gamma)} - \varrho_s^{(\alpha)}] \Gamma_t^{(\alpha\beta)} \right\}$$
$$- \frac{8\pi^2 a^2}{(H^2 + 4Ha)(H+2a)^2} \sum_{s,t} A_{st} \Gamma_s^{(\alpha\gamma)} \Gamma_t^{(\alpha\beta)} . \qquad (8.10)$$

The expression for the body interaction force, $F_{12} = -dW_{12}/dH$, is easily deduced from (8.10) and we will not cite it. To include this section, we note that the main terms of Eqs. (8.9) and (8.10) (the form of which relates to the supposition of the uniformity of the medium and each of the interacting bodies) correspond to the known result by HAMAKER [44] obtained, however, without sufficient theoretical foundation. HAMAKER's results also correspond to the main term of Eq. (8.4) which was rigorously obtained years ago by means of the macroscopic theory [1, 2], as was already stated above.

We also note that taking adsorbed layers into account in the corrective terms of Eqs. (8.4), (8.9), (8.10) was also later considered within the framework of the macroscopic theory [45, 46] for the case of a flat film. The use of the macroscopic theory requires the use of a macroscopic concept, the local dielectric permittivity, within the microscopic scale of adsorbed layers and, moreover, the knowledge of the local dielectric permittivity profile. The model character of such calculations and the absence in their results of such fundamental properties as adsorptions make difficult their comparison with the corrective term in Eq. (8.4) for the disjoining pressure of a flat film.

9. Compressibility and adsorption as parameters for the description of collective phenomena

The applications of the interaction principle demonstrated in Section 3—8 make possible a better evaluation of the meaning of the microscopic theory and of the interaction principle itself.

With the exception of the necessary general apparatus of statistical mechanics presented in Section 1, it may be said that all the microscopic theory developed has asymptotic character. This theory always refers to large distances: the distances between molecular groups, between a molecular group and the boundary surface or, finally, between macroscopic particles. Moreover, the main body of the concrete results is represented in the form of strict asymptotic expansions in powers of the inverse distance. This refers to the results of Sections 3, 5—7, and also to the results on flat films of Section 8.

The asymptotic character of the theory presented corresponds to the asymptotic character of dispersion interactions themselves. The large distances for which the theory is valid correspond to the distances at which dispersion interactions are manifested. Hence, the computational scheme reflects completely the aim of the work pointed out in the title.

The large distances under discussion are only large from the molecular point of view. They are large as compared with molecular dimensions and average intermolecular distances in condensed systems. The may, however, be small from the macroscopic point of view.

As it was already stated, the advantage of the microscopic approach described is the possibility of calculation not only of the main, but also the corrective asymptotic terms. The latter become essential as distances decrease. The possibility of calculation of these terms means the possibility of establishing the theory applicability limits within the framework of the theory itself.

Finding the corrective terms becomes especially important in the case when both the interacting molecular groups have macroscopic dimensions, and the corresponding main term of the asymptotic expansions is, in a way, trivial. We have seen in the disjoining pressure example that interesting physical conclusions may be made only on the level of the corrective terms.

Dealing with a system of strongly interacting molecules, such as condensed matter, the microscopic theory, naturally, should also take into account the short range forces including repulsive intermolecular forces. These forces play a very important part in the microscopic structure. Even more may be said: almost all the properties of condensed systems result from the competition of the short and long range forces. Through this competition, collective phenomena are expressed determining the nature of a condensed system.

Collective phenomena also play the decisive role in the above asymptotic formulation. Indeed, although two molecular groups are removed from each other, the molecules of the surrounding medium, as it was stated in Section 2, remain as it intermediators for the group interaction. The fact that we have managed to take collective phenomena fully into account and to obtain an explicit solution for the asymptotic case, is physically caused by the circumstance that in this case the medium molecules taking part in the interaction transmission may be subdivided into two classes, near and far molecules, the contributions of which may then be calculated separately. The contribution of far molecules is strictly equal to zero in the asymptotic limit for the systems considered with dispersion forces, as it was shown in Section 2. The contribution of near molecules in the asymptotic limit is only determined by integral characteristics of the molecular distribution densities of the surrounding medium in the vicinities of the molecular groups, the total excess molecule number of the medium in a given vicinity. This made it possible in Sections 3 and 5 to express the contributions of the near medium molecules through physically observable quantities, compressibilities and adsorptions.

Let us give a more detailed explanation of these remarks. Using the definition (2.5), we write the compressibility theorem, Eq. (1.7) in the form

$$\int d\vec{r}\, \Delta\varrho_s(\vec{r}/\{n\}) = \chi_s\{n\} , \tag{9.1}$$

where

$$\chi_s\{n\} = \frac{k_B T}{\varrho\{n\}} \frac{\partial\varrho\{n\}}{\partial\mu_s} . \tag{9.2}$$

Equations (9.1) and (9.2) were already used when deriving Eq. (3.1).

In general, we had in view the compressibilities $\chi_s\{n\}$ defined by Eq. (9.2) when speaking about the calculation of the near molecule contribution for the medium. In particular, when the system consists of one component and the complex $\{n\}$ consists of one molecule, Eq. (9.1) becomes

$$\int d\vec{r}\, \Delta\varrho(\vec{r}/\{1\} = \chi , \tag{9.3}$$

where $\chi \equiv (k_B T/\varrho)\, \partial\varrho/\partial\mu$ is the compressibility for this particular case, already introduced in Section 3.

Making the asymptotic expansions (3.3), (3.4), (5.4), (6.1), (7.1) and (7.5) dimensionless by dividing them by the corresponding zero expansion terms, it is easy to see that just the compressibilities $\chi_s\{n\}$ occur in our theory. All the expansions mentioned have the common property that at least one of the interacting molecular groups has microscopic dimensions. The compressibilities $\chi_s\{n\}$ or their products are actually present in the above expansions as coefficients of the force constants. As multipliers in the expansion coefficients, the compressibilities increase when approaching the critical point, which results in increasing the expansion convergence radius and, eventually, in excluding the critical region from the range of applicability of the theory.

Similar explanations may be given for the adsorptions $\Gamma_t^{(\alpha\beta)}$ and $\Gamma_s^{(\alpha\gamma)}$. These adsorptions (or their pair products with one another or with compressibilities) are present as coefficients of the force constant in the asymptotic relationships in which at least one of the interacting molecular groups is a macroscopic body.

Apart from their contribution to the interaction from the near medium molecules, compressibilities and adsorptions, evidently, also include a contribution to the interaction from molecules inside the microscopic molecular groups or the surface layers of macroscopic bodies.

Contributing to the interaction description and also occuring in asymptotic relationships (as coefficients of the force constants), compressibilities and adsorptions are also similar in their physical meaning. As seen from (9.1), compressibilities are the excess molecule numbers in the vicinity of microscopic molecular groups (including the group molecules themselves). And adsorptions are the excess molecule numbers in the surface layers adjacent to macroscopic bodies (including the molecules from the non-uniform regions of the bodies themselves).

Thus, we see that compressibilities and adsorptions play the role of parameters effectively describing collective phenomena. These parameters, in particular, contain the contributions from the short range forces including repulsion. The use of compressibilities and adsorptions as parameters for the collective phenomena description allowed us to obtain rigorous asymptotic results for problems within the framework of the microscopic theory for systems with strongly interacting particles.

The notion of the pair effective potential as a starting point for this microscopic theory refers, strictly speaking, to concrete physical conditions for a system. It was not stipulated in our calculations that the pair effective potential may actually also depend on the chemical potentials, μ_s, of a system and its temperature, T. However, it is easy to see that all of our main agreement virtually admits the presence of such dependences.

Indeed, both the fundamental compressibility theorem, Eq. (1.7), and Eqs. (3.1), (9.1)—(9.3) resulting from it are quite strict results of statistical mechanics. Their derivation, did not include any assumptions about the structure of the total interaction energy, $U(1, \dots, N)$, of a system. The derivatives with respect to the chemical potentials in Eqs. (1.7), (3.1), (9.1)—(9.3) describe true dependences on the chemical potential in a real system.

The notion of the pair effective potential was used in the diagram calculation technique. However, the pair effective potential only referred to fixed values of the parameters μ_S and T, and the effect of changing them was not considered.

This effect does not appear in the functional differentiation technique used in (2.4) when deriving the interaction principle based on the application of the operator $\zeta(i)\, \delta/\delta\zeta(i)$. Though the application of this operator, generally speaking, is connected with a change of the chemical potentials, it implied in its physical meaning that the pair effective potential is kept fixed. Only with this assumption, does the operator $\zeta(i)\, \delta/\delta\zeta(i)$ possess the necessary generating properties and provides a necessary meaning to the operator in (2.4) of the point number density at "the point i" (the diagram points are those to which the operator is applied).

Thus, the interaction principle itself and its direct applications (the computation of the molecular correlations, the distribution functions in a capillary system, the potentials and forces of interaction of macroscopic particles) are also valid when the pair effective potential depends on the conditions in which a system exists.

The matter is more complicated for the derivation of asymptotic expressions for the pressure tensor. Already the starting expressions, Eqs. (1.9) and (1.11), for the pressure tensor are connected with the concept of the pair effective potential .And if this potential depends on the chemical potentials, it is impossible, strictly speaking, to represent the coefficients in asymptotic relationships for the pressure tensor as the derivatives $\partial P^{(\alpha)}/\partial \mu_s$ unidentified with the densities $\varrho_s^{(\alpha)}$ due to the known thermodynamic relationship.

As reviewed in this article, the interaction principl does not exhaust all the possibilities of the microscopic theory. The method of asymptotic summation of diagrams used in the principle derivation, generally speaking, is wider than the principle itself. This is easy to understand already in that the interaction principle refers only to two objects present in a medium. There may be several such objects in more complex situations. The simplest examples are problems on the correlations of several molecular groups located in the medium at large distances from one another (the solution of these problems by the asymptotic summation method may be found in the review article [47]), on the molecular structure of films [6, 14, 18, 19], and so on.

As it was stated above, the microscopic theory and the interaction principle are restricted to regions far from the critical point. The latter demands special study, the use of scaling theory. Because of the universal character (independent of the kind of molecular forces) of correlative and structural properties in the critical point given by scaling theory, the latter is outside the limits of the main problem of this article, which is the study of dispersion interactions.

References

[1] E. M. Lifshitz, Zh. Eksper. Teor. Fiz., **29**, 94 (1955)

[2] I. E. Dzyaloshinsky, E. M. Lifshitz, L. P. Pitaevsky, Adv. Phys. **10**, 165 (1961)

[3] F. M. Kuni, A. I. Rusanov, Dokl. Akad. Nauk USSR, **174**, 406, 1967

[4] F. M. Kuni, A. I. Rusanov, Dokl. Akad. Nauk USSR, **175**, 121, 1967.

[5] F. M. Kuni, A. I. Rusanov, Phys. Letters, **25 A**, 577, 1967.

[6] A. I. Rusanov, F. M. Kuni, in "*Research in Surface Forces*" (B. V. Deryaguin, Ed.), vol. 3, Consultants Bureau, New York-London, 1971, p. III (the Russ. Edition: Nauka, Moscow, 1967, p. 129)

[7] F. M. Kuni, Vestnik Leningrad Univ., No. 4, 30, 1968

[8] F. M. Kuni, Dokl. Akad. Nauk SSSR, **179**, 129, 1968

[9] F. M. Kuni, Phys. Letters, **26 A**, 305, 1968

[10] F. M. Kuni, A. I. Rusanov, Zh. Fiz. Khim. **42**, 849, 1968

[11] A. I. Rusanov, F. M. Kuni, Zh. Fiz. Khim., **42**, 1189, 1968

[12] F. M. Kuni, A. I. Rusanov, Zh. Fiz. Khim., **42**, 1723, 1968

[13] A. I. Rusanov, F. M. Kuni, Zh. Fiz. Khim., **42**, 2563, 1968

[14] F. M. Kuni, A. I. Rusanov, Zh. Fiz. Khim., **42**, 2569, 1968

[15] A. I. Rusanov, F. M. Kuni, Zh. Fiz. Khim., **42**, 2865, 1968

[16] F. M. Kuni, A. I. Rusanov, Phys. Letters, **26 A**, 577, 1968

[17] F. M. Kuni, Vestnik Leningrad Univ., No. 16, 31, 1968

[18] E. N. Trofimova, F. M. Kuni, A. I. Rusanov, Kolloid. Zh., **31**, 578, 1969

[19] F. M. Kuni, A. I. Rusanov, E. N. Brodskaya, Kolloid. Zh. **31**, 860, 1969

[20] A. I. RUSANOV, F. M. KUNI, E. N. BRODSKAYA, Zh. Fiz. Khim. **44**, 756, 1970
[21] A. I. RUSANOV, F. M. KUNI, E. N. BRODSKAYA, Zh. Fiz. Khim., **44**, 766, 1970
[22] F. M. KUNI, A. I. RUSANOV, Kolloid. Zh., **32**, 232, 1970
[23] A. I. RUSANOV, F. M. KUNI, Kolloid. Zh., **33**, 121, 1971
[24] F. M. KUNI, A. I. RUSANOV, Kolloid. Zh., **33**, 238, 1971
[25] A. I. RUSANOV, F. M. KUNI, Zh. Fiz. Khim., **46**, 728, 1972
[26] F. M. KUNI, A. I. RUSANOV, in "*Physical Adsorption from Multicomponent Phases*" (russ.), (M. M. DUBININ, Ed.), Nauka, Moscow, 1972, p. 182
[27] O. SINANOGLU, Chem. Phys. Letters, **1**, 340, 1967
[28] O. SINANOGLU, Adv. Chem. Phys., **12**, 283, 1967
[29] T. HALICIOGLU, O. SINANOGLU, J. Chem. Phys. **49**, 996, 1968
[30] M. P. KEMOKLIDZE, L. P. PITAEVSKY, Zh. Eksper. Teor. Fiz., **59**, 2187, 1970
[31] V. R. BELOSLUDOV, A. M. KOROTKIKH, V. M. NABUTOVSKY, in "*The Centennial of* GIBBS' *Theory of Capillarity*" (russ.), (A. I. RUSANOV, Ed.), Leningrad Univ., 1978, p. 26
[32] B. A. STORONKIN, F. M. KUNI, in "*The Centennial of* GIBBS' *Theory of Capillarity*' (russ.), (A. I. RUSANOV, Ed.), Leningrad Univ., 1978, p. 163
[33] J. ENDERBY, T. GASKELL, N. H. MARCH, Proc. Phys. Soc., **85**, 217, 1965
[34] T. GASKELL, Proc. Phys. Soc., **86**, 693, 1965
[35] N. H. MARCH, in "*Theory of Condensed Matter*", Trieste, 1967
[36] I. F. EFREMOV, "*Periodic Colloid Structures*" (russ.), Khimia. Leningrad, 1971
[37] I. Z. FISHER, Zh. Eksper. Teor. Fiz., **59**, 1336, 1970
[38] E. N. BRODSKAYA, A. I. RUSANOV, Kolloid. Zh. **39**, 636, 646, 1977
[39] B. V. DERYAGUIN, I. I. ABRIKOSOVA, Dokl. Akad. Nauk SSSR, **108**, 214, 1956; Zh. Fiz. Khim., **32**, 442, 1958
[40] D. TABOR, R. H. C. WINTERTON, "Diamond Res., 1969", London, Ind. Diamond, Inform. Bur., p. 13
[41] D. TABOR, J. Colloid Interface Sci., **31**, 364, 1969; Chem. and Ind., No. 25, 969, 1971
[42] F. WITTMAN, H. SPLITTGERBER, K. EBERT, Z. Phys., **245**, 354. 1971
[43] A. I. RUSANOV, "*Phasengleichgewichte und Grenzflächenerscheinungen*", Akademie-Verlag, Berlin, 1978, p. 283
[44] H. C. HAMAKER, Physica, **4**, 1058, 1937
[45] G. H. WEISS, J. E. KIEFER, V. A. PARSEGIAN, J. Colloid Interface Sci., **45**, 615, 1973
[46] B. V. DERYAGUIN, N. V. CHURAEV, Kolloid. Zh. **37**, 1075, 1975
[47] F. M. KUNI, in "*Problems of Theoretical Physics*" (russ.), vol. 2, Leningrad Univ., 1975, p. 72

Thermodynamics of surface phenomena on electrodes

O. A. Petrii

(Faculty of Chemistry, Moscow State University, Moscow/USSR)

Abstract

This contribution considers the main stages in the development of the thermodynamic theory of electrocapillarity. It gives the basic electrocapillarity equations for ideally polarizable and reversible electrodes as well as the relations resulting from these equations. Furthermore, the physical significance of the quantities contained in the electrocapillarity equation of the ideally polarizable and reversible electrodes is discussed. The methods and results of the verification of the thermodynamic electrocapillarity theory for reversible systems are demonstrated as exemplified by electrodes from platinum metals and thallium amalgams. It is shown that the thermodynamic approach to chemisorption phenomena involving transfer of part of the adsorbed particle charge to metal surface allows the determination of the formal charge transfer coefficient, which characterizes both the part of the charge transferred from the adsorbed particle to the electrode and the change in the electric double layer parameters at the electrode-solution interface caused by the adsorption process. The author also discussed the experimental data illustrating the difference between the surface tension and the reversible surface work for solid metals.

1. Introduction

The thermodynamics of surface phenomena on electrodes, or, in other words, the thermodynamic theory of electrocapillarity, deals with the dependence of the reversible surface work σ (equal to the surface tension in the case of liquid metals) [1] on the electrode potential E, the composition of the coexisting phases, pressure, and temperature.

The history of the development of the electrocapillarity theory was given in detail by Frumkin [2]. Therefore in this review we shall only briefly outline the main stages in the development of this branch of the science of surface phenomena.

The beginning of thermodynamic electrocapillarity theory dates back to Lippmann [3, 5], who derived the equation quantitatively relating in a differential form the reversible surface work of a mercury electrode σ to its potential at constant temperature, pressure, and solution composition

$$\partial\sigma/\partial E = -Q\,. \tag{1}$$

In his earlier studies Lippmann defined the quantity Q as the capacity of a unit surface at a constant potential difference, later he [6, 7] equated it with the charge ε of a unit surface of the metal side of the electric double layer at the mercury-electrolyte solution interface.

The next step in the development of electrocapillarity theory were the works of GIBBS [8] and PLANCK [9]. GIBBS applied his wellknown adsorption equation (the pressure p = const)

$$d\sigma = -S^\sigma dT - \sum_i \Gamma_i d\mu_i \tag{2}$$

(Γ_i are the adsorptions, i.e. the surface excesses of the components per unit surface, μ_i are the chemical potentials of the system components, S^σ is the specific excess surface entropy, T is the absolute temperature) to a mercury electrode in sulfuric acid solution and arrived at a relation, which can be written as follows [10] (T = const, p = const):

$$d\sigma = \Gamma_H dE , \tag{3}$$

where Γ_H is the GIBBS' hydrogen adsorption in electrical units. According to GIBBS, the right side of equation (1) should express the surface density of adsorbed electrochemically active substance. The choice of hydrogen as such a substance in the case of mercury was based on the concepts of the nature of mercury polarization generally accepted at that time [2]. GIBBS also pointed out that in the case of a solid there is a difference between the reversible work of surface formation σ and the surface tension γ.

PLANCK [9] showed that equation (1) is valid for a "completely polarizable" electrode, i.e. an electrode whose state is unambiguously determined by the amount of electricity passed through it.

The interest in electrocapillarity theory was significantly stimulated by the extensive, precise studies of electrocapillary phenomena carried out by GOUY [11], who obtained some important thermodynamic relations describing the electrocapillary behaviour of a mercury electrode.

The GIBBS adsorption equation was first used for a detailed analysis of electrocapillary phenomena by FRUMKIN [12—14]. FRUMKIN showed, as exemplified by zinc amalgam in electrolyte solution, that in the general case, the quantity Q cannot be equated with the charge of the metal side of the double layer. In 1927 FRUMKIN formulated the concept of the potential of zero charge as the most important characteristic determining the electrocapillary behaviour of a metal [15]. In [16] he outlined the principles of the thermodynamic theory of the state of the surface of electrodes reversibly adsorbing hydrogen and oxygen. In a certain potential range these electrodes can be considered as "completely polarizable" in PLANCK's sense.

From the middle thirties for a long period of time the development of electrocapillarity theory was associated, primarily, with the analysis of the behaviour of the so-called ideally polarizable electrode (KOENIG [17]), which is characterized by the complete absence of mass transfer between contacting phases [18—33]). Less attention was given to electrodes for which charge transfer across the interface is possible (reversible electrodes) [20, 30, 34—36]. Only in the 60—70's electrocapillary phenomena at reversible electrodes have been subjected to detailed thermodynamic analysis [37—65].

2. The basic electrocapillarity equation

Two approaches were used in deriving the basic electrocapillarity equation. One of them is based on the GIBBS equation [12] and the second on the thermodynamic cycles method [11, 18]. In the first approach, two concepts were used: the GIBBS plane corresponding to zero adsorption of one of the components (usually solvent) [12] and the surface layer of finite thickness [28, 31] separating the phases.

Different methods of deriving the basic electrocapillarity equation lead to virtually the same results. In the thermodynamic approach, an ideally polarizable electrode can be treated as the limiting case of a reversible electrode with an infinitesimal concentration of the potential-determining ions [14]. Particular mention should be made especially in connection with [66], of the derivation of the basic electrocapillarity equation using metal electrons as one of the system components, which was proposed in [67].

Let us consider an ideally polarizable electrode, whose radius of surface curvature is much greater than the surface layer thickness. With the GIBBS adsorption of the solvent $\Gamma_s = 0$ and at constant pressure, temperature and electrode composition, the basic electrocapillarity equation for an ideally polarizable electrode can be written as [64]:

$$d\sigma = -Q\,dE - \frac{Q}{z_j}\,d\mu_j - \sum_i \Gamma_i\,d\mu_i\,, \tag{4}$$

where Q is the specific surface charge, the index i qualifies all the components, except the solvent, including the ion j with the charge z_j, i.e. the component with respect to which the reference electrode is reversible. The physical significance of the quantities Q and Γ will be discussed below, μ_i are the chemical potentials of the system components. All the quantities here and in what follows are expressed in electrical units. With $\mu_j =$ const and at a constant potential drop at the interface between the solution being investigated and the reference electrode solution, instead of the rigorous equation (4), it is possible to use the approximate equation

$$d\sigma = -Q\,dE - \sum_i \Gamma_i\,d\mu_i\,. \tag{5}$$

Often in experiments solutions containing only the salt CA and some organic substance are used. Under the condition that the reference electrode is reversible with respect to the salt cation ($j \equiv C^+$) and using the electroneutrality condition

$$Q = \Gamma_- - \Gamma_+ \tag{6}$$

from equation (4) at $z_j = +1$, we obtain

$$\begin{aligned} d\sigma &= -Q\,dE_+ - \Gamma_-(d\mu_+ + d\mu_-) - \Gamma_{\text{org}}\,d\mu_{\text{org}} \\ &= -Q\,dE_+ - \Gamma_-\,d\mu_{CA} - \Gamma_{\text{org}}\,d\mu_{\text{org}}\,. \end{aligned} \tag{7}$$

Similarly, at $j \equiv A^-$ and $z_j = -1$, we have

$$d\sigma = -Q\,dE - \Gamma_+\,d\mu_{CA} - \Gamma_{\text{org}}\,d\mu_{\text{org}}\,. \tag{8}$$

In these equations E_+ and E_- designate the potentials measured against the reference electrode, reversible to cation or anion, respectively, μ_{CA} and μ_{org} are the chemical potentials of salt and organic substance, Γ_+ and Γ_- are the GIBBS adsorptions of cation and anion, Γ_{org} is the GIBBS adsorption of organic substance. The use of a reference electrode reversible with respect to cation or anion makes it possible to avoid difficulties associated with the existence of a potential difference between the working solution and that of the reference electrode.

For a solution of a constant salt concentration and at not too high organic substance concentrations, it follows from equations (7) and (6) that

$$\Gamma_{\text{org}} = -(\partial\sigma/\partial\mu_{\text{org}})_E \,. \tag{9}$$

In the absence of organic substance or with $\mu_{\text{org}} = \text{const}$, from equations (7) and (8) we obtain

$$\Gamma_+ = -(\partial\sigma/\partial\mu_{\text{CA}})_{E_-}\,; \qquad \Gamma_- = -(\partial\sigma/\partial\mu_{\text{CA}})_{E_+}\,. \tag{10}$$

Using the total differential properties, from equations (7) and (8) it is possible to derive the following four relations [25] ($\mu_{\text{org}} = \text{const}$):

$$(\partial Q/\partial\mu_{\text{CA}})_{E_\mp} = (\partial\Gamma_\pm/\partial E_\mp)_{\mu_{\text{CA}}}\,, \tag{11}$$

$$(\partial E_\mp/\partial\mu_{\text{CA}})_Q = -(\partial\Gamma_\pm/\partial Q)_{\mu_{\text{CA}}}\,, \tag{12}$$

$$(\partial Q/\partial\Gamma_\pm)_{E_\mp} = -(\partial\mu_{\text{CA}}/\partial E_\mp)_{\Gamma_\pm}\,, \tag{13}$$

$$(\partial E_\mp/\partial\Gamma_\pm)_Q = (\partial\mu_{\text{CA}}/\partial Q)_{\Gamma_\pm}\,. \tag{14}$$

These relations are widely used in analyzing the results of measurement of the electrocapillary curves and those of the differential capacity of the electric double layer [25, 29, 64].

Let us assume that a redox system is present in the solution and that equilibrium s established on the electrode

$$0 + n\bar{e} \rightleftarrows R \tag{15}$$

so that the electrode potential is determined by the NERNST equation:

$$dE = d\mu_0 - d\mu_R\,, \tag{16}$$

where μ_0 and μ_R are the chemical potentials of 0 and R. For the reversible electrode under consideration, the GIBBS equation can be written as follows [49, 50]:

$$d\sigma = -\Gamma_0\, d\mu_0 - \Gamma_R\, d\mu_R - \sum_i \Gamma_i\, d\mu_i\,, \tag{17}$$

where Γ_0 and Γ_R are the GIBBS adsorptions of 0 and R, respectively.

At $\mu_0 = \text{const}$ $d\mu_0 = 0$, $dE = -d\mu_R$ and

$$d\sigma = \Gamma_R\, dE_i - \sum_i \Gamma_i\, d\mu_i\,. \tag{18}$$

At $\mu_R = \text{const}$ $d\mu_R = 0$, $dE = d\mu_0$ and

$$d\sigma = -\Gamma_0\, dE - \sum_i \Gamma_i\, d\mu_i\,. \tag{19}$$

Equations (18) and (19), which should be considered as the electrocapillarity equations of a reversible electrode, lead to the relations:

$$(\partial\sigma/\partial E)_{\mu_0, \mu_i} = \Gamma_R\,, \tag{20}$$

$$(\partial\sigma/\partial E)_{\mu_R, \mu_i} = -\Gamma_0\,. \tag{21}$$

In the general case of redox systems, the number of electrocapillarity equations and accordingly that of equations of type (20) and (21) are determined by the number of independent variables in the NERNST equation expressing the system equilibrium conditions. It is evident that from equations (18) and (19) it is possible to obtain systems of equations similar to (11)—(14).

Rusanov generalized the electrocapillarity equations to the case of curved interfaces [31, 32]. A more complex form of the thermodynamic equations corresponding to this case is due to a pressure difference in the coexisting phases arising from curving the interface. The influence of the curvature on the value of σ is apparent at radii of curvature $r < 10^{-5}$ cm, at sufficiently small r the value of σ decreasing with decreasing r. In [68] Rusanov derived electrocapillarity equations for a system containing two phases separated by a thin film being impermeable to all phase components, and also for the case of formation of a thin film between two ideally polarizable electrodes. Thus the thermodynamics of thin films was related to that of electrocapillary phenomena.

3. The concept of the electrode charge

Let us consider the physical significance of the quantities Q and Γ contained in the electrocapillarity equations of ideally polarizable and reversible electrodes.

In defining the value of the electrode charge Q it is usual to apply the concept of the electric double layer at the electrode-solution interface, and to identify the electrode charge with that of the metal side of the double layer ε. Evidently, this definition of Q is a model one, being based on the assumption that the ion adsorption on the electrode surface is of a purely electrostatic nature. Such a model applies only to ions not showing specific adsorbability, such as F^- and Li^+ or Na^+ ions and is not valid when a covalent bond is formed between adsorbed ion and electrode metal. Lorenz [69—71] suggests that the formation of such a bond, which is equivalent to the transfer of part of the adsorbed ion charge to metal, takes place in the case of adsorption of halogen anions and of potassium and cesium cations on mercury and gallium electrodes. The studies of the adsorption phenomena at platinum metals gave conclusive evidence of a charge transfer during the adsorption of most ions [54, 55]. It is clear that in the presence of a partial charge transfer the difference between the concepts of ideally polarizable and reversible electrodes to a certain extent disappears and the free charge of the metal surface ε cannot be equated with Q.

The analysis carried out in [49, 50] shows that the electrode charge Q contained in the thermodynamic equations is equal to the amount of electricity to be supplied to the electrode when its surface is increased by a unit and the concentration of solution components is kept constant. The Gibbs adsorptions Γ of the components are understood to mean the amounts of respective substances to be introduced into the system to keep the bulk phases at constant concentration when the interface increases by a unit.

In the general case, in electrochemical systems the values of Gibbs adsorptions Γ and surface excesses A do not coincide. This situation can be illustrated by the case of a hydrogen-saturated platinum electrode in acidified sodium chloride solution. When the electrode surface increases, a certain amount of hydrogen Γ_H disappears from the gas phase. However, this amount is not equal to the amount of hydrogen A_H per unit electrode surface, so that part of hydrogen undergoes ionization, the electrode surface acquires a negative charge and the hydrogen ions are displaced by sodium ions from the surface layer into solution.

In order to keep constant the solution composition when the ideally polarizable electrode surface increases it is necessary to introduce all solution components in the amounts Γ_i. In the case of a reversible electrode for which equilibrium (15) is established, the solution composition can be kept constant by several methods. For instance, without introducing electricity from outside, it is possible to add the components 0 and R in the amounts Γ_0 and Γ_R, respectively. Another method is to introduce $(\Gamma_0 + \Gamma_R)$ of the substance participating in the redox reaction either entirely in the form 0 or entirely in the form R. In the former case it is necessary to supply to the electrode simultaneously the amount of electricity $Q' = -\Gamma_R$; in the latter case the respective amount of electricity is $Q'' = \Gamma_0$. Thus, the quantities Q' and Q'' (and hence $-\Gamma_R$ and Γ_0) should be considered as the values of the reversible electrode charge at constant chemical potential of oxidant or reductant, respectively.

The foregoing explains why it has been proposed to call the quantities Q' and Q'' the total (or thermodynamic) charges of the electrode [49, 50].

The relation of Q' or Q'' to ε can only be established on the basis of a model of the nature of solution components in the adsorbed state. If no charge transfer other than that corresponding to equation (15) is present, we can write [49, 50]:

$$Q' = -\Gamma_R = \varepsilon - A_R \,, \tag{22}$$

$$Q'' = R_0 = \varepsilon + A_0 \,, \tag{23}$$

where A_R and A_0 are the surface excesses of R and 0, respectively. Hence equations (20) and (21) can be written as:

$$(\partial\delta/\partial E)_{\mu_0,\mu_i} = \Gamma_R = -Q' = \varepsilon + A_R \,, \tag{24}$$

$$(\partial\sigma/\partial E)_{\mu_R,\mu_i} = -\Gamma_0 = -Q'' = -\varepsilon - A_0 \,. \tag{25}$$

The quantities A_R and A_0 can also be called the bound surface charge in distinction to the quantity ε, which reasonably should be called the free charge.

Equations (24) and (25) are the LIPPMANN equations of a reversible electrode. In accordance with these equations, for a reversible electrode whose potential is determined by the equilibrium (15) it is possible to plot two characteristic electrocapillary cuıves. Along one of these curves the electrode potential variation is determined by the change of the chemical potential of reductant (the electrocapillary curves of the 1st kind, equation (24)), and along the second curve by that of the oxidant (the electrocapillary curve of the 2nd kind, equation (25)). It is evident that there exist two potentials of zero charge for a reversible electrode, at which Q' or Q'' vanishes. The derivatives $(\partial Q'/\partial E)_{\mu_0,\mu_i}$ and $(\partial Q''/\partial E)_{\mu_R,\mu_i}$ were named the differential capacities of the 1st and 2nd kind, respectively. They can be obtained by differentiation of the charging curves of the 1st and 2nd kinds, giving the dependence of Q' and Q'' on the electrode potential.

In the case of an ideally polarizable electrode $A_R = A_0 = 0$, we obtain a LIPPMANN equation the right side of which in the absence of partial charge transfer gives the free surface charge.

It should be noted that in some papers (see e.g. [30, 34, 60, 61]) the quantities $-A_R$ and A_0 are designated as $-\Gamma_R$ and Γ_0, and are erroneously treated as the GIBBS adsorptions. The disagreement with the conclusions presented above which arises in this case, is discussed in detail in [2, 49, 50, 62].

4. The experimental verification of the applicability of electrocapillarity theory to reversible systems

Let us consider the form of the basic electrocapillarity equation in the case of some particular reversible electrodes and show how it is possible to verify experimentally electrocapillarity theory for these systems.

To date the surface phenomena at electrodes from platinum metals [37, 55] and at thallium amalgams [72—75] have received the most study.

The following reactions can occur at platinum metals in a certain potential range (the so-called hydrogen region):

$$H_{ads} + H_2O \rightleftarrows H_3O^+ + \bar{e} \qquad \text{(acid solutions)},$$

$$H_{ads} + OH^- \rightleftarrows H_2O + \bar{e} \qquad \text{(alkaline solutions)}$$

The reactions occurring on the surface of platinum metals in the oxygen region are:

$$O_{ads} + 2\,H^+ + 2\bar{e} \rightleftarrows H_2O \qquad \text{(acid solutions)},$$

$$O_{ads} + H_2O + 2\bar{e} \rightleftarrows 2\,OH^- \qquad \text{(alkaline solutions)}.$$

Thus, platinum electrodes are not ideally polarizable, but in a certain potential range, their state is completely determined by the amount of electricity passed through them, which is characteristic of "completely polarizable" electrodes [39, 40].

In the thermodynamic analysis of surface phenomena at platinum metals in acid solutions it is convenient to use as components in building up the surface layer the hydrogen atoms H and HA, CA and solvent molecules, i.e. uncharged species [41] (naturally, the final results do not depend on the choice of the surface layer components). Then the system state is determined by the chemical potentials of atomic hydrogen μ_H, acid, μ_{HA}, CA salt and solvent. At $\Gamma_s = 0$ for the total differential of the reversible surface work we have the expression:

$$d\sigma = -\Gamma_H\, d\mu_H - \Gamma_{HA}\, d\mu_{HA} - \Gamma_{CA}\, d\mu_{CA}\,, \tag{26}$$

where Γ_H, Γ_{HA}, Γ_{CA} are the GIBBS adsorptions of the respective components. Using the electroneutrality condition

$$\Gamma_{HA} = \Gamma_{H^+};\quad \Gamma_{CA} = \Gamma_{C^+};\quad \Gamma_{A^-} = \Gamma_{HA} + \Gamma_{CA};\quad \Gamma_{H^+} = \Gamma_{A^-} - \Gamma_{C^+} \tag{27}$$

and the relation $d\mu_H = -dE_r$, we obtain

$$d\sigma = \Gamma_H\, dE_r - \Gamma_{H^+}\, d\mu_{HA} - \Gamma_{C^+}\, d\mu_{CA}\,, \tag{28}$$

where E_r is the potential measured against the reversible hydrogen electrode in the same solution. In the case of a platinum-hydrogen electrode, equations (22) and (23) assume the form

$$Q' = -\Gamma_H = \varepsilon - A_H\,, \tag{29}$$

$$Q'' = R_{H^+} = \varepsilon + A_{H^+} \tag{30}$$

and the LIPPMANN equations can be written as follows

$$(\partial\sigma/\partial E_-)_{\mu_{HA},\mu_{CA}} = \Gamma_H = -Q' = -\varepsilon + A_H\,, \tag{31}$$

$$(\partial\sigma/\partial E_-)_{\mu_H,\mu_{CA}} = -\Gamma_{H^+} = -Q'' = -\varepsilon - A_{H^+}\,, \tag{32}$$

where E_- is the hydrogen electrode potential measured against the electrode reversible with respect to the A^- anion in the same solution.

Equation (28) can be rewritten as

$$d\sigma = -Q'\,dE_r - \Gamma_{H^+}\,d\mu_{HA} - \Gamma_{C^+}\,d\mu_{CA} \tag{33}$$

and carrying out conventional operations on the partial derivatives of the total differential, it is possible to obtain a set of thermodynamic equations similar to (11)—(14):

$$(\partial Q'/\partial\mu_{HA})_{E_r,\mu_{CA}} = (\partial\Gamma_{H^+}/\partial E_r)_{\mu_{HA},\mu_{CA}}\,, \tag{34}$$

$$(\partial E_r/\partial\mu_{HA})_{Q',\mu_{CA}} = -(\partial\Gamma_{H^+}/\partial Q')_{\mu_{HA},\mu_{CA}}\,, \tag{35}$$

$$(\partial E_r/\partial\mu_{HA})_{\Gamma_{H^+},\mu_{CA}} = -(\partial\Gamma_{H^+}/\partial Q')_{E_r,\mu_{CA}}\,, \tag{36}$$

$$(\partial\Gamma_{H^+}/\partial E_r)_{Q',\mu_{CA}} = (\partial Q'/\partial\mu_{HA})_{\Gamma_{H^+},\mu_{CA}}\,. \tag{37}$$

Let us consider equation (35) more in detail. The quantity $(\partial E_r/\partial\mu_{HA})_{Q',\mu_{CA}}$ contained in this equation was named the isoelectric potential shift. The following particular cases of this equation are of special interest.

1. Pure acid solution (salt concentration is zero: $[CA] = 0$). It is evident that

$$(\partial E_r/\partial\mu^{\pm}_{HA})_{Q'} = -2(\partial\Gamma_{H^+}/\partial E_r)_{\mu^{\pm}_{HA}}/(\partial Q'/\partial E_r)_{\mu^{\pm}_{HA}}\,, \tag{38}$$

where $\mu^{\pm}_{HA}$ is the mean chemical potential of acid ions. For the acid H_2A we have:

$$(\partial E_r/\partial\mu^{\pm}_{H_2A})_{Q'} = -3(\partial\Gamma_{H_2A}/\partial E_r)_{\mu^{\pm}_{H_2A}}/(\partial Q'/\partial E_r)_{\mu^{\pm}_{H_2A}}\,, \tag{39}$$

where $\mu^{\pm}_{H_2A}$ is also the mean chemical potential of acid ions.

2. Acidified neutral salt solution in which $[CA] \gg [HA]$. In this case

$$d\mu_{HA} = d\mu_{H^+} + d\mu_{A^-} = d\mu_{H^+} \tag{40}$$

and

$$(\partial E_r/\partial\mu_{H^+})_{Q',\mu_{CA}} = -(\partial\Gamma_{H^+}/\partial E_r)_{\mu_{H^+},\mu_{CA}}/(\partial Q'/\partial E_r)_{\mu_{H^+},\mu_{CA}}\,. \tag{41}$$

This equation is also valid for the C_nA solution acidified with H_nA at $[C_nA] \gg [H_nA]$. If the electrode potential is measured against a constant reference electrode, then instead of (41), we have

$$(\partial E/\partial\mu_{H^+})_{Q',\mu_{CA}} = 1 - (\partial\Gamma_{H^+}/\partial E)_{\mu_{H^+},\mu_{CA}}/(\partial Q'/\partial E)_{\mu_{H^+},\mu_{CA}} \tag{42}$$

and instead of (38)

$$\left(\frac{\partial E}{\partial\mu^{\pm}_{HA}}\right)_{Q'} = 1 - 2\,\frac{(\partial\Gamma_{H^+}/\partial E)_{\mu^{\pm}_{HA}}}{(\partial Q'/\partial E)_{\mu^{\pm}_{HA}}}\,. \tag{43}$$

It was shown in [41, 43] that the field of application of the above relations can be extended to alkaline solutions and to the potentials at which adsorbed hydrogen on the electrode surface is substituted by adsorbed oxygen, i.e. to the oxygen section of the charging curve, if the system can be treated as a reversible one also in the case of oxygen adsorption. In [39, 40] relations are given for electrodes dissolving hydrogen.

The derived thermodynamic equations allow a verification of the reversible electrode theory. In fact, the quantities $(\partial E_r/\partial\mu^{\pm}_{HA})_{Q'}$ and $(\partial E_r/\partial\mu_{H^+})_{Q',\mu_{CA}}$ can be determined by the isoelectric potential shift method [76]. The values of $\partial Q'/\partial E_r$ can be found from the equilibrium charging curves of the electrode [41]. Then by means of appropriate equations, it is possible to calculate $\partial\Gamma_{H^+}/\partial E_r$ and, integrating the $(\partial\Gamma_{H^+}/\partial E_r)$, E_r curve, to find the dependence of $\Delta\Gamma_{H^+}$ on E_r. The coincidence of the calculated $\Delta\Gamma_{H^+}$, E_r

curve with the experimentally obtained dependence of Γ_{H^+} on E_r proves the validity of the thermodynamic theory and allows the determination of the potential range in which the system behaviour is reversible. The integration constant being unknown, some experimental value of Γ_{H^+} is used as such in comparing calculation and experiment.

The verification was carried out for many systems on electrodes from platinum, palladium, rhodium, ruthenium and platinum-ruthenium alloy (for a review see [54, 55]).

Fig. 1 illustrates the verification of the theory for the case of a platinized platinum electrode in 10^{-2} N H_2SO_4, HCl and HBr solutions. For 10^{-2} N H_2SO_4 the points corresponding to the experimental data were obtained by two methods: by titration and

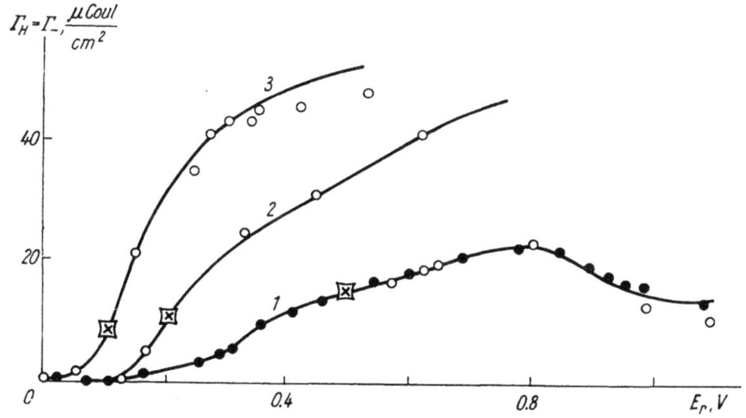

Fig. 1: Comparison of calculated (solid curves) and experimental (points) dependences of hydrogen ion adsorption on the potential of a platinized platinum electrode in the solutions.

1 – 0.01 N H_2SO_4, 2 – 0.01 N HCl, 3 – 0.01 N HBr. Circles – the results obtained by titration, black dots – the results obtained by the radioactive tracer method.

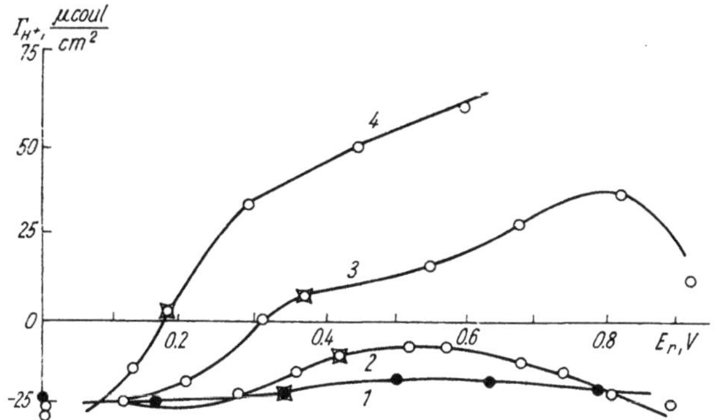

Fig. 2: Comparison of calculated (solid curves) and experimental (points) dependences of hydrogen ion adsorption on the potential of a platinized platinum electrode in the solutions.

1 – 0.01 N KOH, 2 – 0.01 N KOH + 1 N KCl, 3 – 0.01 N KOH + 1 N KBr, 4 – 0.01 N KOH + + 1 N KJ.

by the radioactive tracer method [77, 78]. Fig. 2 gives the calculated and experimental Γ_{H^+}, E_r curves for a platinized platinum electrode in 10^{-2} N KOH and in the presence of 1 N KCl, KBr and KJ [45]. In Figs 1. and 2 the points taken as integration constants in comparing calculation and experiment are specially marked.

The examples given above as well as the results of verification for other systems show that the thermodynamic theory of the reversible electrode is applicable in a wide potential range, including the initial section of the oxygen region. The pronounced ion chemisorption phenomena in the presence of which surface layer formation can evidently not be treated as a completely reversible process, narrow down the practical applicability range of the theory.

The theory was also verified for systems in which the concentration of a certain ion i, specifically adsorbable on the electrode, was low, as compared to that of a surface-inactive electrode, so that the chemical potentials of the electrolyte ions remained virtually constant when the ion i concentration varied. Under these conditions

$$d\sigma = -Q'\,dE_r - \Gamma_i\,d\mu_i\,, \tag{44}$$

where Γ_i is the GIBBS adsorption of the surface-active ion. It follows from equation (44) that

$$(\partial Q'/\partial\mu_i)_{E_r} = (\partial\Gamma_i/\partial E_r)_{\mu_i}\,, \tag{45}$$

$$(\partial E_r/\partial\mu_i)_{Q'} = -\,(\partial\Gamma_i/\partial Q')_{\mu_i}\,, \tag{46}$$

$$(\partial E_r/\partial\mu_i)_{\Gamma_i} = -\,(\partial\Gamma_i/\partial Q')_{E_r}\,, \tag{47}$$

$$(\partial\Gamma_i/\partial E_r)_{Q'} = (\partial Q'/\partial\mu_i)_{\Gamma_i}\,. \tag{48}$$

If equation (46) is rewritten as

$$\left(\frac{\partial E_r}{\partial\mu_i}\right)_{Q'} = -\,\frac{(\partial\Gamma_i/E_r)_{\mu_i}}{(\partial Q'/\partial E_r)_{\mu_i}} \tag{46a}$$

it becomes clear that the dependence of the ion i adsorption on potential (to the accuracy of the integration constant) can be found from the charging curve slope and the electrode potential shift $(\partial E_r/\partial\mu_i)_{Q'}$ when the concentration of the specifically adsorbable ion varies. This shift became known as the adsorption potential shift. By means of equation (48) bromine ion adsorption was determined on platinum, rhodium and iridium electrodes in 10^{-2} N HBr + 1 N H_2SO_4 [79]. The comparison of the calculated Γ_i values with those obtained by direct analytical methods (from the change in the bromine ion concentration in solution) confirmed the validity of the thermodynamic theory of the reversible electrode. Due to the excess of supporting electrolyte in the systems under consideration, $d\mu_i = RT/F\ d\ln c_i$ and hence equations (45)—(48) can be rewritten substituting μ_i by $RT/F\ d\ln c_i$, and the condition μ_i = const by the condition c_i = const.

If a change in the concentration of specifically adsorbable ions involves a change not only of their chemical potential, but also of that of the oppositely charged ions, the following relations are valid [39, 40]:

$$\left(\frac{\partial Q'}{\partial\mu_{CA}}\right)_{\mu_{H^+},E_r} = \left(\frac{\partial\Gamma_{H^+}}{\partial E_r}\right)_{\mu_{H^+},\mu_{CA}} + \left(\frac{\partial\Gamma_-}{\partial E_r}\right)_{\mu_{H^+},\mu_{CA}}, \tag{69}$$

$$\left(\frac{\partial E_r}{\partial\mu_{CA}}\right)_{Q',\mu_{H^+}} = -\left(\frac{\partial\Gamma_{H^+}}{\partial Q'}\right)_{\mu_{H^+},\mu_{CA}} - \left(\frac{\partial\Gamma_-}{\partial Q'}\right)_{\mu_{H^+},\mu_{CA}}. \tag{50}$$

Equation (41) can be reduced to the form

$$\left(\frac{\partial E_r}{\partial \mu_{H^+}}\right)_{Q',\mu_{CA}} = -\left(\frac{\partial \Gamma_-}{\partial Q'}\right)_{\mu_{H^+},\mu_{CA}} - \left(\frac{\partial \Gamma_+}{\partial Q'}\right)_{\mu_{H^+},\mu_{CA}} . \qquad (51)$$

Thus, measurement of the quantities $(\partial E_r/\partial \mu_{CA})_{Q',\mu_{H^+}}$ and $(\partial E_r/\partial \mu_{H^+})_{Q',\mu_{CA}}$ makes it possible to determine by means of equations (50) and (51) the dependence of Γ_- and Γ_+ on potential, i.e. to determine separately the anion and cation adsorptions. The validity of this conclusion was confirmed in [80].

The theory was verified with the use of data on ion coadsorption on platinum. In the case of coadsorption of ions 1 and 2 at constant electrode potential in the presence of a large foreign electrolyte excess, the Gibbs equation reduces to the form

$$d\sigma = -\Gamma_1\, d\mu_1 - \Gamma_2\, d\mu_2 . \qquad (52)$$

Hence it follows that

$$(\partial \Gamma_1/\partial \mu_2)_{\mu_1} = (\partial \Gamma_2/\partial \mu_1)_{\mu_2} . \qquad (53)$$

Thus, the slopes of the isotherms of the displacement from the electrode surface of a particular ion by another ion at constant potential should coincide. In our case, due to the presence of a foreign salt excess, μ_i can be substituted by $RT/F\ d \ln c_i$.

In Fig. 3 the dependence of the cesium ion adsorption on the concentration of sodium ions is plotted alongside the dependence of the sodium ion adsorption on the concentration of cesium ions. Actually, these dependences are parallel [81]. A similar phenomenon was observed in the case of coadsorption of sulfate- and chloride anions in acid solutions, and of bromide- and iodide anions in alkaline solutions on platinized platinum [82]. The coincidence of the slopes of the displacement isotherms may be considered as confirming the validity of the thermodynamic theory of a reversible electrode

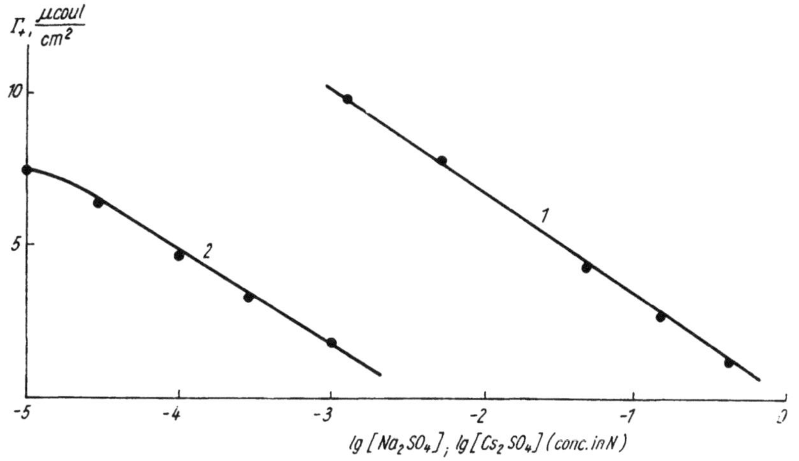

Fig. 3: Isotherms of the displacement of cesium cations by sodium cations (1) and of sodium cations by cesium cations (2) at $E_r = 0$ on a platinized platinum electrode in the solutions.

1 – 10^{-3} N H_2SO_4 + 2 × 10^{-3} N Cs_2SO_4 + x N Na_2SO_4; 2 – 10^{-3} N H_2SO_4 + 2 × 10^{-3} N Na_2SO_4 + x N Cs_2SO_4.

For an electrode on which reaction (15) occurs, the reversible surface work σ in a solution with a constant supporting electrolyte concentration is a function of μ_0 and μ_R, and when plotted against σ, μ_0, μ_R can be represented by a paraboloid surface. The section of this paraboloid by the plane $\mu_0 = \text{const}$ gives an electrocapillary curve of the 1st kind and its section by the plane $\mu_R = \text{const}$, an electrocapillary curve of the 2nd kind. The maximum of the paraboloid corresponds to the condition $Q' = Q'' = 0$ (Fig. 4) [64]. For platinum metals σ is a function of μ_H and μ_{H^+}. The electrocapillary curves of the 1st and 2nd kind are none other than the section of the paraboloid surface by the planes $\mu_{H^+} = \text{const}$ or $\mu_H = \text{const}$. The electrocapillary curves of a platinum electrode (to the accuracy of the integration constant), obtained by integration of equation (31) and (32) are shown in Fig. 5 [2]. In their shape, these curves resemble the electrocapillary curves of a mercury electrode, but point to a much greater dependence of σ on potential than for mercury. This is due to the fact that on platinum σ is affected not only be the adsorption of solution ions but also by that of hydrogen and oxygen atoms. Another specific feature of the curves of the 1st kind is the dependence of the potential of the electrocapillary maximum on solution pH, which in accordance with the electrocapillarity equation, results from the influence of pH on hydrogen and oxygen adsorption. The thermodynamic equations describing the influence of pH on the potentials of zero total charge of platinum metals are given and discussed in [2, 54, 56].

On the basis of equation (33) were derived the relations for the dependence of hydrogen and oxygen adsorption on solution pH and electrolyte concentration [36, 47, 83].

In [84] electrocapillarity theory was extended to the reversible adsorption phenomena of organic substances on hydrogen- and oxygen-adsorbing metals. Assuming

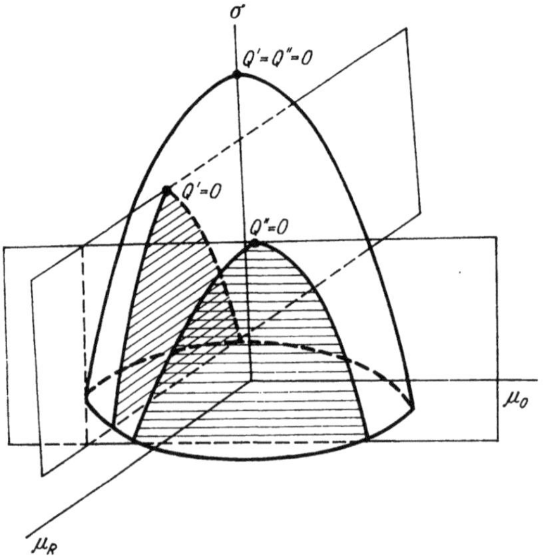

Fig. 4: Electrocapillary curves of the 1st and 2nd kinds of a reversible electrode represented as sections of the surface σ by the planes $\mu_0 = \text{const}$ and $\mu_R = \text{const}$; from [64].

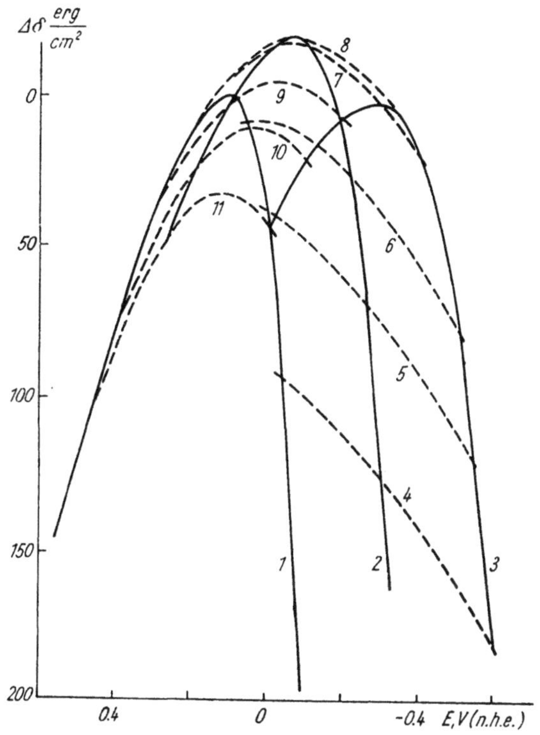

Fig. 5: Electrocapillary curves of the 1st (1—3) and 2nd (4—11) kinds for a platinized platinum electrode in 0.1 N KCl at pH.

1 – 2.3, 2 – 7.0, 3 – 12.0; and at $E_r(V)$: 4 – 0.1, 5 – 0.15, 6 – 0.2, 7 – 0.3, 8 – 0.4, 9 – 0.5, 10 – 0.6, 11 – 0.7.

the solution pH and electrolyte concentration to be constant and under the condition that a change of the organic substance concentration does not affect the chemical potentials of the other system components, we can write

$$d\sigma = -Q'\, dE_r - \Gamma_{org}\, d\mu_{org}\,, \tag{54}$$

whence it follows, in particular, that

$$(\partial \mu_{org}/\partial E_r)_{\mu_{org}} = -(\partial Q'/\partial \Gamma_{org})_{E_r} \tag{55}$$

or

$$(\partial \Delta G_{org}/\partial E_r)_{\Gamma_{org}} = -(\partial Q'/\partial \Gamma_{org})_{E_r}\,, \tag{56}$$

where ΔG_{org} is the standard free adsorption energy of organic substance. Analysis of equation (56) leads to the conclusion if on electrodes reversibly adsorbing hydrogen, oxygen and organic substance appear the maximum of Γ_{org} should occur in the potential range in which hydrogen and oxygen adsorption has a minimum [84]. Unfortunately, so far this conclusion has not been experimentally checked since no substance has been found which would be reversibly adsorbed on platinum metals.

Frumkin and coworkers [72—74] studied in detail the electrocapillary behaviour of the system Hg, Tl, Tl^+, H_2O. In this case it is convenient to define the interface by the condition $\Gamma_{Hg} = 0$. If the concentrations of the salts TlA and CA in solutions are low, so that μ_{H_2O} can be assumed to be constant, the system may be treated in the same way as a platinum-hydrogen electrode and we can write

$$d\sigma = -\Gamma_{Tl}\, d\mu_{Tl} - \Gamma_{TlA}\, d\mu_{TlA} - \Gamma_{CA}\, d\mu_{CA} \,. \tag{57}$$

It follows from equation (57) that

$$(\partial\delta/\partial E_-)_{\mu_{TlA},\mu_{CA}} = -Q' = \Gamma_{Tl} = A_{Tl} - \varepsilon \,, \tag{58}$$

$$(\partial\sigma/\partial E_-)_{\mu_{Tl},\mu_{CA}} = -Q'' = \Gamma_{Tl^+} = -\varepsilon - A_{Tl^+} \,, \tag{59}$$

where A_{Tl} is the amount of Tl present in the surface layer as atoms, A_{Tl^+} is the extent of adsorption of Tl^+ ions remaining in the ionic side of the double layer. Sets of electrocapillary curves of the 1st and 2nd kind for the system Hg, Tl, Tl^+, H_2O are given in [74], where it is also shown under what conditions the quantities A_{Tl} or A_{Tl^+} may be neglected and when this assumption is inadmissible. Some of the above-mentioned curves are reproduced in Fig. 6.

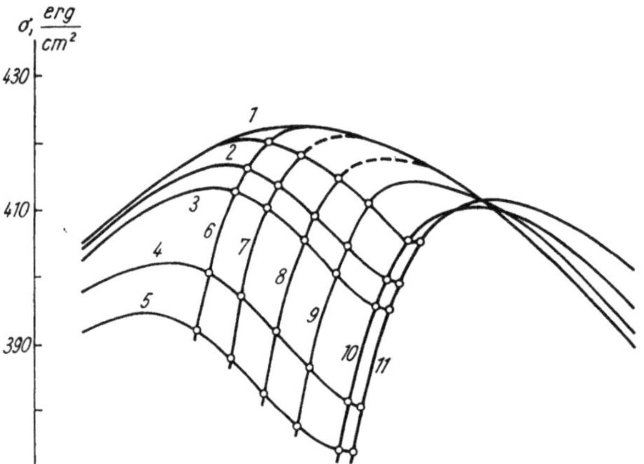

Fig. 6: Electrocapillary curves of the 1st and 2nd kinds in the system Hg, Tl, Tl^+, H_2O (from the data of [73]).

Tl^+ concentration (x) in x M $TlNO_3$ + $(1 - x)$ M KNO_3 solutions: 1 — 0.001, 2 — 0.005, 3 — 0.01, 4 — 0.05. Tl concentration in amalgams: 6 — 0.002, 7 — 0.014, 8 — 0.12, 9 — 0.7, 10 — 4.9, 11 — 7.8 atom %. Reference electrode —9.9 % Tl amalgam in 0.1 M $TlNO_3$ + 0.9 M KNO_3.

Proceeding from the concepts of total and free charges, the authors of [49, 50] considered the electrocapillary behaviour resulting when the condition

$$[0] + \gamma[R] = \text{const} \tag{60}$$

(where $\gamma = \text{const}$) is satisfied. At $\gamma = (D_R/D_0)^{1/2}$, where D_R and D_0 are the diffusion coefficients of the red and ox forms, equation (60) corresponds to the concentration change of 0 and R at the electrode surface under polarographic conditions. A generalized electrode charge concept allows a more consistent treatment of electrocapillary curves in redox systems than was done in [85]—87], as well as interpretation of the polarographic "electrocapillary curves" and the electrocapillary curves for inorganic and organic redox systems obtained by means of a capillary electrometer [49, 50, 88].

5. The thermodynamic approach to charge-transfer chemisorption

As it has been pointed out above, up to the present quite a number of chemisorption processes involving transfer to metal surface of a part of the charge of particles undergoing adsorption have some to light. A thermodynamic approach to these processes was already suggested in [16], where consideration was given to the adsorption of hydrogen atoms on a platinum surface involving formation of hydrogen dipoles and decrease of the electric double layer capacity. In recent years, on the basis of the generalized electrode charge concept, a consistent phenomenological treatment of the charge transfer during chemisorption has been developed [89—93].

Let us assume the process

$$0 + \bar{e} \rightleftarrows R \tag{61}$$

to occur through an intermediate adsorption state A:

$$0 + n_1\bar{e} \rightleftarrows A\,, \tag{62a}$$

$$A + n_1\bar{e} \rightleftarrows R\,. \tag{62b}$$

It is evident that $n_1 + n_2 = 1$. The quantities n_1 and n_2 were called the formal charge transfer coefficients [92]. In the case of chemisorption of 0 the process may stop at the stage (62a) and in the case of chemisorption of R at the stage (62b). It can be shown that the formal charge transfer coefficients are determined by the relations

$$n_1 = -(\partial Q'/\partial \Gamma_\Sigma)_E = (\partial \Gamma_R/\partial \Gamma_\Sigma)_E\,, \tag{63}$$

$$n_2 = (\partial Q''/\partial \Gamma_\Sigma)_E = (\partial \Gamma_0/\partial \Gamma_\Sigma)_E\,, \tag{64}$$

where $\Gamma_\Sigma = \Gamma_0 + \Gamma_R$. Using the electrocapillarity equation of a reversible electrode, it is possible to transform equations (63) and (64) as follows

$$n_1 = (\partial \Gamma_R/\partial \Gamma_\Sigma)_E = \left[\left(\frac{\partial \Gamma_R}{\partial \mu_0}\right)_{\mu_R} + \left(\frac{\partial \Gamma_R}{\partial \mu_R}\right)_{\mu_0}\right] \Big/ \left[\left(\frac{\partial \Gamma_\Sigma}{\partial \mu_R}\right)_{\mu_0} + \left(\frac{\partial \Gamma_\Sigma}{\partial \mu_0}\right)_{\mu_R}\right], \tag{65}$$

$$n_2 = (\partial \Gamma_0/\partial \Gamma_\Sigma)_E = \left[\left(\frac{\partial \Gamma_0}{\partial \mu_R}\right)_{\mu_0} + \left(\frac{\partial \Gamma_0}{\partial \mu_0}\right)_{\mu_R}\right] \Big/ \left[\left(\frac{\partial \Gamma_\Sigma}{\partial \mu_R}\right)_{\mu_0} + \left(\frac{\partial \Gamma_\Sigma}{\partial \mu_0}\right)_{\mu_R}\right]. \tag{66}$$

In the general case, the coefficients n_1 and n_2 represent not only the part of the charge transferred from adsorbed particle to electrode (characterized by the so-called true, or microscopic, charge transfer coefficient λ [69]), but also the change in the double layer parameters caused by the adsorption process. It is for this reason that n_1 and n_2 should be called formal (thermodynamic) charge transfer coefficients. The relations for n_1 and n_2 in the case of some particular redox systems are given in [2, 92].

VETTER and SCHULTZE [94—97] introduced the quantity γ_i, similar to n_i, for which they proposed the name "the electrosorption valence". This term is less fitting than "the formal charge transfer coefficient" since values γ_i different from zero are obtained, for example, in the case of adsorption of saturated organic compounds at the mercury-solution interface. VETTER and SCHULTZE found the values of γ_i for a large

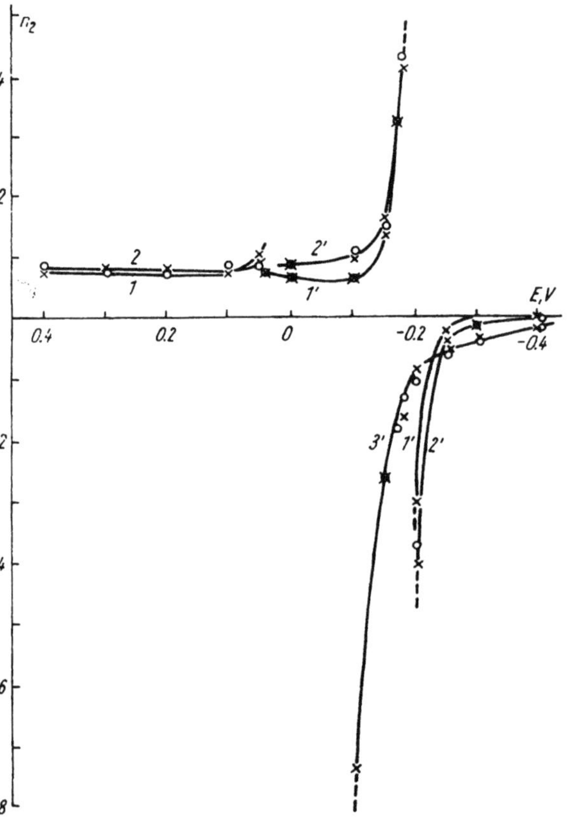

Fig. 7: Dependences of n_2 on potential in the system Hg, Tl, Tl^+, H_2O, calculated by different methods from the electrocapillary curves of the 1st (1, 2) and 2nd (1′, 2′, 3′) kinds.

1 – C_{Tl} = 0.01 M, 2 – C_{Tl^+} = 0.04 M, 1′ – C_{Tl} = 0.12, 2′ – C_{Tl} = 0.68, 3′ – C_{Tl} = 4.86 atom % (from the data of [74]). Reference electrode as in Fig. 6.

number of adsorption processes. They attempted with the use of a model to separate out from γ_i the true charge transfer coefficient λ [97], this attempt, naturally, being outside the thermodynamic approach.

The formal character of n_1 and n_2, as well as their difference from the true charge transfer coefficients λ is evidenced by the results of the calculation of these quantities for the systems Hg, Tl, Tl^+, H_2O carried out in [74] by means of the formulae

$$n_1 = (\partial \Gamma_{Tl}/\partial \Gamma_{\Sigma})_E \,, \tag{67}$$

$$n_2 = (\partial \Gamma_{Tl^+}/\partial \Gamma_{\Sigma})_E \,. \tag{68}$$

The obtained results are illustrated in Fig. 7. It is clear from this figure that even if at positive potentials n_2 has a quite reasonable value, close to 0.75—0.80, in the vicinity of $E = -0.15$ V the dependence of n_2 on E is disturbed, i.e. $n_2 \to \pm\infty$ [74].

6. The dependence of the surface tension of solid metals on potential

The thermodynamic relations used earlier contained the reversible surface work σ, which coincides with the surface tension γ only in the case of liquid surfaces. The difference between σ and γ in the case of solid surfaces, which was first pointed out by GIBBS [8], was given by SHUTTLEWORTH [98] for the case of an isotopic surface as

$$\gamma = \sigma + s(d\sigma/ds)_T \,, \tag{69}$$

where s is the surface area of a solid. For a long time the difference between γ and σ was not taken into account in electrocapillarity theory, evidently for lack of experimental data on the surface tension of solid electrodes.

This difference was demonstrated by A. GOKHSHTEIN who developed the "estance" method — a method of determining the dependence of the quantity $|\partial\gamma/\partial Q'|$ on the electrode potential [98—103].

Since

$$\frac{\partial \gamma}{\partial Q'} = \frac{\partial \gamma}{\partial E} \cdot \frac{\partial E}{\partial Q'} = \frac{1}{C} \cdot \frac{\partial \gamma}{\partial E}, \tag{70}$$

where C is the electrode capacity, then by means of $|\partial\gamma/\partial Q'|$ it is easy to calculate $|\partial\gamma/\partial E|$.

GOKSHTEIN showed that $\partial\gamma/\partial E$ obeys the equation

$$\partial\gamma/\partial E = -Q' - (\partial Q'/\partial \ln s)_E \,. \tag{71}$$

The validity of this equation was confirmed by comparing $(\partial\gamma/\partial E)_s$ and $(\partial E/\partial \ln s)_{Q'}$, i.e. by the similarity between the change of the surface tension with varying potential and the change of potential upon elastic deformation of the electrode.

The relative contribution of the terms on the right hand side of equation (71) to $\partial\gamma/\partial E$ depends on the metal nature, solution composition and electrode potential. Thus, in the case of lead in sodium fluoride solution $|Q| \gg |(\partial Q/\partial \ln s)_E|$. At the same time, for cadmium and, especially, platinum, the role of the second term in the right hand side of equation (71) is significant.

The derivatives $\partial\gamma/\partial E$ and $\partial\sigma/\partial E$ for platinum in 1 N H_2SO_4 are compared in Fig. 8 [102, 103]. The values of $\partial\gamma/\partial E$ were obtained by the"estance"method, $\partial\sigma/\partial E$ was equated with the total charge $-Q'$. It is clear that there is a significant difference between the quantities being compared and that the shape of the $\partial\gamma/\partial E$, E curve is mainly determined by the derivative $\partial Q'/\partial \ln s$.

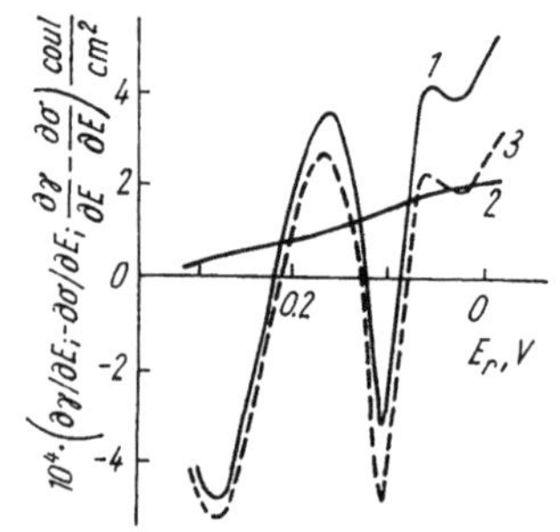

Fig. 8: Difference between the surface tension γ and the reversible surface work σ given as the derivatives of these quantities for platinum in 1 N H_2SO_4.

1 — $\partial\gamma/\partial E$, 2 — $\partial\sigma/\partial E$, 3 — $(\partial\gamma/\partial E) - (\partial\sigma/\partial E)$ (from the data of [102, 103]).

If a solid electrode is immersed in a solution containing m electroneutral components, then at $\Gamma_s = 0$ for the total differential γ the following relation is valid [102]:

$$d\gamma = -\left[Q + \left(\frac{Q}{\ln s}\right)_{E,\mu_k}\right] dE - \sum_{l=1}^{m}\left[\Gamma + \left(\frac{\partial \Gamma}{\partial \ln s}\right)_{E,\mu_k}\right] d\mu_k\,, \tag{72}$$

where $k, l = 1, \dots, m$.

Some problems of the influence of shear deformation on the surface tension and the contact angle of solid electrodes were recently considered by Rusanov [104].

7. Conclusion

To date considerable advances have been made in the thermodynamic description of electrocapillary phenomena both on ideally polarizable and on reversible electrodes. An important role in this has been played by the formulation of the generalized concept of electrode charge, leading to development of a consistent electrocapillarity theory for various electrochemical systems. Some methods and suitable relations have been found for experimental verification of the thermodynamic theory of reversible electrodes. A very careful verification has confirmed the validity of the thermodynamic theory and demonstrated the advantages of the methods of investigation of the electrode-solution interface based on thermodynamic analysis.

The development of electrocapillarity theory proves once more how varied the problems in surface thermodynamics are which can be solved on the basis of the GIBBS adsorption equation.

References

[1] A. N. FRUMKIN, O. A. PETRII, B. B. DAMASKIN, J. Electroanal. Chem., 1972, V. 35, p. 439—440

[2] A. N. FRUMKIN, "*Zero Charge Potentials*", Nauka, Moscow, 1979

[3] G. LIPPMANN, Pogg. Ann. Phys., 1873, V. 149, p. 546—561

[4] G. LIPPMANN, C.R., 1873, V. 76, p. 1407

[5] G. LIPPMANN, Ann. chim. phys., (5), 1875, V. 5, p. 494; 1877, V. 12, p. 265

[6] G. LIPPMANN, C.R., 1882, V. 95, p. 686

[7] G. LIPPMANN, J. Phys. (2), 1883, V. 2, p. 116

[8] W. GIBBS, "*On the Equilibrium of Heterogeneous Substances* (1877—1878)", Coll. Works, Longmans and Green, 1928, V. 1

[9] M. PLANCK, Ann. Phys., 1891, V. 44, p. 385

[10] A. N. FRUMKIN, J. Electroanal. Chem., 1975, V. 64, p. 247

[11] G. GOUY, Ann. chim. phys. (7), 1903, V. 29, p. 145; (8), 1906, V. 8, p. 291; 1906, V. 9, p. 75; C.R., 1908, V. 146, p. 612; Ann. phys. (9), 1917, V. 7, pp. 129, 176

[12] A. N. FRUMKIN, "*Electrocapillary Phenomena and Electrode Potentials*", Odessa, 1919

[13] A. N. FRUMKIN, Phil. Mag., 1920, V. 40, pp. 363, 375

[14] A. N. FRUMKIN, Z. physik. Chem., 1923, V. 103, pp. 43—55

[15] A. N. FRUMKIN, A. GORODETZKAJA, Z. physik. Chem., 1928, V. 136, p. 451

[16] A. N. FRUMKIN, A. I. SHLYGIN, Acta physicochim. URSS, 1936, V. 5, p. 819

[17] F. KOENIG, J. Phys. Chem., 1934, V. 38, pp. 111, 339

[18] S. CRAXFORD, O. GATTY, Philpot J. Phil. Mag. (7), 1935, V. 19, p. 965

[19] A. N. FRUMKIN, Z. A. JOFA, Zh. Fiz. Khim., 1939, V. 13, p. 931; Z. JOFA, A. FRUMKIN, Acta physicochim. URSS, 1939, V. 10, p. 473

[20] D. GRAHAME, R. WHITNEY, J. Amer. Chem. Soc., 1942, V. 64, p. 1548

[21] D. GRAHAME, Chem. Rev., 1947, V. 41, p. 441

[22] R. PARSONS, M. DEVANATHAN, Trans. Faraday Soc., 1953, V. 49, p. 404

[23] R. PARSONS, Modern Aspects of Electrochemistry, V. 1, ed. by J. BOCKRIS, B. CONWAY, N.Y., Acad. Press, 1954, p. 103

[24] A. N. FRUMKIN, Zh. Fiz. Khim., 1956, V. 30, p. 2066

[25] R. PARSONS, Proc. Second Int. Congress Surface Activity, Butterworths, London, 1957, V. 3, p. 38

[26] D. GRAHAME, B. SODERBERG, J. Chem. Phys., 1954, V. 22, p. 449

[27] D. GRAHAME, J. Chem. Phys., 1948, V. 16, p. 1117

[28] R. PARSONS, Can. J. Chem., 1959, V. 37, p. 308

[29] P. DELAHAY, Double Layer and Electrode Kinetics, N.Y., Inters. 1965

[30] D. MOHILNER, Electroanalytical Chem., V. 1, ed. by A. BARD, N.Y., Marcel Dekker, 1966, p. 231

[31] A. I. RUSANOV, Zh. Fiz. Khim., 1962, V. 36, p. 549

[32] A. I. RUSANOV, "*Phase Equilibria and Interfacial Phenomena*" (russ.), "Khimija", Leningrad, 1967, p. 239—258. German Ed.: "*Phasengleichgewichte und Grenzflächenerscheinungen*", Akademie-Verlag, Berlin 1978

[33] J. Newman, "*Electrochemical Systems*", Prentice Hall, Inc., Englewood Cliffs, N.Y., 1973
[34] D. Mohilner, J. Phys. Chem., 1962, V. 66, p. 724
[35] A. I. Rusanov, Zh. Fiz. Khim., 1962, V. 36, p. 690
[36] A. N. Frumkin, "*Adv. Electrochemistry and Electrochem. Eng.*", V. 3, ed. P. Delahay, Intersc., N.Y., 1963, p. 287
[37] A. N. Frumkin, "*Elektrokhimiya*", 1966, V. 2, p. 387
[38] A. N. Frumkin, N. A. Balashova, V. E. Kazarinov, J. Electrochem. Soc., 1966, V. 113, p. 1011
[39] A. N. Frumkin, O. A. Petrii, R. V. Marvet, J. Electroanal. Chem. 1966, V. 12, p. 504
[40] A. N. Frumkin, O. A. Petrii, R. V. Marvet, "*Elektrokhimiya*", 1967, V. 3, p. 1311
[41] A. N. Frumkin, O. A. Petrii, A. M. Kossaya, V. S. Entina, V. V. Topolev, J. Electroanal. Chem., 1968, V. 16, p. 175
[42] A. N. Frumkin, O. A. Petrii, Abhandl. Sächs. Akad. Wissenschaft., Leipzig, Math.-Naturwiss. Kl., 1968, V. 49, p. 17
[43] A. N. Frumkin, O. A. Petrii, A. M. Kossaya, "*Elektrokhimiya*", 1968, V. 4, p. 475
[44] O. A. Petrii, Yu. G. Kotlov, "*Elektrokhimiya*", 1968, V. 4, p. 1256
[45] O. A. Petrii, A. N. Frumkin, Yu. G. Kotlov, J. Res. Inst. Catal. Hokk. Univer., 1968, V. 16, p. 367
[46] O. A. Petrii, A. N. Frumkin, Yu. G. Kotlov, "*Elektrokhimiya*", 1969, V. 5, p. 476, 735
[47] O. A. Petrii, Yu. G. Kotlov, "*Elektrokhimiya*", 1970, V. 6, p. 404
[48] A. N. Frumkin, O. A. Petrii, Electrochim. Acta, 1970, V. 15, p. 391
[49] A. N. Frumkin, O. A. Petrii, B. B. Damaskin, J. Electroanal. Chem., 1970, V. 27, p. 81
[50] A. N. Frumkin, O. A. Petri, B. B. Damaskin, "*Dvoinoi sloi i adsorbtsija na tverdikh elektrodakh*", Tartu, Tartuskii Gos. Universitet, 1970, V. 2, p. 5
[51] A. N. Frumkin, O. A. Petrii, Dokl. Akad. Nauk SSSR, 1971, V. 196, p. 1387
[52] O. A. Petrii, B. M. Grafov, "*Elektrokhimiya*", 1973, V. 9, p. 852
[53] B. B. Damaskin, O. A. Petrii, V. V. Batrakov, "*Adsorption of Organic Compounds on Electrodes*", Plenum Press, N.Y.-London, 1971, Ch. 9
[54] O. A. Petrii, Usp. Khim., 1975, V. 44, p. 2048
[55] O. A. Petrii, "*Itogi nauki i tekhniki*", in "*Elektrokhimiya*", V. 12, VINITI, M., 1977, p. 56
[56] A. N. Frumkin, J. Electroanal. Chem., 1968, V. 18, p. 328
[57] A. N. Frumkin, O. A. Petrii, Yu. G. Kotlov, "*Elektrokhimiya*", V. 5, p. 1214
[58] A. N. Frumkin, O. A. Petrii, T. Ya. Kolotyrkina, "*Elektrokhimiya*", 1974, V. 10, p. 1741
[59] A. N. Frumkin, O. A. Petrii, T. Ya. Kolotyrkina-Safonova, Dokl. Akad. Nauk SSSR, V. 222, p. 1159
[60] W. Plieth, K. Vetter, Ber. Bunsenges. physikal. Chem., 1968, Bd. 72, S. 673; 1969, Bd. 73, S. 79
[61] W. Plieth, J. Electroanal. Chem., 1969, V. 23, p. 305
[62] A. N. Frumkin, O. A. Petrii, Electrochim. Acta, 1975, V. 20, p. 347
[63] A. Soffer, J. Electroanal. Chem., 1972, V. 40, p. 153
[64] B. B. Damaskin, O. A. Petrii, "*Vvedenie v elektrokhimicheskuju kinetiku*", "Visshaya shkola", M., 1975
[65] E. Schmidt, Helv. Chim. Acta, 1969, V. 52, p. 1763
[66] A. N. Frumkin, B. B. Damaskin, Dokl. Akad. Nauk SSSR, 1975, V. 221, p. 395
[67] R. Schofield, Phil. Mag., (7), 1926, V. 1, p. 641

[68] A. I. Rusanov, Dokl. Akad. Nauk SSSR, 1972, V. 203, p. 387
[69] W. Lorenz, Z. physik. Chem., 1961, Bd. 218, S. 272; 1962, Bd. 219, S. 421; 1963, Bd. 224, S. 145; 1966, Bd. 232, S. 176
[70] G. Salie, W. Lorenz, Z. physik. Chem. (N.F.), 1961, Bd. 29, S. 408; Z. Elektrochem., 1964, Bd. 68, S. 197; Z. physik. Chem., 1975, Bd. 256, S. 386
[71] W. Lorenz, Z. physik. Chem., 1970, Bd. 244, S. 65; 1973, Bd. 254, S. 123
[72] A. N. Frumkin, N. S. Polianovskaya, B. B. Damaskin, J. Electroanal. Chem., 1972, V. 39, p. 123
[73] A. N. Frumkin, N. S. Polianovskaya, B. B. Damaskin, O. Os'kina, J. Electroanal. Chem., 1973, V. 49, p. 7
[74] A. N. Frumkin, N. S. Polianovskaya, B. B. Damaskin, J. Electroanal. Chem., 1976, V. 73, pp. 267, 273
[75] G. Salié, W. Lorenz, Z. physik. Chem., 1975, Bd. 256, S. 386; Electrochim. Acta, 1975, V. 20, p. 309
[76] O. A. Petrii, R. V. Marvet, A. N. Frumkin, "*Elektrokhimiya*", 1967, V. 3, p. 116
[77] O. A. Petrii, A. N. Frumkin, V. V. Topolev, "*Elektrokhimiya*", 1969, V. 5, p. 1104
[78] V. E. Kazarinov, O. A. Petrii, V. V. Topolev, A. V. Losev, "*Elektrokhimiya*", 1971, V. 7, p. 1365
[79] O. A. Petrii, Nguen van Tue, Yu. G. Kotlov, "*Elektrokhimiya*", 1969, V. 5, p. 1108
[80] A. N. Frumkin, O. A. Petrii, Yu. G. Kotlov, "*Elektrokhimiya*", 1969, V. 5, p. 1214
[81] B. B. Damaskin, O. A. Petrii, V. E. Kazarinov, "*Elektrokhimiya*", 1972, V. 8, p. 1373
[82] D. A. Sveshnikova, V. E. Kazarinov, O. A. Petrii, "*Elektrokhimiya*", 1977, V. 13, p. 1505
[83] O. A. Petrii, J. N. Malisheva, V. E. Kazarinov, "*Elektrokhimiya*", 1971, V. 7, pp. 1574, 1842
[84] A. N. Frumkin, Dokl. Akad. Nauk SSSR, 1964, V. 154, p. 1432
[85] P. Delahay, G. Susbielles, J. Phys. Chem., 1966, V. 70, p. 315
[86] B. Timmer, M. Sluyters-Rehbach, J. Sluyters, J. Electroanal. Chem., 1967, V. 15, p. 343
[87] R. De Levie, J. Electroanal. Chem., 1969, V. 20, p. 332
[88] G. P. Girina, A. N. Frumkin, L. G. Feoktistov, "*Elektrokhimiya*", 1975, V. 11, p. 762
[89] B. Grafov, E. Pekar, J. Electroanal. Chem., 1971, V. 31, p. 137
[90] B. Grafov, E. Pekar, O. Petrii, J. Electroanal. Chem., 1972, V. 40, p. 179
[91] B. M. Grafov, E. A. Ukshe, "*Elektrokhimicheskie tsepi peremennogo toka*", "Nauka", M., 1973
[92] A. N. Frumkin, B. B. Damaskin, O. A. Petrii, "*Elektrokhimiya*", 1976, V. 12, p. 3; Z. physik. Chem., 1975, Bd. 256, S. 728
[93] A. N. Frumkin, B. B. Damaskin, O. A. Petrii, J. Electroanal. Chem., 1974, V. 53, p. 57
[94] K. Vetter, J. Schultze, Ber. Bunsenges. physikal. Chem., 1972, Bd. 76, S. 920, 927
[95] J. Schultze, K. Vetter, J. Electroanal. Chem., 1973, V. 44, p. 63
[96] K. Vetter, J. Schultze, J. Electroanal. Chem., 1974, V. 53, p. 67
[97] J. Schultze, F. Kappitz, Electrochim. Acta, 1976, V. 21, pp. 327, 337
[98] R. Shuttleworth, Proc. Phys. Soc. (London), 1950, V. 63, p. 444
[99] A. Ya. Gokhshtein, Elektrokhimiya, 1966, V. 2, p. 1318; 1968, V. 4, p. 886; 1969, V. 5, p. 637; 1970, V. 6, p. 979; 1971, V. 7, pp. 3, 594
[100] A. Ya. Gokhshtein, Dokl. Akad. Nauk SSSR, 1967, V. 174, p. 395; 1958, V. 183, p. 859; 1971, V. 200, p. 620

[101] A. Ya. Gokhshtein, Electrochim. Acta, 1970, V. 15, p. 219
[102] A. Ya. Gokhshtein, Usp. Khim., 1975, V. 44, p. 1956
[103] A. Ya. Gokhshtein, "*Surface Tension of Solids and Adsorption*" (russ.), "Nauka", M., 1976, 399 p
[104] A. I. Rusanov, Kolloid. Zh., 1977, V. 39, vip. 4, p. 711—717

On the mechanics and thermodynamics of three phase contact line systems

A. Scheludko, B. V. Toshev and D. Platikanov

(Faculty of Chemistry, University of Sofia, Sofia/Bulgaria)

Abstract

Gibbs' idea of line tension for a system with a three-phase contact line is discussed at length. Several cases are examined. The dependence of the equilibrium contact angle on the size of the three-phase contact junction when line tension exists is recognized. The effect of line tension on phase formation on a substratum is discussed and it is shown, on the basis of Gretz's formula, that the work of nucleus formation is strongly influenced by the line tension. The formation of a new phase surface when joining two liquid phases or a liquid one to a solid one is considered as well. The line tension in that case, if it is positive, influences both the force and energy barrier of merging of the phases, the barriers being measurable when the joining phases (or one of them) are small.

In conclusion various experiments for measurement of line tension are proposed and discussed. The case of a line tension along the perimeter of the Newton black film, formed when two bubbles join in a solution of an appropriate surfactant, is considered in detail as a basis for two different methods for line tension measurement: the method of the decreasing bubble and the method of critical bubble size.

1. The problem of small phases. Homogeneous and heterogeneous nucleation. Line tension and heterogeneous nucleation

A hundred years ago Gibbs made the thermodynamic theory of capillarity an essential part of his third work: "On the Equilibrium of Heterogeneous Substances" (Gibbs, 1906). The motto from Clausius "*The energy of the Universe is constant. The entropy of the Universe strives to a maximum*" underscores the great scale of this work. The method, however, used to describe heterogeneous systems by macroscopic parameters, determining phase equilibrium, creates difficulties when applied to very small phases.

In the sections "On the Possibility of the Formation of a Fluid of different Phase within any Homogeneous Fluid" (p. 252–258)[1] and "Liquid Films" (p. 300–314) where the subjects of investigation are small droplets and thin liquid films, Gibbs intentionally confines himself to those phase-dimensions where there still exists substance which can be regarded as substance of a certain phase in bulk. For such sufficiently large drops and sufficiently thick films Gibbs offers definitive solutions. For smaller phases he only restricts himself to advice about the way they should be treated thermodynamically. These recommendations have been more or less realised, e.g. by means of the concept of disjoining pressures in thin liquid films, as introduced by

[1] All references to pages here and hereafter apply to Gibbs (1906).

DERJAGUIN (1955). Hence we shall not consider them here. GIBBS also deals with the problem of transition to "zero" phases (to phases of molecular dimensions in our concept) and for such a case he thinks that "*it will hardly be profitable to seek to interpret*..." the equations obtained with large phases (p. 255). It seems as if it is impossible with such limitations to solve thermodynamically the problem of the genesis of a new phase (e.g. from supersaturated vapours), for such a phase forms by growth from molecular dimensions. The important contribution of GIBBS is that he demonstrated the inconsistency of such a complete limitation. As a result of a detailed analysis of systems in the supersaturated state (e.g. a supersaturated gas) GIBBS showed that the droplets determining system stability (and phase nucleation kinetics) are those for which the corresponding vapour pressure p_r equals the pressure of the supersaturated vapours in the system. Hence, they are droplets/nuclei of a new phase whose sizes, at not very high supersaturations $(p_r - p_\infty)/p_\infty$, are substantially greater than molecular dimensions.

The relation between p_r and p_∞ is given by THOMSON-GIBBS' equation

$$p_r/p_\infty = \exp \frac{2\sigma\bar{v}}{RkT}, \tag{1}$$

where p_∞ is the vapour pressure of a very large drop, σ the surface tension, $\bar{v}$ the molecular volume of the liquid, R the drop radius and k and T are BOLTZMANN's constant and the KELVIN temperature, respectively. Equation (1) holds for ideal gases and incompressible liquids. The work of formation of such a nucleus is

$$W = \sigma S - P_\sigma v = \tfrac{1}{3}\sigma S\,, \tag{2}$$

where S is the surface area of the spherical nucleus, v its volume and

$$P_\sigma = \frac{2\sigma}{R}, \tag{3}$$

the capillary pressure in it. This work is the energy barrier to nucleation of a new phase and appears in this form in all kinetic theories of the rate of nucleation developed after GIBBS.

The experiments of VOLMER and FLOOD (1934) on the formation of a cloud in WILSON's chamber have shown that the theory of cloudformation agrees satisfactorily with experiment although the sizes of the nuclei calculated by equation (1) at the respective supersaturations are very small. They correspond to 72 molecules for water, 27 for methanol, 144 for n-propanol, 119 for 2-propanol, 74 for n-butanol, 64 for nitromethane and 41 for ethyl acetate. It follows that a macroscopic thermodynamical description of small droplets is admissible for aggregates of a few hundreds and even a few tens of molecules. Hence the problem of very small phases, though in an approximate manner, can be avoided in the thermodynamic part of the theory of homogeneous nucleation of a new phase at sufficiently small supersaturations.

It is different, however, in the case of the heterogeneous nucleation of a new phase if one follows the theory of GIBBS as outlined in "On the Possible Formation of the Surface where two different Homogeneous Fluids meet of a Fluid of different Phase from either" (p. 258–264) but does not take into account his critical remarks. In this theory a liquid drop, a lens C is considered in thermodynamic equilibrium between liquid phases A and B which meet at a plane surface. As in the previous case the surfaces of the drop are spherical. GIBBS obtains the angles of contact with respect to the plane surface AB by a balance of forces, thus deriving a rule, named later the rule of NEUMANN.

The work of formation of such a droplet-nucleus, i.e. a droplet the vapour pressure of which is equal to the pressure in the supersaturated phases A or B, is thus of the same type as in the homogeneous nucleation of a new phase, namely

$$W_1 = \sum \sigma_i S_i - P_\sigma v = \tfrac{1}{3} \sum \sigma_i S_i \,. \tag{4}$$

Here $\sigma_i S_i$ is the change of free surface energy in the system due to the appearance in it of a lens C of volume v.

For the spherical case of a lens on a solid surface, equation (4) yields

$$W_2 = \tfrac{4}{3} \pi R^2 \sigma \tfrac{1}{4} (2 - 3 \cos \theta + \cos^3 \theta) \,. \tag{5}$$

Here the rule of YOUNG is applied as a partial case of the rule of NEUMANN for a drop wetting a plane surface at an angle θ. This formula, first derived by VOLMER (1939), is employed below for the description of the thermodynamic part of our theory of heterogeneous nucleation on a solid substratum and nuclei of condensation.

Actually, GIBBS has seriously criticised formula (4) as follows: "*It should however be observed that in the immediate vicinity of the circle in which the three surfaces of discontinuity intersect, the physical state of each of these surfaces must be affected by the vicinity of the others. We cannot, therefore, rely upon the formula* (4) *except when the dimensions of the lentiform mass are of sensible magnitude*", (p. 261). As we shall demonstrate later, this remark of GIBBS also applies to formula (5) as well as to NEUMANN's and YOUNG's rules.

It is worth noting that the lower limit of applicability of equation (4) with respect to size has not been confined by GIBBS either to molecular dimensions or to the presence in the drop of a substance with bulk properties, although such a stipulation is made in all other cases where the talks of very small phases. GIBBS has obviously not meant the specific properties of small phases in this sense, but that in a three-phase system it is necessary to account for the line parameters which are absent in a two-phase system. In this sense the concept of linear thermodynamics, linked to two- and three-dimensional thermodynamics for a three-phase system is developed in detail in the remark on p. 288: "*We may remark that a nearer approximation in the theory of equilibrium and stability might be attained by taking special account, in our general equations, of the lines in which surfaces of discontinuity meet. These lines might be treated in a manner entirely analogous to that in which we have treated surfaces of discontinuity. We might recognise linear densities of energy, of entropy, and of the several substances which occur about the line, also a certain linear tension.*"

The introduction of the concept of line tension $\varkappa$ results as is well known, in the appearance of a two-dimensional pressure (or tension) $\sigma_\varkappa$, when the line of three-phase contact is curved, analogous to the appearance of a capillary pressure P_σ from the concept of surface tension when the interface is curved. This is readily appreciated by reference to Figure 1. Here points a and b are symmetrically situated with respect to the intersection of a normal N with a three-phase contact line of radius r, and the tension $\varkappa$ acts along the tangents to these points. This tension yields a force $2\varkappa \sin \dfrac{\delta\alpha}{2}$

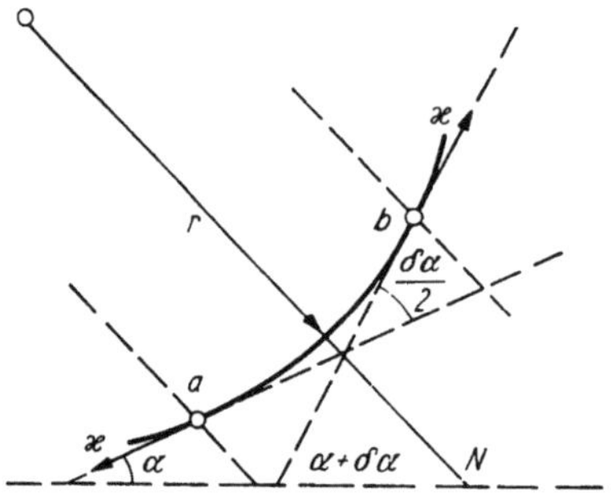

Fig. 1: Line tension and a curved three phase contact line.

in a direction parallel to N. Thus, a force $\frac{2\varkappa}{\breve{ab}} \sin \frac{\delta\alpha}{2}$ will act upon a unit arc $\breve{ab}$ The two-dimensional pressure or tension for the points of crossing with N is obtained by $\breve{ab} \to 0$, i.e.

$$\sigma_\varkappa = \lim_{\breve{ab}\to 0} \frac{2\varkappa \sin \frac{\delta\alpha}{2}}{\breve{ab}} = \varkappa \lim_{\breve{ab}\to 0} \frac{\delta\alpha}{\breve{ab}} = \varkappa K_L = \frac{\varkappa}{r}, \tag{6}$$

since the $\lim \frac{\delta\alpha}{\breve{ab}}$ at $\breve{ab} \to 0$ equals the curvature of the line $K_L = 1/r$ in this point.[1] The resulting expression is a complete two-dimensional analog of the capillary pressure P_σ, while formula (6) is an analog to LAPLACE's equation $P_\sigma = \sigma K_s$, where K_s is the surface curvature.

In the derivation of NEUMANN's (and YOUNG's) rule by force balance, the tension $\sigma_\varkappa$ must now be added to the balance of tensions along the perimeter of the drop C between phases A and B. In the general case of a curved surface for the solid body, a balance along the tangent of the surface gives (Figure 2):

$$\sigma_{AB} - \sigma_{AC} \cos\theta - \sigma_{BC} - \sigma_\varkappa \cos\psi = 0\,, \tag{7}$$

where ψ denotes the angle between the plane of the three-phase contact line and the solid-liquid interface.

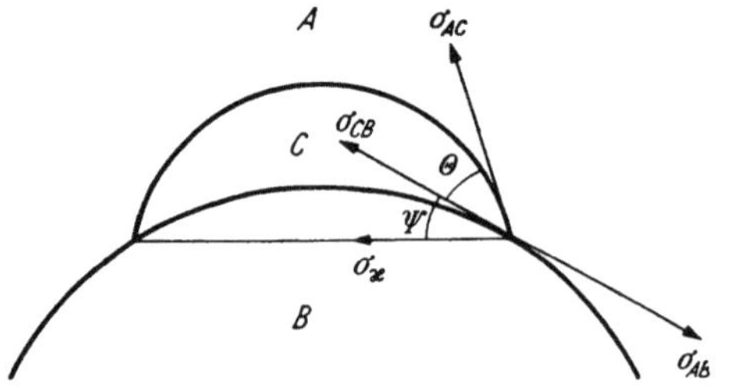

Fig. 2: Profile of a drop wetting a curved surface of a solid body.

[1] A similar analysis has already been carried out by WESSELOVSKII and PERTZOV (1936).

The thermodynamic derivation of equation (7) has recently been given by one of us (TOSHEV (1974/1975)). For a plane surface of the substrate B, when $\psi = 0$, we have from equations (6) and (7)

$$\sigma_{AB} - \sigma_{AC} \cos \theta - \sigma_{BC} - \frac{\varkappa}{r} = 0 , \tag{7'}$$

which turns into YOUNG's rule at a sufficiently large radius of the wetting perimeter when $\theta = \theta_\infty = \text{const}$, namely

$$\sigma_{AB} - \sigma_{AC} \cos \theta_\infty - \sigma_{BC} = 0 . \tag{8}$$

Equation (7′) can be expressed through the angle of wetting θ_∞ in the form

$$\sigma_{AC}(\cos \theta_\infty - \cos \theta) = \varkappa/r . \tag{7''}$$

Formulae (7), (7′) and (7″) can be regarded as describing the dependence of the wetting angle on the magnitude of the wetting perimeter. Formula (7″) is not exact because the vapour pressure in a system with a large drop is different from that with a small drop and thus the σ_{AB} are also different, for example because of the adsorption of vapours of C on the surface AB. Since, however, the change of σ_{AB} is related to the adsorption ability of the surface, a particularly simple case of $\sigma_{AB} = \text{const}$ can be singled out.

In the above equations, obtained by making use of the force balance method, the quantities denoted by σ have a different meaning in the different systems. If the system is fluid then the σ are surface tensions whose interpretation as forces is well-known. If a solid body exists in the system then it is necessary to distinguish between the works of stretching of the solid interfaces and phase lines and the works of formation of these boundaries (p. 315, also TOSHEV and ERIKSSON (1975/1976)). The latter appear in the NEUMANN-YOUNG kind of equation (p. 326) but these quantities can not be interpreted as forces. Nevertheless this does not mean that it is impossible to apply the force balance method of mechanical equilibrium to such systems; there exists a suitable interpretation in terms of forces as applied to the differences of these quantities (e.g., $\sigma_{AB} - \sigma_{BC}$ in equation (8)), (p. 328)[1].

In the derivation of the work of formation of a nucleus on a substratum, $\varkappa$ is accounted for in the energy balance in equation (4) by adding the line energy $\varkappa L$, where L is the length of the wetting perimeter. Thus for a drop on a plane solid surface we get

$$W_2' = \frac{4}{3} \pi R^2 \sigma \frac{1}{4} (2 - 3 \cos \theta + \cos^3 \theta) + \frac{\varkappa L}{2} . \tag{9}$$

The last term represents the work of formation of a two-dimensional nucleus W^*. It can be obtained separately by transforming formula (4) from a two/three-dimensional system into a one/two-dimensional system:

$$W^* = \varkappa L - \sigma_\varkappa S - \tfrac{1}{2} \varkappa L . \tag{10}$$

Here $\varkappa$ corresponds to σ, L to S, $\sigma_\varkappa$ to $\dot{P}_\sigma$ and S to v. The angle θ in equation (9) is the real angle of wetting of the nucleus/drop and does not equal θ_∞ in expression (5). Formula (9) has recently been derived by GRETZ (1966) in another form.

[1] For more details see TOSHEV and ERIKSSON (1975/1976).

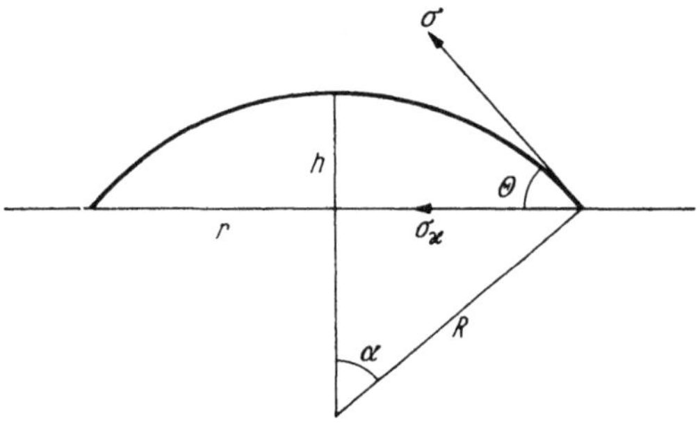

Fig. 3: Profile of a drop wetting a plane surface of a solid body.

Here we shall draw attention to some important corollaries of this formula.
In equation (7'') we shall introduce the dimensionless curvature

$$\tilde{K} = \frac{\varkappa/\sigma}{R}, \tag{11}$$

which in place of equation (7'') yields

$$(\cos \theta_\infty - \cos \theta) \sin \theta = \tilde{K} . \tag{7'''}$$

Herein $r = R \sin \alpha$ and $\alpha = \theta$ (Figure 3).

Formula (7''') shows that a wetting drop cannot exist at each R and, therefore, at each respective supersaturation, whilst preserving at the same time the spherical form of its surface. The equality (7''') is only fulfilled up to certain limiting values $\tilde{K}_{\max}$ corresponding to wetting angles θ_m. The solution for the extremum of (7''') yields

$$\cos \theta_m = \frac{\cos \theta_\infty}{4} \left[1 \pm \sqrt{1 + \frac{8}{\cos^2 \theta_\infty}} \right]. \tag{12}$$

The angle θ_m alters from 120° at $\theta_\infty = 0$ to 180° at $\theta_\infty = 180°$. Correspondingly, the maximum value of K runs from 1.299 at $\theta_\infty = 0$, through 0.5 at $\theta_\infty = 90°$, to zero at $\theta_\infty = 180°$. In investigations of lens nucleation, account must therefore be taken of $\varkappa$, when θ lies between θ_∞ and θ_m or $\tilde{K}$ lies between 1.299 and zero. It is useful to compare the results obtained with and without taking into account the effect of $\varkappa$ on heterogeneous nucleation. This is readily done by examining the ratio

$$\frac{W_2'}{W_2} = \frac{2 - 3 \cos \theta + \cos^3 \theta + 3\varkappa \sin \theta/\sigma R}{2 - 3 \cos \theta_\infty + \cos^3 \theta_\infty}. \tag{13}$$

Here W_2 is written according to formula (5) with $\theta = \theta_\infty$.

Combining equations (13) and (7''') we get W_2'/W_2 vs. $\tilde{K}$. This dimensionless dependence is shown in Figure 4 with the three values of θ_∞, 20°, 45° and 90°. This figure demonstrates that neglecting $\varkappa$ can strongly underestimate the work of nucleus formation, especially in the case of small wetting angles θ_∞ on a large drop, even when $\tilde{K} \ll 1$.

The introduction of $\varkappa$ also gives very substantial corrections when calculating the nucleus volume from the supersaturation data, or the curvature of the wetting nucleus/drop by formula (1).

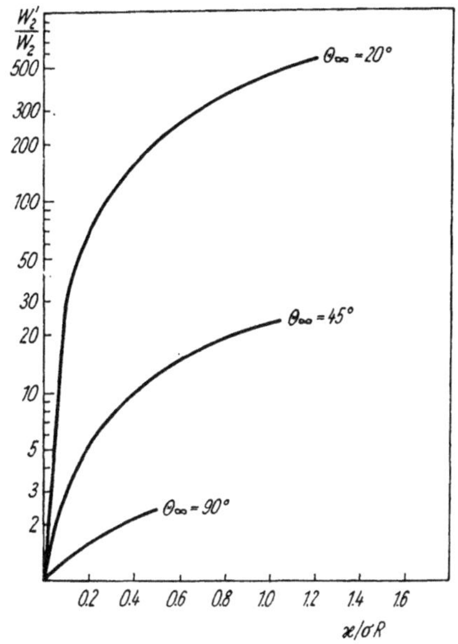

Fig. 4: Plot of the theoretically calculated ratio work W_2' of formation of a nucleus on a substratum accounting for line tension to "work W_2 of formation of the same nucleus without line tension" vs. dimensionless drop curvature $\tilde{K} = |\varkappa/\sigma R|$ for three wetting angles θ_∞ of a large drop.

At equal curvatures, the ratio of the real drop volume v', to the volume v calculated without allowing for $\varkappa$, i.e. at $\theta = \theta_\infty$, gives

$$\frac{v'}{v} = \frac{(1 - \cos\theta)^2 (2 + \cos\theta)}{(1 - \cos\theta_\infty)^2 (2 + \cos\theta_\infty)}. \tag{14}$$

Here θ is again expressed through $\tilde{K}$, according to (7'''), and the dependence v'/v vs. $\tilde{K}$ shows the accuracy of interpretation of the volume of the nucleus when $\varkappa$ is ignored.

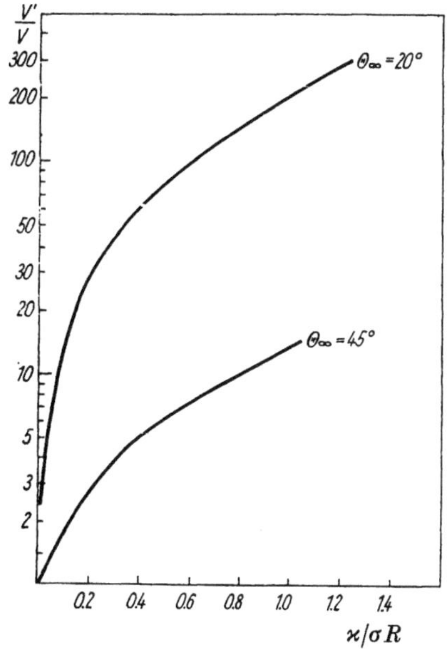

Fig. 5: Plot of the theoretically calculated ratio of wetting drop volume v' obtained with line tension to drop volume v without line tension vs. dimensionless drop curvature $\tilde{K} = |\varkappa/\sigma R|$ for two wetting angles θ_∞ of a large drop.

In Figure 5 the result of the solution of formulae (14) and (7‴) is illustrated by the curves for the two θ_∞, 20° and 45°. As it can be seen, the deviation from VOLMER's dependence with respect to nucleus volume increases very strongly (almost exponentially) with $\tilde{K}$, and especially strongly at small angles of wetting θ_∞. Thus, to remain within the range of the macroscopic thermodynamic theory of nucleation on a substratum, it is obligatory to incorporate line tension. This remark obviously also holds for the deposition of a new phase on condensation nuclei at incomplete wetting. The remark of GIBBS in this sense (p. 261) has been realised by KRASTANOV (1957) but, unfortunately, without taking into account the line tension along the perimeter of wetting of the nucleus by the drop.

GIBBS also admits the possibility of $\varkappa$ being negative (p. 296). In the third part of the present paper, data (PLATIKANOV, NEDYALDOV and SCHELUDKO (1978), PLATIKANOV, NEDYALKOV and NASTEVA (1978)) are presented for the special case of a NEWTON film where $\varkappa$ on its perimeter can change sign depending on electrolyte concentration. Hence, it is necessary to consider also the effect of negative $\varkappa$ on the process of heterogeneous nucleation.

The effect of such a possibility can be seen from the derivation of formulae (7) and (9) for a drop on a plane solid substratum. According to equation (7‴) the wetting angle θ diminishes when the drop radius r diminishes whereas according to equation (9) the work of formation of a nucleus W_2' can fall to zero even before the angle θ becomes zero. This means that at $\varkappa < 0$, at certain supersaturations from equation (1), corresponding to given values of $\tilde{K}_0 = \left|\frac{\varkappa/\sigma}{R}\right|_0$ for $W_2' = 0$, heterogeneous nucleation will occur without a barrier and its velocity will only be limited by kinetic circumstances (mass transfer at the drop). We have calculated these values of $\tilde{K}_0$ from (9) and (7‴) as a function of the wetting angle θ_∞ of the large drop. It can be seen in Figure 6 that for small θ_∞ these values are much lower than unity.

At $\varkappa < 0$ our reasonings about $\tilde{K}_0$ are limited because in this case $\theta < \theta_\infty$ and $v' < v$, which at small angles (θ_∞) are very small themselves. We have calculated v'/v from equation (14) with θ for $W_2' = 0$ in a wide interval of values of θ_∞ (from 20° to 150°) and have established here that v' is only approximately two times smaller than v (a factor of 0.49 for $\theta_\infty = 20°$ and 0.58 for $\theta_\infty = 150°$).

If the line tension is negative the equilibrium of the two-dimensional phases α and β (with surface tensions σ_α and σ_β), divided by a circular contour (Figure 7), is stable (p. 296). Indeed, from the extremum of the free energy F at constant area of the two-

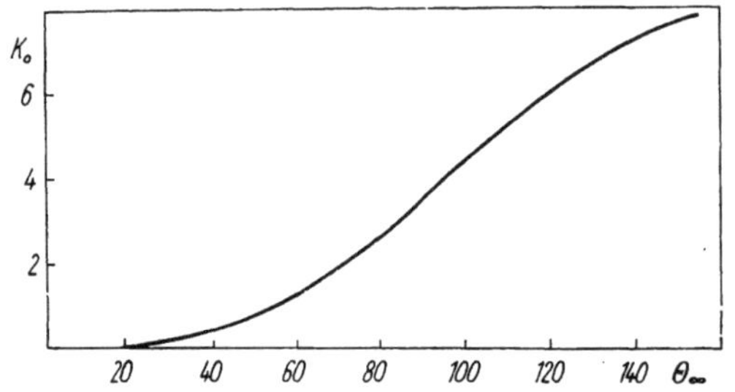

Fig. 6: Plot of the theoretically calculated dimensionless curvature $\tilde{K}_0$ of wetting drop, when the work of its formation is zero vs. wetting angle θ_∞ of a large drop.

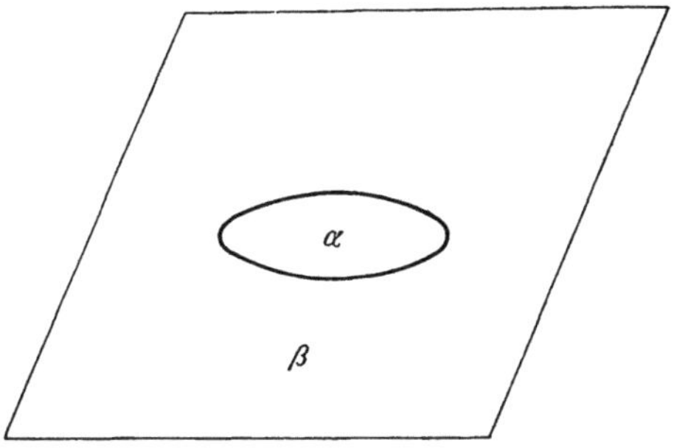

Fig. 7: Two dimensional phases.

dimensional system

$$dF = [2\pi r(\sigma_\alpha - \sigma_\beta) + 2\pi\varkappa]\, dr = 0 \tag{15}$$

follows

$$\sigma_\varkappa = \sigma_\beta - \sigma_\alpha = \varkappa/r \tag{6'}$$

and

$$\frac{\partial^2 F}{\partial r^2} = -2\pi\varkappa/r > 0\,. \tag{16}$$

The equilibrium of the lentiform mass of phase C with its surroundings as treated earlier, is unstable (p. 261–262). Allowing for line tension does not change this conclusion if $\varkappa > 0$. At the transition to $\varkappa < 0$, in view of equation (16) obtained for two-dimensional phases, a special case may be expected in which the liquid formation (for example, on a solid substratum) will be in stable equilibrium with the neighbouring phases, i.e. it will not represent a nucleus for a new phase. Indeed (Figure 3), from

$$dF = [-\tfrac{1}{2}\pi(r^2 + h^2)\, P_\sigma + 2\pi h\sigma]\, dh + [-\pi r h P_\sigma + 2\pi r\sigma + 2\pi r(\sigma_{BC} - \sigma_{AB}) + 2\pi\varkappa]\, dr = 0 \tag{15'}$$

equations (3) and (7′) follow, and using them, together with

$$\frac{\partial^2 F}{\partial h^2} > 0; \qquad \frac{\partial^2 F}{\partial h^2}\,\frac{\partial^2 F}{\partial r^2} - \left(\frac{\partial^2 F}{\partial h\, \partial r}\right)^2 > 0$$

we find that at

$$\tilde{K} > \frac{\sin^3\theta}{\cos\theta} \tag{16'}$$

the drop on the substratum will be in stable equilibrium with the neighbouring phases. This does not coincide with the case $W_2' = 0$ which has been considered above.

2. Two-dimensional nucleation. Contact between liquid phases. Adhesion of a solid sphere to a plane liquid surface

We have demonstrated the effect of line tension on the process of heterogeneous nucleation. Now we shall consider the formation of a new surface phase, without the formation of a new volume phase, when the line tension is a determining factor.

The formation of a new interface between two volume phases always takes place when a fluid phase is pressed to or collides with another fluid or solid phase. Such contacts appear at the interaction of a drop or a bubble with a solid surface (wetting/dewetting); when a solid particle touches the boundary of two fluid phases (e.g. in flotation); at the collision of drops of two immiscible liquids in a third liquid. In addition, a thin liquid film, including the NEWTON black film, can form under certain conditions on the boundary of two identical liquid or gas phases. The latter represents a new two-dimensional phase confined by the perimeter of wetting. At the rupture of such a film, or when it does not form at all, coalescence occurs.

All these phenomena are characterized by a deformation of the merging liquid phases as a result of which an extended contact forms. This differs from the adhering of solid bodies which occurs without such a spontaneous deformation. The hardening of contact between solid phases by recrystallisation and bridge formation between them corresponds to the strengthening by a deformation of the contacts between liquid phases. Contact formation between liquids or between liquid and solid phases is associated with a certain force barrier, determined by the action of the line tension when the latter is positive.

To illustrate the involvement of line tension in contact formation we now examine in detail the example of a solid sphere adhering to a liquid/liquid or liquid/gas interface (as in Figure 8) which has formed the basis of a recent capillary theory of flotation (SCHELUDKO, TOSHEV and BOJADJIEV (1976)). The curving of a plane surface by a pressing force, characterized by an angle $\gamma > 0$, creates a capillary force of reaction

$$F_\sigma = 2\pi r \sigma \sin \gamma \tag{17}$$

by which the surface tension is measured through balance methods, including the sphere method (SCHELUDKO and NIKOLOV (1975); FIEBER and SONNTAG (1975); HUH and MASON (1976)). At sufficiently large r all tensions along the perimeter of wetting are compensated, according to equation (7), with $\psi = \pi - \alpha$ determining the position of this perimeter on the sphere. If r, however, is very small, then the condition (7) cannot be fulfilled even when $\theta = 0$. In such a case a certain part $\sigma_\varkappa$ remains uncompensated, namely,

$$\Delta\sigma_\varkappa = \varkappa/r - \varkappa/r_c\,. \tag{18}$$

Here r_c is the critical radius of that perimeter of wetting which corresponds to the ultimate possibility of a compensation of tensions along the surface of the sphere as θ converges to zero, given by

$$\sigma_{AB} - \sigma_{BC} - \sigma_{AC} + \varkappa \cos\alpha_c/r_c = 0\,. \tag{19}$$

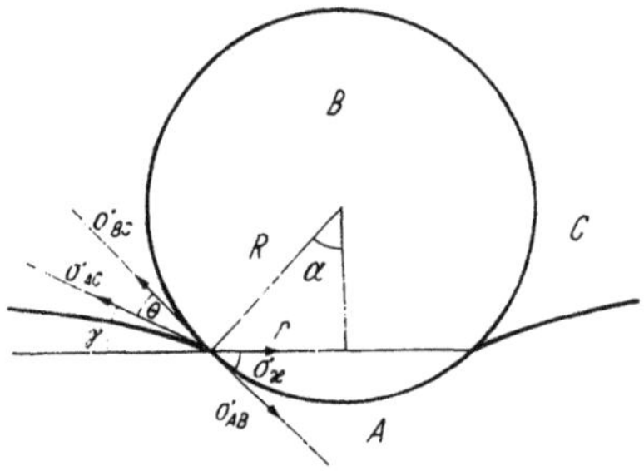

Fig. 8: Scheme of solid sphere profile, on a lower pendant drop surface

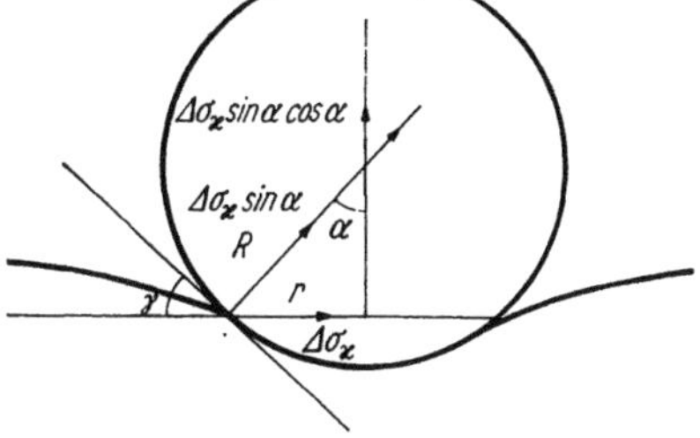

Fig. 9: Scheme illustrating the correlation of the uncompensated part $\Delta\sigma_\varkappa$ and the pressing force.

This condition together with equation (8) and $\sin\alpha = r/R$ yields for small α_c (i.e. when $\cos\alpha_c \simeq 1$):

$$r_c = \frac{\varkappa/\sigma}{1 - \cos\theta_\infty}, \tag{19'}$$

where $\sigma = \sigma_{AC}$.

In Figure 9 the uncompensated part of the tension, $\Delta\sigma_\varkappa$, (directed toward the centre of the perimeter of wetting when $\varkappa > 0$) is resolved along the normal to the sphere and the resultant transferred to the centre of the sphere is resolved along the vertical. The force $\Delta\sigma_\varkappa \sin\alpha\cos\alpha$, from a unit length of the wetting perimeter multiplied by the perimeter length $2\pi r$ yields,

$$F_\varkappa = 2\pi r\varkappa \sin\alpha\cos\alpha \left(\frac{1}{r} - \frac{1}{r_c}\right) = 2\pi\varkappa \sin^2\alpha\cos\alpha \left(\frac{1}{\sin\alpha} - \frac{1}{\sin\alpha_c}\right). \tag{20}$$

The sum of this force and F_σ must be counterbalanced by the external pressing force G so that the equilibrium of the system is preserved.

Thus

$$G = 2\pi R\sigma \sin^2\alpha \left[1 + \frac{\varkappa/\sigma}{R}\left(\frac{1}{\sin\alpha} - \frac{1}{\sin\alpha_c}\right)\cos\alpha\right]. \tag{21}$$

Here r is expressed in terms of the central angle α.

For the region of small G, where $r < r_c$, the angle of wetting of the sphere by the liquid, θ, equals zero and thus, γ in equation (17) equals α and $F_\sigma = 2\pi R\sigma \sin^2\alpha$. Expression (21) can be rewritten for this case as

$$G = 2\pi R\sigma_1 \sin^2\alpha\,, \tag{22}$$

where the effective tension is

$$\sigma_1 = \sigma + \frac{\varkappa}{R}\left(\frac{1}{\sin\alpha} - \frac{1}{\sin\alpha_c}\right)\cos\alpha\,. \tag{23}$$

That is to say, the system behaves as if the liquid tension had risen by the component $\Delta\sigma_\varkappa \cos\alpha$.

Expression (21) demonstrates that at $G = 0, \alpha = 0$ and the wetting perimeter gradually increases with G up to $r = r_c$ at $R = R_C$ with

$$G_c = 2\pi R_c\sigma \sin^2\alpha_c = F_{\sigma,c}\,, \tag{24}$$

when $\Delta\sigma_\varkappa$, and together with it $F_\varkappa$, converges to zero. If, however, G is a little higher than G_c and $r > r_c$ the wetting angle θ is allowed to increase, then r increases simultaneously. Such a situation must satisfy the conditions

$$G = 2\pi R\sigma \sin\alpha \sin(\alpha - \theta) \tag{25}$$

and

$$\frac{\varkappa/\sigma}{R} = (\cos\theta - \cos\theta_\infty)\tan\alpha \tag{26}$$

arising from equation (17) with $\gamma = \alpha - \theta$, equation (7) with $\psi = \pi - \alpha$ and equation (8) with $\sigma_{AC} = \sigma$. Thus for $G = G_c$ there are two solutions: one for $\theta = 0$ and $r = r_1 = r_c$ and the second for $\theta < 0$ and $r = r_2$. At sufficiently large R the perimeter of wetting increases so much that θ becomes almost equal to θ_∞ because the teım with $\varkappa$ vanishes.

Furthermore, at the transition $G = G_c$, it can be seen that as r increases from r_1 to $r_2 \gg r_1$ the angle $\gamma_2 = \alpha - \theta$ strongly diminishes with respect to $\gamma_1 = \gamma_c$ because $G_c = 2\pi r_1\sigma \sin^2\gamma_1 = 2\pi r_2\sigma \sin^2\gamma_2$. In most cases, even with $G = G_c$, γ may be assumed zero i.e. $\alpha \simeq \theta$, so that r_2 can be estimated by the approximate formula $r_2 = R_c \sin\theta_\infty$.

We shall now illustrate the above feature using the specific example of a small solid spheıe which is pressed by its weight onto the lower surface of a pendant drop.

Let us assume for simplicity that the drop is large, so that its surface is planar. The pressing force, equal to the weight of the sphere in the liquid can be written in the form

$$G = \tfrac{4}{3}\pi R^3 \varrho g\,, \tag{27}$$

where g is the accelleration of gravity and ϱ is the difference in density between the sphere and the liquid. Expression (27) is not exact, because the change in weight of the portion of the sphere protruding from the liquid is neglected. It has, however, been shown on the basis of precise calculations (SCHELUDKO and NIKOLOV (1975)), that formula (27) is sufficiently exact for small spheres and small angles of wetting.

For purposes of calculation we assume $\varkappa = 10^{-5}$ dyn[1] $(= 10^{-10}$ N), $\sigma = 50$ dyn cm^{-1}, $\varrho g = 10^3$ dyn cm^{-3} and $\theta_\infty = 20°$. With an assumed contact angle of 25° a large sphere would have no thin film on the AB interface and for a small sphere with the same solid/liquid combination the thin film must also be absent, i.e. the sphere must form a real contact. The inherent instability of a film on the protruding part of the sphere with this contact is heightened by a capillary pressure of $2\sigma/R$.

Expressing G as a function of R (equation 27), we can calculate from equation (21) the dependence of r on R in the region $0 < r < r_c$ as shown in curve 1 of Figure 10 using the expanded scale on the right hand ordinate for r. The last point of the curve has co-ordinates $r_c = 3.32 \times 10^{-6}$ cm and $R_c = 9.53 \times 10^{-4}$ cm. At $R > R_c$, the angle θ differs from zero and both angles θ and α begin to rise spontaneously until the system of equations (25) and (26) is satisfied. The solution of the latter gives a second value $r = r_2$ at $R = R_c$, namely 3.2×10^{-4} cm. Thus at the transition through $R = R_c$ a hundredfold increase in the perimeter of wetting must occur. Prior to the transition, r is too small to be observed, but after it the wetting perimeter radius becomes comparable to R and should be microscopically observable. This value of r, naturally, depends on θ_∞, to which θ becomes almost equal after the transition.

[1] $\varkappa$ throughout is quoted in dyn to conform with earlier work.

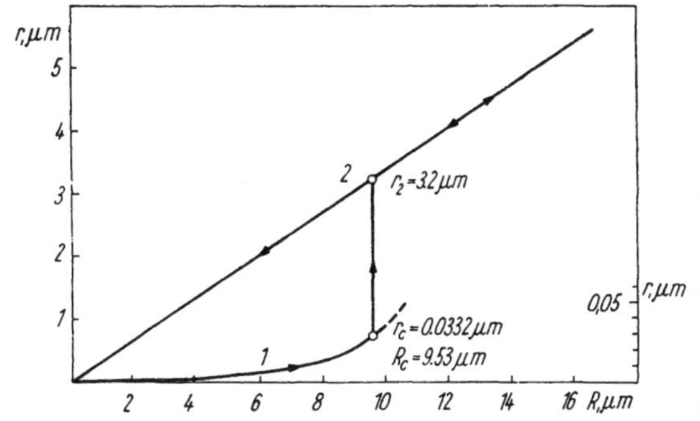

Fig. 10: Plot of the dependence of the wetting perimeter radius r on the radius R of a solid sphere in equilibrium at a liquid interface, calculated with $\theta_\infty = 20°$, $\varkappa = 10^{-5}$ dyn, $\sigma = 50$ dyn cm^{-1} and $\varrho g = 10^3$ dyn cm^{-3}.

For curve 2 (Figure 10), $\alpha \simeq \theta$ and is very close to θ_∞ for R equal to or even larger than R_c. When decreasing R on curve 2 we return to $R = 0$ at $r = 0$ and in so doing θ decreases with respect to θ_∞.

In curve 1 of Figure 11 a section of the dependence of θ on R for very small and decreasing R is shown. As can be seen, θ diminishes down to $\theta = \theta_m = 11.6°$ at $R = R_m = 2.44 \times 10^{-5}$ cm after which it jumps down to zero and brings us on to curve 1 of Figure 10 again. Here, r is not yet zero and continues to diminish according to equation (21). This section cannot be observed on the rough scale of R in Figure 10. The jump in θ to zero at θ_m, when R becomes equal to R_m corresponds to the vanishing of the derivative $dR/d\theta$ from the relation $(\varkappa/\sigma)/R = (\cos\theta - \cos\theta_\infty)\tan\theta$ describing curve 1 in Figure 10, which arises from equation (26) putting α equal to θ. The condition $dR/d\theta = 0$ gives

$$\cos\theta_m = (\cos\theta_\infty)^{1/3}\,. \tag{28}$$

In the case of negative $\varkappa$, as R decreases the angle θ gives the reverse behaviour, increasing smoothly from $\theta = \theta_\infty$ at large R, and approaching $\theta = 90°$ when R tends to zero (Figure 11, curve 2 for $\varkappa = -10^{-5}$ dyn). With the increase of R, the radius of the wetting perimeter gradually increases (without a jump) along the curve describing equations (25), (26) and (27).

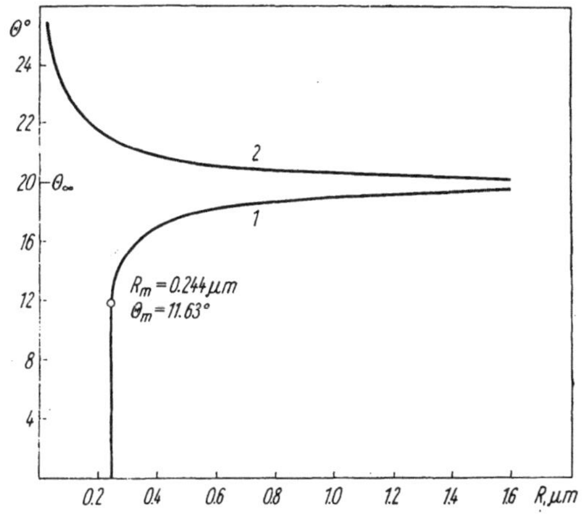

Fig. 11: Plot of wetting angle θ^f of a small solid sphere, in equilibrium at a liquid interface, as a function of its radius, calculated with $\theta_\infty = 20°$, $\varkappa = 10^{-5}$ dyn, $\sigma = 50$ dyn cm^{-1} and $\varrho g = 10^3$ dyn cm^{-3}.

3. Measurement of the line tension. Method of the diminishing bubble. Method of the critical bubble

In the theory of crystal growth STRANSKI and KAISCHEW (1934) have put forward a mechanism of genesis of a new monolayer on a crystal facet the work of formation of which $\varkappa L/2$ is determined by the free energy $\varkappa L$ of its perimeter L but no measurement of this term was effected.

During the compression of an insoluble monolayer on a LANGMUIR trough a region of constant two-dimensional pressure can often be observed, indicating a transition from a homogeneous gaseous state of the monolayer to a heterogeneous state with condensed regions. Unfortunately, the conditions for the onset of measurable supersaturation at this transition have so far not been found. Nor are there reliable ways to determine the number of appearing condensed spots, although their presence has been described in the literature as an experimentally established fact (ADAM and HARDING (1932), BOUHET (1931), ZOCHER and STIEBEL (1930)). As a consequence this case cannot, for the present, be employed to detect and measure the line tension of the perimeter of the condensed spots in the monolayer.

The first attempt to measure line tension was recently made by TORZA and MASON (1971). In a specially prepared emulsion the sticking together of microscopic droplets of liquids A and B dispersed in liquid C was photographed. Tensions σ_{AC} and σ_{BC} are known and the tension σ_{AB} between the drops is obtained from σ_{AC}, σ_{BC} and the curvature of the three surfaces determined from the photographs of the double drops. Furthermore, from the shape of the droplets the angles along the perimeter of contact between them were determined. The line tension of the contact of the phases A and B in C was then calculated from the difference between the angles determined above and the angles which would have appeared at larger contact perimeters (large drops), when $\sigma_{\varkappa} = 0$, and the latter angles are determined by the three surface tensions only.

In the quoted paper five photographs of double droplets are used. The calculated values of $\sigma_{\varkappa}$ differ strongly from each other. The respective values of $\varkappa$ are also dispersed more than an order of magnitude and in one case $\varkappa$ is even negative. Thus, neither the value of $\varkappa$ nor its sign have been reliably established.

The reason for the setback seems to be the fact that the axis of symmetry of the double droplets in the photographs has not been parallel to the photographic plane as assumed in the calculations. Such disorientation of the freely swimming double droplets from the assumed orientation must appear as an imaginary deformation of the surface of contact between the droplets and will lead to faulty values of σ_{AB}. In addition the values of $\varkappa$ have been erroneously calculated because of the incorrectly determined curvatures of the surfaces and radii of contact as well as the angles at the perimeters. If these authors had measured σ_{AB} separately and chosen from all the photographs of double drops those whose surface curvature gave σ_{AB} which corresponded to the measured values of σ_{AB}, then the error from disorientation of the droplets would probably have been excluded.

The second attempt to measure the line tension was made in connection with the capillary theory of flotation (SCHELUDKO, TOSHEV and BOJADJIEV (1976), MINGINS and SCHELUDKO (1978)). In this case the second possibility pointed out in the previous section was used. Microscopic glass spheres (ballotini), treated with a suitable solution of surfactant so that the angle of contact was approximately $\theta_\infty = 20°$, were allowed to deposit on the lower surface of a pendant drop of the same surfactant solution.

Spheres of different diameter behaved in a different way: the bigger spheres formed a clearly visible perimeter of wetting when observed from beneath with a microscope in reflected light, while the smaller ones stayed on the liquid surface without forming a detectable perimeter of wetting. From this the radius of the smallest particles forming a visual perimeter of wetting was determined (R_c) from which the line tension $\varkappa$ was evaluated using the formula

$$\varkappa = \sqrt{\tfrac{2}{3}\,\varrho g \sigma}\,(1 - \cos\theta)\,R_c^2\,, \tag{29}$$

which follows from equations (19), (27) and (24) at $r = R\sin\alpha$.

Although the value of $\varkappa \simeq 10^{-5}$ dyn obtained by this method agrees well with the mean value of $\varkappa$ calculated from flotation data, assuming that the lower limit of flotability corresponds to a particle size of the order of a micron and is determined by the work of formation of a two-dimensional nucleus of contact, this result also cannot be considered indisputable. As one of the referees of that paper has rightly noted, although unlikely, it is not impossible that the ballotini used may have been of two kinds: smaller, with smaller angle of contact and bigger, with larger angle of contact. Thus the lack of adherence of small particles may not have been governed by line tension but by their greater wettability. The verification of such an interpretation has not been possible because the angles of contact have only been measurable with big particles forming a visual three phase contact.

It is only recently that unambiguous results on the measurement of line tension have been obtained (PLATIKANOV, NEDYALKOV and SCHELUDKO (1978); PLATIKANOV, NEDYALKOV and NASTEVA (1978)). Here the perimeter of a NEWTON film was measured, the film being formed on the protruding part of a bubble emerging on the surface of sodium lauryl sulphate (NaLS) + NaCl aqueous solution. The bubbles used in these investigations were very small, of radius $R_b < 100\mu$, and because the capillary pressure in them is three to four orders of magnitude higher than the ambient pressure, such bubbles must behave as solid spheres like those considered in the previous section. The difference from the latter case is that the region of contact now is formed by a NEWTON film of a radius of curvature R_f which divides two identical phases. The tension of the film Δ is approximately only one per cent lower than the doubled tension of the surfaces meeting at the film perimeter (KOLAROV, SCHELUDKO and EXEROWA (1968); PRINCEN (1968); HUISMAN and MYSELS (1969); PRINS (1969); CLINT et al. (1969); PLATIKANOV and NEDYALKOV (1969/1970)). Because the capillary pressure in the bubble is $P_\sigma = 2\sigma/R_b = 2\Delta/R_f$, R_f is almost double R_b. The result of the symmetry of the system, i.e. the equal tensions of the bulk phases, is that the tangent to the film divides the angle between liquid surface and bubble surface into two exactly equal halves. This half θ_∞^f is accepted as the angle of contact film/bulk liquid. On very small bubbles (and films) a measurable component $\sigma_\varkappa$ arises. Although $\sigma_\varkappa$ acts not along the film but along its perimeter, i.e. it forms a certain angle with the tangent to the film at its perimeter, the symmetry of the system is almost preserved since $\sigma_\varkappa$ is much lower than σ even for the smallest bubbles under investigation. All these considerations, as has been shown (PLATIKANOV, NEDYALKOV and SCHELUDKO (1978)), lead to the following sufficiently exact formula for the tangential equilibrium

$$\varkappa^f = 4\sigma R_b(\cos\theta^f - \cos\theta_\infty^f)\sin\theta^f \tag{30}$$

instead of equation (26) for the solid sphere. Here the central angle $\alpha = 2\theta^f$, $r = R_b \sin 2\theta^f$ and

$$\Delta = 2\sigma \cos \theta^f_\infty \, . \tag{31}$$

Expression (30) leads to a similar dependence of the angle θ^f on R_b as for a solid sphere (Figure 11).

When small bubbles emerge, the formation of a Newton film (as with the formation of a bare surface on a solid sphere) is limited by the action of the line tension (if positive). Only when the force $G = \frac{4}{3}\pi R_b^3 \varrho g$ (27) pressing the bubble is greater than a certain critical value G_c, does a visible Newton film form. According to condition (19) where $\sigma_{AB} = \Delta$ and $\sigma_{BC} = \sigma_{AC} = \sigma$ with equation (31) and for $\theta^f = 0$ we get

$$r_c = \varkappa^f / 2\sigma(1 - \cos \theta^f_\infty) \tag{32}$$

in place of equation (19′). If the radius of the critical bubble is R_c, then to the extent that $G = G_c$, $F_\varkappa = 0$ and $F_\sigma = 2\pi R_c \sigma \sin^2 \alpha_c$, where $\sin \alpha_c = r_c / R_b$, we obtain for $F_\sigma = G_c$

$$\varkappa^f = 2 \sqrt{\tfrac{2}{3} \varrho g \sigma}\, (1 - \cos \theta^f_\infty)\, R_c^2 \tag{33}$$

in place of equation (29) for a solid sphere.

Formulae (30) and (33) offer the possibility of measuring the line tension by the change of the contact angle of a Newton film on the top of a diminishing bubble (according to equation (30)) as well as by the smallest (critical) bubble, ($R_b = R_c$), which forms a visible Newton film when emerging (according to equation (33)).

The thin-film aspects of the system under investigation here (Platikanov, Nedyalkov and Scheludko (1978); Platikanov, Nedyalkov and Nasteva (1978)) have been studied in many other works (Kolarov, Scheludko and Exerowa (1968); Princen (1968); Huisman and Mysels (1969); Prins (1969); Clint et al. (1969); Platikanov and Nedyalkov (1969/1970); de Feijter (1973)), so that Δ, σ and θ^f_∞ are reliably established. The parameters of the film/contact, in contrast to the contact on a solid sphere, are well reproduced. The bubble, in which the gas is compressed by additional capillary pressure, diminishes spontaneously on account of the gas diffusion out of it. This makes measurements by formula (30) especially convenient. Results from both methods of measurement are now described and have been obtained with the same concentrations of NaLS (0.05%) and different concentrations of NaCl at a temperature of 25 °C.

Relatively large bubbles ($R_b > R_c$) emerging at the solution surface are observed in a suitable manner (see Platikanov, Nedyalkov and Nasteva (1978)) and the process of diminution followed. Eventually, a set of values for R_b and r are obtained, which through calculation by the formula, $\sin 2\theta^f = r/R_b$, yields θ^f vs. R_b. These data, according to equation (30), allow a determination of $\varkappa$ and θ^f. The agreement of the latter with the latest data of de Feijter (1973), confirms the correctness of the usage of formula (30).

Regardless of the excellent reproducibility of the system parameters, the value of θ^f vs. R_b are considerably dispersed, mainly because of the not very high precision of measurement of R_b and especially of r. For this reason, measurements with several bubbles have been made for each concentration and the obtained values (of the order of a hundred) have been treated statistically (Platikanov, Nedyalkov and Nasteva

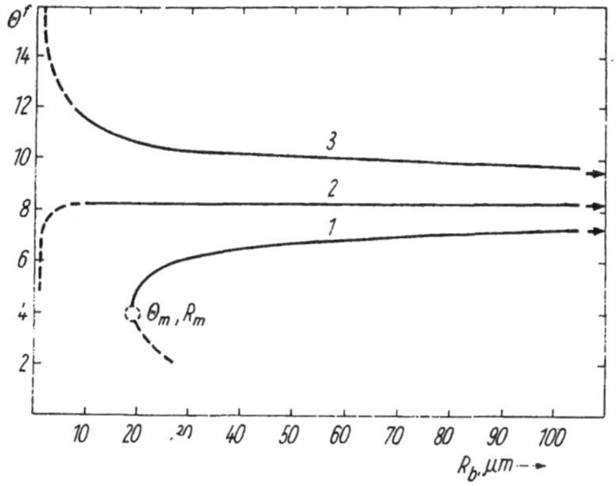

Fig. 12: Plot of the contact angle θ^f (NEWTON film/bulk liquid interface) as a function of bubble radius R_b; curve 1 is for 0.320, curve 2 is for 0.360 and curve 3 is for 0.459 mol/l NaCl (PLATIKANOV, NEDYALKOV and NASTEVA (1978)).

(1978)). The averaged results are presented in Figure 12 for three concentrations of NaCl. As it can be seen, at concentration 0.320 mole/l NaCl (curve 1), the angle θ^f diminishes with diminishing R_b which corresponds to $\varkappa^f = 7.6 \times 10^{-5}$ dyn; at concentration 0.360/mole/l (curve 2), θ^f is almost constant i.e. $\varkappa^f \simeq 0$ and, finally, at concentration 0.450 mole/l (curve 3) θ^f increases, i.e. the tension of the film perimeter is negative with $\varkappa^f$ -12.9×10^{-5} dyn.

The corresponding values of θ^f_∞, indicated by arrows, coincide very well with the values measured by DE FEIJTER (1973) using a different method with the same solutions. In Figure 12 the parts of the curves covered by the measurements are given with continuous lines and those parts not covered are dotted. The lower limit of R_b of the measured bubbles has been determined by the almost instant extinction of the very small bubbles on account of the high capillary pressure in them. It is interesting to mention that we managed to approach very closely on curve 1 up to the point of discontinuity, predicted in the case of a solid sphere at $\varkappa^f > 0$. The value of θ^f_m, corresponding to $d\theta^f/dR_b = \infty$, is obtained from equation (30) in the form of equation (12) with $\theta_\infty = \theta^f_\infty$, instead of equation (28) for a solid surface.

The second method, the method of the critical bubble (PLATIKANOV, NEDYALKOV and SCHELUDKO (1978)), has been applied for measuring $\varkappa^f$ at concentrations of NaCl lower than 0.360 mol/l, where $\varkappa^f > 0$. Obviously, for $\varkappa^f < 0$ formation of a NEWTON film must always take place, which has been observed for high NaCl concentrations, so that the application of the method of the critical bubble in this region is impossible.

By a special device, described in the article by PLATIKANOV, NEDYALKOV and SCHELUDKO (1978), hydrogen bubbles of different size, including very small ones, have been produced by electrolysis and by squeezing. Starting from a certain size NEWTON films are formed on the protruding surface of the bubble after its emergence. This phenomenon has been very distinctly observed as an instant expansion of a black spot on the bubble and the disappearance of NEWTONian interference fringes which before that (as well as on smaller bubbles) framed the dark central part of the bubble dome. The frequency of formation of visible NEWTON films increased when R_b increased over R_c, and on sufficiently big bubbles visible films always formed. Taking into account that according to the considerations presented in section II the state with an existing

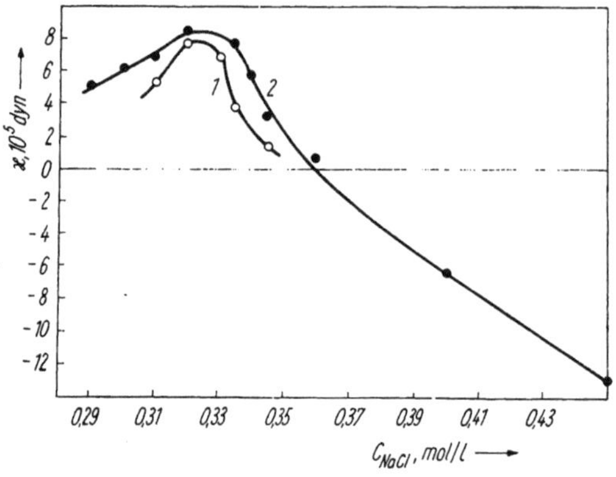

Fig. 13: Plot of NEWTON film line tension $\varkappa^f$ as a function of NaCl concentration: curve 1 is obtained by the critical bubble method (PLATIKANOV, NEDYALKOV and SCHELUDKO (1978)), curve 2 is obtained by the diminishing bubble method (PLATIKANOV, NEDYALKOV and NASTEVA (1978)).

submicroscopic contact is stable at $R_b \leqq R_c$, R_c has been determined from the beginning of the curve of frequency of film appearance versus bubble size. With the values of R_b found by extrapolation from the dependence of probability on R_c, (i.e. again statistically) $\varkappa^f$ was calculated by means of equation (33) for each concentration.

All values of $\varkappa^f$ obtained as a function of NaCl concentration are shown in Figure 13, where curve 2 corresponds to the method of the diminishing bubble (PLATIKANOV, NEDYALKOV and NASTEVA (1978)) and curve 1 to the method of the critical bubble (PLATIKANOV, NEDYALKOV and SCHELUDKO (1978)). As it can be seen, the region of $\varkappa^f > 0$ overlapped by both methods yields close values for $\varkappa^f$ vs. NaCl concentration (C_{NaCl}). Since in both cases the values of $\varkappa^f$ are obtained statistically from a large number of single measurements, the systematically lower values of $\varkappa^f$ obtained by the critical bubble method can hardly be attributed to the insufficient accuracy of the measurement of R_b and r with the first method and of $R_b = R_c$ with the second. The reason lies in the fact that the conditions of each type of experiment were at variance in different aspects with those in the model by which they were interpreted. It was assumed in the model that the bubble tension equals the tension of the plane liquid surface. Actually, the newly formed bubble which rises immediately up to the surface (the critical bubble method) can hardly manage to lower, by adsorption, the surface tension to its equilibrium value at the moment of attachment to the solution surface. On the contrary, in the diminishing bubble method when the bubbles dimensions sharply diminish to values where θ^f begins notably to deviate from θ^f_∞ and when the rate of reduction of the surface is considerable, it is possible that the density of the adsorption layer on the bubbles is higher than the equilibrium value, and the tension is correspondingly lower. If these considerations will be confirmed later, the correct values of $\varkappa^f$ must lie between the two curves, i.e. the accuracy of determination of $\varkappa^f$ is about 10^{-5} dyn.

The order of magnitude of the obtained values of the line tension as well as the fact that it changes its sign under the influence of increasing electrolyte concentration is unquestionable. The latter contradicts the conclusion of DE FEIJTER and VRIJ (1972) (derived on the basis of theoretical calculations) that the film must always possess negative line tension along the line with the bulk liquid.

In conclusion, returning to the initial problem of this review, the problem of a small phase, a remark may be added. From the point of view of formal thermodynamics the introduction of line tension (and of respective excess quantities) recommended by GIBBS (p. 288) solves the problem of the description of a three phase contact in the case of sufficiently large phases. In the molecular statistical treatment of this case, however, the problem is much more complicated than in the case of a two phase system. In the latter case (e.g. a liquid drop in a vapour phase) the parameters for bulk phases (e.g. surface tension) can be used without much loss of accuracy, taking into account only the closest neighbours. In the case of line tension, particularly for large angles of contact θ_∞, the distances in the regions where three surfaces meet without formation of a film of noticeable thickness (or NEWTON film) approach molecular dimensions. Besides, it becomes obligatory to take into account the interactions between the molecules of all three phases in the zone. Thus the effect of a type of disjoining pressure in a thin film is powerfully felt here. The simple reduction of the problem, however, to the problem of the thin film is hardly admissible when the distances between the surfaces of all three phases in the zone of their contact are of the same order of magnitude.

The methods of measurement of line tension described above are universal by nature and, at present, attempts are made to apply them to other systems, viz. a drop or bubble at a solid surface or the interface of two different fluids.

One may hope that similar investigations will enable line tension to be used in describing the behaviour of microheterogeneous dispersion systems and allow us to approach more closely a complete molecular interpretation of this parameter.

References

N. K. ADAM and J. B. HARDING, Proc. Roy. Soc. A **138**, 1932, 419
C. BOUHET, Ann. Physique **15**, 1931, 1102
J. H. CLINT, J. S. CLUNIE, J. R. TATE and J. F. GOODMAN, Nature **223**, 1969, 291
B. V. DERYAGUIN, Kolloid. Zh. **17**, 1955, 207
J. A. DE FEIJTER, Thesis, Univ. Utrecht, 1973
J. A. DE FEIJTER and A. VRIJ, J. Electroanal. Chem. **37**, 1972, 9
CH. FIEBER and H. SONNTAG, Colloid Polymer Sci. **253**, 1975, 32
J. W. GIBBS, 1906, "*The Scientific Papers of J. Willard Gibbs*", vol. 1: "*Thermodynamics*", New York: Longmans, Green, and Company; 1961, New York: Dover
R. D. GRETZ, Surface Sci. **5**, 1966, 239
C. HUH and S. G. MASON, Canadian J. Chem. **54**, 1976, 969
F. HUISMAN and K. J. MYSELS, J. Phys. Chem. **73**, 1969, 489
T. KOLAROV, A. SCHELUDKO and D. EXEROWA, Trans. Faraday Soc. **64**, 1968, 2864
L. KRASTANOV, Meteorologia i gidrologia **12**, 1957, 16
J. MINGINS and A. SCHELUDKO, 1978, J. Chem. Soc. Faraday I, in press
D. PLATIKANOV and M. NEDYALKOV, Ann. Univ. Sofia, Fac. Chimie **64**, 1969/1970, 353
D. PLATIKANOV, M. NEDYALKOV and V. NASTEVA, 1978 (to be published)
D. PLATIKANOV, M. NEDYALKOV and A. SCHELUDKO, 1978 (to be published)
H. M. PRINCEN, J. Phys. Chem. **72**, 1968, 3342
A. PRINS, J. Colloid Interface Sci. **29**, 1969, 177
A. SCHELUDKO and A. NIKOLOV, Colloid Polymer Sci. **253**, 1975, 396

A. Scheludko, B. V. Toshev and D. T. Bojadjiev, J.C.S. Faraday I **72**, 1976, 2815
J. N. Stranski and R. Kaischew, Z. phys. Chem. **B26**, 1934, 100
S. Torza and S. G. Mason, Kolloid-Z. u. Z. Polymere **246**, 1971, 593
B. V. Toshev, Ann. Univ. Sofia, Fac. Chimie **69**, 1974/1975, 25
B. V. Toshev and J. C. Eriksson, Ann. Univ. Sofia, Fac. Chimie **70**, 1975/1976, 75
M. Volmer, "*Kinetik der Phasenbildung*", Dresden: Th. Steinkopff, 1939
M. Volmer and H. Flood, Z. phys. Chem. **A190**, 1934, 273
H. Zocher and F. Stiebel, Z. phys. Chem. **A147**, 1930, 401
V. S. Wesselovskii and V. N. Pertzov, Zhur. Fiz. Khim. **8**, 1936, 245

Thermodynamic stability of charged surfaces

A. Steinchen and A. Sanfeld

(Chimie Physique II, Université Libre de Bruxelles, Bruxelles/Belgium)

Abstract

A purely thermodynamical stability analysis of charged liquid-liquid interfaces is developed, on the basis of the Glansdorff-Prigogine theory. Destabilizing terms accounting or experimentally observed surface instabilities appear clearly.

1. Introduction

During the last ten years, there has been an extraordinary impetus in the study of nonlinear nonequilibrium phenomena. Prigogine was the first to give a thermodynamic framework to the completely new situations observed in far from equilibrium conditions. He introduced the concept of "dissipative structures", i.e. the new ordering of matter, maintained by external fluxes, occurig after an instability threshold. In 1971, Glansdorff and Prigogine published their famous monograph [1] devoted to the thermodynamic theory of nonequilibrium continuous systems in the entire range of macroscopic description, starting from equilibrium and including nonlinear situations and instabilities. This monograph contains the thermodynamic criterion for the occurrence of dissipative structures, to which we will refer here. More recently, Nicolis and Prigogine [2] showed the important role played by the fluctuations near the bifurcation points. This stochastic approach stressed the "nucleation" process of the dissipative structures, involving fluctuations beyond a critical size. The meaning of the concept of "order through the fluctuations" is there clearly emphasized.

The Glansdorff and Prigogine thermodynamic stability criteria were extended a few years ago by the authors of the present paper to continuous charged and polarized systems [3]. Later on, Takeyama, Kitahara [4] and Matsushita [5] applied the same approach to analyze some examples of instabilities in the electric field.

With regard to the nonequilibrium thermodynamics of capillary systems, the domain of investigations was only recently extended beyond the linear region [6]. Previously Defay and coworkers [7], [8] devoted a great amount of work to the "dynamical surface tension" and showed the difficulties arising when treating surface phenomena out of equilibrium due to the non-autonomy of the surface in the Gibbs surface model. Recently, Bedeaux and al. [9], Kovac [10], and Kehlen and Baranowski [11] gave the entropy production for interfacial systems. Haase [12] calculated the entropy production for multiphase electrochemical systems in the linear region of phenomenological laws.

For nonlinear nonequilibrium capillary systems, the authors of the present paper obtained the excess balance of surface entropy [13] and they stressed the new types of instabilities that can be observed in interfaces. We will give here an extension of the same thermodynamic stability analysis for charged capillary systems in far from equilibrium conditions. We will show how electrochemical constraints applied to an interface are able to induce motion and deformations of the interface. This thermodynamic approach is complementary to the kinetic stability analysis developed a few years ago by our group to determine the threshold conditions for surface instabilities due to matter transfer [14, 15], to surface reactions [16, 17, 18], and to electrical constraints [19, 20].

2. Thermodynamics of irreversible processes and thermodynamic stability analysis for continuous systems

The basic assumption of irreversible thermodynamics for continuous systems is that the local entropy of the system out of equilibrium depends on the same local macroscopic variables as at equilibrium. The above assumption, often called "local equilibrium" assumption, reads

$$\delta s = T^{-1}\,\delta u + pT^{-1}\,\delta v - \sum_{\gamma} \mu_\gamma T^{-1}\,\delta x_\gamma\,, \tag{1}$$

s is the entropy per unit mass
u is the internal energy per unit mass
v is the volume of the mass unit
x_γ is the mass fraction of constituent
μ_γ is the chemical potential of constituent
p is the hydrostatic pressure

Any finite variation of entropy around a reference state can be developed in a series expansion of the type

$$\Delta s = \delta s + \tfrac{1}{2}\,\delta^2 s + \text{(higher order)}. \tag{2}$$

By an extension of the DUHEM stability criterion of equilibrium for small perturbations, GLANSDORFF and PRIGOGINE [1], showed that the stability criterion of equilibrium was related to the negative sign of the second derivative of entropy:

$$\delta^2 s = \delta T^{-1}\,\delta u + \delta(pT^{-1})\,\delta v - \sum_{\gamma} \delta(\mu_\gamma T^{-1})\,\delta x_\gamma \leqq 0 \tag{3}$$

(equilibrium stability condition).

Moreover, they wrote the above quantity as the quadratic form

$$\delta^2 s = -\frac{\varrho}{T}\left[\frac{C_v}{T}\,(\delta T)^2 + \frac{\varrho}{\chi}\,(\delta v)^2_{x_\gamma} + \sum_{\gamma\gamma'}\left(\frac{\partial \mu_\gamma}{\partial x_{\gamma'}}\right)\delta x_\gamma\,\delta x_{\gamma'}\right], \tag{4}$$

ϱ is the density
C_v the heat capacity at constant volume
χ the isothermal compressibility

This quadratic form permits us to obtain three separate stability conditions for equilibrium that read:

$$C_v > 0 \quad \text{thermal stability}\,, \tag{5}$$

$$\chi > 0 \quad \text{mechanical stability}\,, \tag{6}$$

$$\Sigma \left(\frac{\partial \mu_\gamma}{\partial x_{\gamma'}}\right) \delta x_\gamma\, \delta x_{\gamma'} > 0 \quad \text{stability with respect to diffusion}\,. \tag{7}$$

For polarized continuous systems, we have already shown a few years ago [3] that the equilibrium stability criterion can usually be read

$$\delta^2 s = -\frac{\varrho}{T}\left[\frac{C_v}{T}(\delta T)^2 + \frac{\varrho}{\chi}(\delta v)^2_{x_\gamma} + \Sigma\left(\frac{\partial \mu_\gamma}{\partial x_{\gamma'}}\right)\delta x_\gamma\, \delta x_{\gamma'} + \frac{\varrho}{k}(\delta \vec{P})^2\right] \leqq 0\,, \tag{8}$$

where $\vec{P}$ is the polarization vector per unit mass and $k = \frac{\varepsilon - 1}{4\pi}$ the electric susceptibility. The dipolar stability condition thus reads

$$\varepsilon \geqq 1\,. \tag{9}$$

Glansdorff and Prigogine have extended the thermodynamic theory of stability to nonequilibrium conditions. They have assumed for systems far from phase transitions, in the complete range where a macroscopic description is possible and where the basic hypothesis of local equilibrium is valid that the inequalities (5), (6), (7) and (8) are fulfilled. They considered then the negative quantity $\delta^2 s$ as a Lyapounov function. If macroscopic perturbations of the state variables arise around a nonequilibrium state corresponding to the above conditions, they give rise to the negative quantity $\delta^2 s$. The condition for the system to be stable, in the sense of Lyapounov, is that the time variation of that quantity should be positive i.e. that the stability condition for the nonequilibrium system reads

$$\partial_t \delta^2 s \geqq 0\,. \tag{10}$$

The choice of $\delta^2 s$ as Lyapounov function instead of any other quadratic function finds its justification in its physical significance in terms of fluctuations in the Einstein theory [1].

To take into account convective processes in continuous media, Glansdorff and Prigogine introduced a new Lyapounov function:

$$\delta^2 z = \delta^2 \left(s - T_0^{-1}\frac{v^2}{2}\right) \leqq 0\,, \tag{11}$$

where v is the barycentric velocity modulus, and they derived the explicit form of the generalized balance of excess entropy. The stability condition obtained then reads

$$\begin{aligned} \tfrac{1}{2}\,\partial_t \delta^2 \varrho z = &- \{\delta J_j\, \delta T^{-1} - \Sigma\, \delta(\varrho_\gamma \Delta_{\gamma j})\, \delta(\mu_\gamma T^{-1}) \\ &- T^{-1}(\delta \hat{P}^{ij}\, \delta v^i + \tfrac{1}{2}\varrho v_j (\delta v)^2 - v_j\, \delta^2(\varrho s))\}_{,j} \quad \text{[flux terms]}\,, \\ &+ \sum_k \delta J^k\, \delta X^k \quad \delta\ \text{Fluxes} \times \delta\ \text{Forces [excess entropy production]}\,, \end{aligned}$$

$$
\begin{aligned}
&\text{BÉNARD effect} \quad && - [\delta(\varrho u)\,\delta v_j + \delta \hat{P}^{ij}\,\delta V_i + \tfrac{1}{2}\,\varrho v_j (\delta v)^2]\,T^{-1}_{,j} \\
& && - \Sigma\,\delta\varrho_\gamma\,\delta v_j [F_{\gamma j} T^{-1} - (\mu_\gamma T^{-1})_{,j}]\,, \\
&\text{convective terms} && - [\delta \hat{P}^{ij}\,\delta T^{-1} - T^{-1}\,\delta(\varrho v_j)\,\delta v_i]\,v_{i,j} \\
& && + \tfrac{1}{2}\,T^{-1}(\delta v)^2\,(\varrho v_j)_{,j} + v_j^0[\delta(\varrho u)\,\delta T^{-1}_{,j} - \Sigma\,\delta\varrho_\gamma\,\delta(\mu_\gamma T^{-1})_{,j} \geqq 0\,, \quad (12)
\end{aligned}
$$

where $\hat{P}^{ij}$ is the pressure tensor, $F_{\gamma j}$ the external force acting on component γ, J_j the heat flux, $\Delta_{\gamma j}$ the diffusion velocity of component γ, and the superscript 0 refers to the reference stationary state.

The second term of the r.h.s. of (12) is the excess entropy production related to dissipative effects (heat conduction, diffusion, viscous dissipation and chemical reactions). The subsequent terms concern the dissipative phenomena together with mechanical effects. If the relations between the fluxes J_k and the forces X_k are non-linear, for example in chemical autocatalytic processes, the term $\delta J_k\,\delta X_k$ can become $\leqq 0$ and may lead to a violation of the inequality (12), giving rise to a "chemical instability". The system may leave the so-called thermodynamic branch to follow another branch, which may correspond to a quite different structure. The terms in $T^{-1}_{,j}$ and $F_{\gamma j}T^{-1} - (\mu_\gamma T^{-1})_{,j}$ account for the onset of convection in multicomponent systems due to a temperature gradient. If these terms become < 0, they may lead to an instability for a reference state at rest, that is the onset of convection cells (the wellknown BÉNARD effect [21]). The convection cells observed after a critical threshold of the temperature gradient are stationary patterns corresponding to a self-organization of the system maintained by a thermal flux.

For charged and polarized continuous systems undergoing convection, we have derived in a previous paper [3] the explicit form of the nonequilibrium stability condition:

$$
\begin{aligned}
\tfrac{1}{2}\,\partial_t \delta^2(\varrho\zeta) = - \Big\{ & \delta J_j\,\delta T^{-1} - T^{-1}\,\delta\hat{P}^{ij}\,\delta v^i + T^{-1}\vec{P}^i\,\delta\vec{E}^i\,\delta v^j - T^{-1}\vec{M}^i\,\delta\vec{B}^i\,\delta v^j \\
& - \sum_\gamma \delta(\mu_\gamma T^{-1})\,\delta(\varrho_\gamma \Delta_\gamma^j) - T^{-1}\,\frac{\varrho v_j}{2}\,(\delta v)^2 - v^j\,\delta T^{-1}\Big[\delta(\vec{E}^i\vec{P}^i) \\
& + \delta\vec{B}^i\vec{M}^i) - \frac{cT_0^{-1}}{2\pi}\,(\delta\vec{E}\times\delta\vec{H})^j\Big] \\
& + v_j[\delta T^{-1}\,\delta(\varrho u) - \Sigma\,\delta(\mu_\gamma T^{-1})\,\delta\varrho_\gamma]\Big\}_{,j} + \sigma(\delta Z) \geqq 0\,, \quad (13)
\end{aligned}
$$

where Z is the global generalized entropy including the convective velocity and where the LYAPOUNOV function is

$$
\delta^2(\varrho\zeta) = \delta^2(\varrho z) - 4T_0^{-1}\left\{\frac{(\delta\vec{E})^2}{8\pi} + \frac{(\delta\vec{B})^2}{8\pi} - \delta\vec{B}\,\delta\vec{M}\right\} \quad (14)
$$

$\vec{E}$ the electric field
$\vec{B}$ the magnetic field
$\vec{H}$ the magnetic induction
$\vec{M}$ the magnetization
$\vec{P}$ the electric polarization

The generalized excess entropy production reads

$$\sigma(\delta Z) = \sum_\alpha \delta J^\alpha \, \delta X^\alpha - T_{0,j}^{-1} \Big[\delta v^j \, \delta(\varrho u - \vec{E}^i \vec{P}^i - \vec{B}^i \vec{M}^i) + \frac{c}{2\pi} (\delta \vec{E} \times \delta \vec{H})^j$$
$$+ \delta \hat{P}^{ij} \, \delta v^i + \frac{1}{2} \varrho v_j (\delta v)^2 \Big]$$
$$- \sum \delta \varrho_\gamma \, \delta v^j [g^j T^{-1} + z_\gamma \vec{E}^j T^{-1} - (\mu_\gamma T^{-1})_{,j}]$$
$$- v^i_{,j} \Big[\delta \hat{P}^{ij} \, \delta T^{-1} - T^{-1} \, \delta(\varrho v^j) \, \delta v^i + \frac{1}{2} T^{-1} (\delta v)^2 \, (\varrho v^j)_{,j} \Big]$$
$$+ v^j [\delta T_{,j}^{-1} \, \delta(\varrho u - \vec{E}^i \vec{P}^i - \vec{B}^i \vec{M}^i)$$
$$+ \sum \delta \varrho_\gamma \, \delta(z_\gamma \vec{E}^j T^{-1}) - \sum \delta \varrho_\gamma \, \delta(\mu_\gamma T^{-1})_{,j}$$
$$- v^j \, \delta T^{-1} [\delta(\varrho z \vec{E}^j) + \delta(\vec{E}^i_{,j} \vec{P}^i) + \delta(\vec{B}^i_{,j} \vec{M}^i)] \qquad (15)$$

with the j component of the gravity vector and with

$$\sum \delta J^\alpha \, \delta X^\alpha = \delta J^j \, \delta T_{,j}^{-1} + \sum \delta w_\varrho \, \delta(A_\varrho T^{-1})$$
$$+ \sum \delta(\varrho_\gamma \Delta^j_\gamma) \, \delta(\tilde{A}^j_\gamma T^{-1}) - \delta(\hat{P}^{ij} T^{-1}) \, \delta v^i_{,j}$$
$$+ \delta(\vec{I}^j_T T^{-1}) \, \delta \vec{E}^j + \delta(\partial_\gamma \vec{M}^j T^{-1}) \, \delta \vec{B}^j \, , \qquad (16)$$

where w_ϱ and A_ϱ are respectively the chemical rate and the affinity of reaction ϱ, $\tilde{A}^j$ is the total affinity including gravity effects (g^j), electrochemical diffusion $(-T(\mu_\gamma T^{-1})_{,j} + z_\gamma \vec{E}^j)$ and magnetic contributions related to the time derivative of the vector potential $\left(-\frac{z}{\partial} \frac{cA^j}{\partial t}\right)$ and where $\vec{I}_T$ is the total electric current including diffusion-migration current, convection current and displacement current. The additional electric term $\delta(\vec{I}^j_T T^{-1}) \, \delta \vec{E}^j$ in the excess entropy production may have a destabilizing contribution for systems exhibiting a negative resistance as has already been stressed by many electrochemists [3, 22, 23].

3. Equilibrium stability of a plane charged surface, in a Gibbs model

The electrochemical system considered here contains all the relevant charged species (i.e. it is not subjected to an electrical field of external origin). We assume that the Helmholtz electrochemical free energy $\tilde{F}$ is a function

$$\tilde{F} = \tilde{F}(T, V', V'', A, n'_\gamma, n''_\gamma, n^\sigma_\gamma) \qquad (17)$$

of the volumes of the bulk phases V' and V'', of the area A of the surface and of the number of moles of all the components γ in the bulk phases λ' and λ'', and in the surface $^\sigma$.

As it was done classically by Defay et al. [7] the variation of the surface entropy perunit area reads, at equilibrium

$$\delta\tilde{s}^{\sigma} = T^{-1}\,\delta\tilde{u}^{\sigma} - \Sigma\,\tilde{\mu}_{\gamma}T^{-1}\,\delta\Gamma_{\gamma} \tag{18}$$

with $\tilde{\mu}_{\gamma}$ the electrochemical potential of γ in both phases and in the surface ($\tilde{\mu}_{\gamma}' = \tilde{\mu}_{\gamma}'' = \tilde{\mu}_{\gamma}^{\sigma}$ at equilibrium), and $\tilde{u}^{\sigma}$ the total surface energy but unit area [24]

$$\tilde{u}^{\sigma} = u^{\sigma} + \frac{1}{A}\int\limits_{V}\frac{\vec{E}^2}{8\pi}\,\delta V \tag{19}$$

with $\vec{E}$ the electrical field created by the charge distribution in the system.

In the set of variables $\tilde{u}^{\sigma}$ and Γ_{γ}, the second derivative of the surface entropy reads

$$\delta^2\tilde{s} = \delta T^{-1}\,\delta\tilde{u}^{\sigma} - \Sigma\;(\tilde{\mu}_{\gamma}T^{-1})\,\Gamma_{\gamma} \tag{20}$$

or in the quadratic form

$$\delta^2\tilde{s}^{\sigma} = -\frac{1}{T}\left[\frac{c^{\sigma}}{T}\,\delta(T)^2 + \sum_{\gamma\gamma'}\left(\frac{\partial\tilde{\mu}_{\gamma}}{\partial\Gamma_{\gamma'}}\right)\delta\Gamma_{\gamma}\,\delta\Gamma_{\gamma'}\right] \tag{21}$$

where

$$c^{\sigma} = \frac{\partial\tilde{u}^{\sigma}}{\partial T} = T\,\frac{\partial\tilde{s}^{\sigma}}{\partial T} \tag{22}$$

is the surface heat capacity per unit area.

The same relationship can be obtained for the second derivative of the surface electrochemical entropy per unit mass, $\tilde{s}_m$. The equilibrium stability condition then reads

$$\delta^2\tilde{s}_m^{\sigma} = -\frac{1}{T}\left[\frac{c_m^{\sigma}}{T}\,(\delta T)^2 + \frac{\Gamma}{E}\,(\delta\omega)^2_{x_{\gamma}} + \sum_{\gamma\gamma'}\left(\frac{\partial\tilde{\mu}_{\gamma}}{\partial x_{\gamma'}}\right)\delta\chi_{\gamma}^{\sigma}\,\delta\chi_{\gamma'}^{\sigma}\right] \leq 0\,. \tag{23}$$

c_m^{σ} is the surface heat capacity per unit mass, > 0

$E = A\,\dfrac{\partial\sigma}{\partial A}$ the Gibbs elasticity, >0

χ_{γ} the surface mass fraction of component γ

Γ the total surface density

ω the area per unit mass

At equilibrium, and far from phase transitions the r.h.s. of eq. (23) is a negative definite quantity.[1] The G.P. criterion for equilibrium stability can be extended to multiphase systems. The Lyapounov function describing the stability of the total system then reads as a volume integral of the second derivatives of the entropy densities over the volumes of the bulk phases in equilibrium plus a surface integral of the second derivatives of the surface entropy densities over the area of all the interfaces of the system. Since the volumes of the bulk phases can be reduced to zero for non vanishing area of the interface, the equilibrium stability of the interface will be described by the < 0 sign of $\delta^2\tilde{s}^{\sigma}$ or of $\delta^2\tilde{s}_m^{\sigma}$ independently of the sign of the second derivatives of entropy in bulk.

[1]) To assign a negative sign to $\delta^2 s_m^{\sigma}$, the dividing surface has to be chosen in such a way that $\Gamma \geqq 0$.

4. Nonequilibrium stability of a plane charged surface

As we have seen in Section 3 the LYAPOUNOV function $\delta^2\tilde{s}^\sigma$ is a 2d order function in the variables $\delta\tilde{u}^\sigma$ and $\delta\Gamma_\gamma$. We will assume that out of equilibrium, (20) still remains valid. In fact, when adsorption is not in equilibrium, DEFAY et al. [6], [7], [8] showed that $\tilde{s}^\sigma$ depends not only on the local properties of the interface but also on the properties of the neighbouring phases. We will, however, write (20) even out of equilibrium, and we will control the system in such a way that the concentrations c'_γ and c''_γ in the two neighbouring bulk phases will be fixed. The difference between these concentrations and the equilibrium concentrations is a chemical constraint imposed open the system.

Let us now write down the excess entropy balance: by application of the EULER theorem we have

$$\tfrac{1}{2}\,\delta_t\delta^2\tilde{s}^\sigma = \delta(T^{-1})^\sigma\,\partial_t\delta\tilde{u}^\sigma - \sum_\gamma \delta(\tilde{\mu}_\gamma T^{-1})^\sigma\,\partial_t\delta\Gamma_\gamma\,, \tag{24}$$

where T and $\tilde{\mu}/T^{-1}$ are no longer uniform throughout the system, but are defined locally in the surface. If we take into account the onset of convection we have to introduce a new LYAPOUNOV function

$$\delta^2 z^\sigma = \delta^2\,(\tilde{s}^\sigma - \tfrac{1}{2}\,\Gamma T_0^{-1\sigma} v^{\sigma 2}) < 0 \tag{25}$$

and the stability criterion for nonequilibrium surfaces reads

$$\tfrac{1}{2}\,\partial_t\delta^2 z^\sigma = \delta T^{-1\sigma}\,\partial_t\delta\tilde{u}^\sigma - \sum_\gamma \delta(\tilde{\mu}^\sigma_\gamma T^{-1\sigma} - \tfrac{1}{2}\,T_0^{-1\sigma} v^{\sigma 2})\,\partial_t\delta\Gamma_\gamma$$
$$- T^{-1\sigma}\,\delta v^\sigma_i\,\partial_t\delta(T v^\sigma_i) \geqq 0\,, \tag{26}$$

where v^σ is the barycentric velocity taken at the surface.

The excess balances of surface momentum read

$$\Gamma\,\frac{dv^\sigma_z}{dt} = \Delta_s\delta P^{zz} + \sigma\,\delta C\,, \tag{27}$$

$$\Gamma\,\frac{dv^\sigma_\alpha}{dt} = \Delta_s\delta P^{\alpha z} + \delta\sigma^{\alpha\beta}_{,\beta}\,, \tag{28}$$

$\Delta/\delta P^{zz}$ is the discontinuity of the normal component z of the perturbation of the pressure tensor across the surface.

δC is the perturbation of the mean curvature of the surface, $\Delta_s\delta P^{\alpha z}$ is the discontinuity of the perturbation of the tangential viscous stress across the surface and $\delta\sigma^{\alpha\beta}_{,\beta}$ is the perturbation of the surface divergence of the surface tension tensor, including the surface viscosity, α and β are the coordinates in the deformed surface.

These momentum balances do not include any external force. In the system considered here, all the relevant charges of the system are distributed in the interfacial region and create the electrical potential difference between the two bulk phases. The electrical effects due to the potential difference through the surface are included in the surface tension [24]. The balance of excess surface kinetic energy may be deduced from the momentum balances and it reads

$$v^\sigma_z\,\partial_t\delta(\Gamma v^\sigma_z) = v^\sigma_z\Delta_s\delta P^{zz} + v^\sigma_z\,\delta(\sigma C)\,, \tag{29}$$

$$v^\sigma_\alpha\,\partial_t\delta(\Gamma v^\sigma_\alpha) = v^\sigma_\alpha\Delta_s\delta P^{\alpha z} + v^\sigma_\alpha\,\delta\sigma^{\alpha\beta}_{,\beta} \tag{30}$$

and the excess balance of surface internal energy is given by

$$\partial_t \delta \tilde{u}^\sigma = -\delta[j^\alpha + \tilde{u}^\sigma v^\alpha]_{,\alpha} - \Delta_s \delta J^z + \delta[\sigma^{\alpha\beta} v^\alpha_{,\beta}] , \tag{31}$$

where j^α is the surface heat flux and $\Delta_s \delta J^z$ the discontinuity of the perturbation of heat flux normal to the surface.

The excess balance of matter in the surface reads for each species

$$\partial_t \delta \Gamma_\gamma = -\delta[\Gamma_\gamma v^\alpha]_{,\alpha} - \delta[\Gamma_\gamma \Delta_{\gamma\alpha}]_{,\alpha} - \Delta_s[\delta(\Delta_{\gamma z} - v_z^\sigma)\,\varrho_\gamma] + \sum_\varrho \nu_\varrho M_\gamma \,\delta w_\varrho . \tag{32}$$

The first term of the r.h.s. is the two-dimensional divergence of the convective flow of γ in the surface, the second is the surface divergence of the tangential diffusion flow, the term $\Delta_s[\delta(\Delta_{\gamma z} - v_z^\sigma)\,\varrho_\gamma]$ is the discontinuity of the normal fluxes of γ from the two neighbouring bulk phases and the last term is the source term due to chemical reactions in the surface, M_γ is the molar weight of γ.

Combining equations (26) and (32) we obtain by classical, but tedious, transformations (see [1]) the explicit form of the nonequilibrium stability criterion:

$$\begin{aligned}
\tfrac{1}{2}\,\partial_t \delta^2 z^\sigma = & -[\delta T^{-1\sigma}\,\delta j^\alpha - \delta(\tilde{\mu}_\gamma T^{-1})_\sigma\,\delta(\Gamma_\gamma \Delta_{\gamma\alpha})_{,\alpha} \qquad \text{Flux terms} \\
& + \Delta_s[\delta T^{-1\sigma}\,\delta J^z + \delta T^{-1\sigma}\,u\,\delta v_z - \delta(\varrho_\gamma \Delta_{\gamma z})\,\delta(\mu_\gamma T^{-1})^\sigma] \\
\delta_{\text{Flux}} \cdot \delta_{\text{Force}} & \left\{ \begin{aligned}
& + \delta T^{-1}_{\sigma,\alpha}\,\delta j^\alpha - \sum_\gamma \delta(\tilde{\mu}^\sigma T^{-1\sigma})_{,\alpha}\,\delta(\Gamma_\gamma \Delta_{\gamma\alpha}) + \sum_\varrho \delta(A_\varrho^\sigma T_\sigma^{-1})\,\delta w_\varrho \\
& - \delta J^z_+\,\delta(T_+^{-1} - T_\sigma^{-1}) + \delta J^z_-\,\delta(T_-^{-1} - T_\sigma^{-1}) \\
& + \sum_\gamma \delta\varrho_\gamma(\Delta^+_{\gamma z} - v_\sigma^z)\,\delta[\tilde{\mu}_\gamma^+ T_+^{-1} - \tilde{\mu}_\gamma^\sigma T_\sigma^{-1}] \\
& - \sum_\gamma \delta\varrho_\gamma(\Delta^-_{\gamma z} - v_z^\sigma)\,\delta[\tilde{\mu}_\gamma^- T_-^{-1} - \tilde{\mu}_\gamma^\sigma T_\sigma^{-1}] \\
& + v^\alpha_{,\beta}\,\delta(\Pi^{\alpha\beta} T_\sigma^{-1}) .
\end{aligned} \right. \\
\text{LAPLACE effect} \quad & -T_\sigma^{-1}\,\delta v_z^\sigma\,\Delta_s \delta p^{zz} - T_\sigma^{-1}\,\delta v_z^\sigma\,\delta(\sigma C) , \\
\text{MARANGONI effect} \quad & -T_\sigma^{-1}\,\delta v^\alpha\,\Delta_s \delta P^{\alpha z} T_\sigma^{-1}\,\delta v^\alpha\,\delta\sigma^{\alpha\beta}_{,\beta} - T_\sigma^{-1}\,\delta v^\alpha_{,\beta}\,\delta\sigma^{\alpha\beta} , \\
\text{Surface BÉNARD effect} \quad & -v^\alpha\,\delta\tilde{u}^\sigma\,T^{-1}_{\sigma,\alpha} + \delta v^\alpha \sum \delta\Gamma_\gamma[\tilde{\mu}_\gamma^\sigma T_\sigma^{-1}]_{,\alpha} \geqq 0 .
\end{aligned} \tag{33}$$

Let us now analyze the various terms of (33) and see what the terms are which are able to destabilize the surface. The first bracket of the r.h.s. is the surface divergence of the excess flux term $\delta T^{-1}\,\delta\bar{j} - \delta(\tilde{\mu} T^{-1})^\sigma\,\delta(\Gamma_\gamma \bar{\Delta}_\gamma)$.

At the boundary of the surface (limited by a closed line) the fluctuations of fluxes $\delta\bar{j}$ and $\delta(\Gamma_\gamma \bar{\Delta}_\gamma)$ can be maintained equal to zero or the temperature and the electrochemical potential can be maintained constant.

The second term of the r.h.s. of (33) is the discontinuity of an excess of normal flux term from the bulk phases. By application of the phenomenological laws in both bulk phases for heat and matter diffusion, it can be shown that the sign of the terms $\Delta_s\,\delta T_\sigma^{-1}\,\delta J^z$ and $\Delta_s\,\delta(\varrho_\gamma \Delta_\gamma^z)\,\delta(\mu_\gamma T^{-1})_\sigma$ is always > 0 for fixed temperature and chemical potential in the bulk phases only for no slip boundary conditions.

As was already shown by BEDEAUX [9], for the surface entropy production, the excess entropy production includes an additional term due to the temperature slip boundary condition i.e.

$$-\delta J^z_+ \,\delta(T_+^{-1} - T_\sigma^{-1}) + \delta J^z_- \,\delta(T_-^{-1} - T_\sigma^{-1}) \,,$$

where J^z_+ and J^z_- are the normal heat fluxes on both sides of the interface, T_+ and T_- the temperatures on both sides of the interface, and T_σ the temperature in the surface. The analog of these terms exists also for the diffusion of matter when there is a slip boundary condition for the electrochemical potentials, i.e.

$$\sum_\gamma \delta\varrho_\gamma(\Delta^+_{\gamma z} - v^\sigma_z)\,\delta[\tilde{\mu}^+_\gamma T_+^{-1} - \tilde{\mu}^\sigma_\gamma T_\sigma^{-1}] - \sum_\gamma \delta[\varrho_\gamma(\Delta^-_{\gamma z} - v^\sigma_z)]\,\delta[\tilde{\mu}^-_\gamma T_-^{-1} - \tilde{\mu}^\sigma_\gamma T_\sigma^{-1}]\,.$$

These terms may have a destabilizing effect if the electroadsorption-desorption processes are nonlinear [19], [25]. It is assumed here that there is a no slip boundary condition for the barycentric velocity i.e.

$$v_z^+|_{z=0} = v_z^-|_{z=0} = v_z^\sigma\,.$$

The terms $\delta T^{-1}_{\sigma,\alpha}\,\delta j^\alpha$ and $-\sum_\gamma \delta(\tilde{\mu}^\sigma T_\sigma^{-1})_{,\alpha}\,\delta(\Gamma_\gamma \Delta_{\gamma\alpha})$ are the classical two dimensional excess entropy sources for thermal diffusion and diffusion-migration. They are always > 0. The term $\sum_\varrho \delta(A^\sigma_\varrho T_\sigma^{-1})\,\delta w_\varrho$ is the excess surface entropy production term due to chemical reactions, it has a destabilizing contribution for nonlinear chemical reactions occurring at the interface [13, 16, 17, 18, 25, 26]. The term $v^\alpha_{,\beta}\delta(\Pi^{\alpha\beta}T_\sigma^{-1})$ is the viscous dissipation in the surface [9, 15, 27, 28], always > 8. The terms

$$-v^\alpha\,\delta\tilde{u}^\sigma T^{-1}_{\sigma,\alpha} + \delta v^\alpha \sum_\gamma \delta T_\gamma[\tilde{\mu}^\sigma_\gamma T_\sigma^{-1}]_{,\alpha}$$

are the twodimensional analogs of the extended BÉNARD effect for multicomponent systems. The terms arising from the LAPLACE and MARANGONI effect may also exhibit destabilizing properties [17, 25]. The coupling of the chemical composition of the surface, the surface thermal flux and the surface convection through the term $-T_\sigma^{-1}\,\delta v^\alpha\,\delta\sigma_{,\alpha}$ together with a state equation $\sigma(T, \Gamma_\gamma ..)$ gives rise to nonlinearities and thus to destabilizing terms. The dependence of the surface tension σ on the electric field through the intertace may also have a destabilizing effect [19, 24].

Conclusion

An extension of the GLANSDORFF and PRIGOGINE thermodynamic stability criteria to charged surfaces separating two immiscible fluids is given. The surface stability for a GIBBS surface, out of equilibrium, is characterized by the positive sign of the excess surface entropy production. The explicit form of this quantity reveals new destabilizing terms accounting for a surface type of the BÉNARD effect, for RAYLEIGH-TAYLOR and MARANGONI effects, and destabilizing terms due to nonlinear electrochemical surface reactions. The negative sign of the excess surface entropy production gives a sufficient condition for the occurrence of surface dissipative structures.

References

[1] P. Glansdorff and I. Prigogine, "*Thermodynamic Theory of Structure, Stability and Fluctuations*", Wiley-Interscience, New York 1971
[2] G. Nicolis and I. Prigogine, "*Self-organization in Nonequilibrium Systems*", Wiley-Interscience 1977
[3] A. Sanfeld and A. Steinchen-Sanfeld, Bull. Ac. Roy. Belg. Cl. Sc. 1971, **57**, (684)
[4] K. Takeyama and K. Kitahara, Journ. Phys. Soc. Japan 1975, **39** (125)
[5] M. Matsushita, Journ. Phys. Soc. Japan 1976, **41** (674)
[6] R. Defay, I. Prigogine and A. Sanfeld, Journ. Coll. Interf. Sci. 1977, **58** (498)
[7] R. Defay, I. Prigogine, A. Bellemans and D. H. Everett, "*Surface Tension and Adsorption*", Longmans Green, London 1966
[8] R. Defay, "*Sortir de l'équilibre*", Collection "*Discours de la Methode*" Gauthier-Villars, 1979
[9] D. Bedeaux, A. M. Albano and P. Mazur, Physica 1976, **82A** (438)
[10] J. Kovac, Physica **1977**, **86A** (1)
[11] H. Kehlen and B. Baranowski, J. Non-Equilib. Thermodyn. 1977, **2** (169)
[12] R. Haase, Z. Phys. Chem. Neue Folge 1977, **106**, (113)
[13] A. Steinchen and A. Sanfeld, in "*Colloid and Interface Science*", V. III, Ed. M. Kerker, Ac. Press. New York, 1976
[14] M. Hennenberg, T. S. Sørensen and A. Sanfeld, J. Chem. Soc. Faraday Trans. II, 1977, **73**, (48); ibid., J. Coll. Interf. Sci. 1977, **61**, (62)
[15] M. Hennenberg, P. M. Bisch, M. Vignes-Adler and A. Sanfeld, "Symposium on Dynamics and Instability of Fluid Interface", Technical University of Denmark, Lyngby, May 1978
[16] A. Sanfeld and A. Steinchen, Biophys. Chem. 1975, **3** (99)
[17] M. Hennenberg, T. S. Sørensen, A. Steinchen and A. Sanfeld, J. Chim. Phys., 1975, **72** (1202); ibid., J. Coll. Interf. Sci. 1976, **56**, (191)
[18] M. G. Velarde, J. L. Ibanez, T. S. Sørensen, A. Sanfeld and M. Hennenberg, Proc. Intern. Conference on Physical Chemistry and Hydrodynamics, Oxford, July 1977, Hemisphere Publ.
[19] P. M. Bisch and A. Sanfeld, in "*Lectures Notes in Physics*", Ed. J. C. Legros and J. Platten, Springer Verlag, 1978, **72**, Berlin
[20] D. Van Lamsweerde-Gallez, P. M. Bisch, A. Sanfeld, Bioelectrochem. and Bioenerg. 1978, **5**, (1)
[21] C. Normand, Y. Pomeau and M. G. Velarde, Rev. on Modern Physics 1977, **49**, (581)
[22] J. Wojtovicz in "*Modern Aspects of Electrochemistry*", V. 8, Plenum Press, New York 1972
[23] P. Poncet, M. Braizaz, B. Pointu et J. Rousseau, J. Chim. Phys. 1977, **74** (452); ibid., J. Chim. Phys. 1968, **75**, (287)
[24] A. Sanfeld, "*Thermodynamics of Charged and Polarized Layers*", Wiley. London 1968
[25] A. Sanfeld, A. Steinchen, M. Hennenberg, P. M. Bisch, D. Van Lamsweerde-Gallez and W. Dalle-Vedove, "Symposium on Dynamics and Instability of Fluid Interface", Technical University of Denmark, Lyngby, May 1978
[26] W. Dalle-Vedove, P. M. Bisch, A. Sanfeld and A. Steinchen, C.R.Acad. Sci. Paris 1978
[27] F. Goodrich, The theory of capillary excess viscosities, cf. this volume, p. 19
[28] H. C. Maru, V. Mohan and D. T. Wasan, submitted to Chem. Eng. Sci.
[29] J. W. Gardner, R. Schechter in "*Colloid and Interface Science*", Ed. M. Kerker, Ac. Press, 1976

Thermodynamics of bilayer lipid membranes

J. Ch. Eriksson

(Department of Physical Chemistry, The Royal Institute of Technology, Stockholm/Sweden)

Introduction

Lipid bilayers are widely used as models for biomembranes and have been studied extensively during recent years with a variety of experimental techniques (cf. e.g. Refs. [1, 2]). In particular, methods to measure the bilayer membrane tension have been devised. Thus interesting information concerning the thermodynamic properties of lipid bilayers can be reached. The purpose of the present contribution is to forward the development of a rigorous and complete thermodynamic formalism for bilayer lipid membranes.

Previous works related specifically to the thermodynamics of lipid bilayers include papers by Tien [3], Good [4], Evans and Simons [5], and Evans and Waugh [6]. Toshev and Ivanov have recently treated the closely related subject of thin liquid films between fluid phases [7, 8]. The famous (but still, in all details, not always well understood) Gibbs Theory of Capillarity [9] and earlier appears in the author's series on surface thermodynamics, in particular papers V [10], VI [11], VII [12], furnish an appropriate general basis from which we shall proceed.

Depending on such details in the experimental arrangement which determine the mass transfer possibilities and on the ambient temperature a bilayer lipid membrane is actually either a partially closed or an open thermodynamic system. As was indicated by Gibbs (loc. cit.) and has been shown explicitly for solid surfaces [10, 12, 13] and insoluble surface films [11], the thermodynamics of closed or partially closed surface phase systems has to be developed along lines that differ from those ordinarily followed for completely open surface phase systems. The resulting thermodynamic fundamental equations are also of a different nature. In essence this is due to the need to account explicitly for changes of state caused by surface strain variations.

As it will be demonstrated below, the (lateral) membrane strain may be taken as an independent state variable for a partially closed bilayer membrane system leading to a Shuttleworth type of fundamental equation in a manner similar to that for insoluble surface films. Alternatively, the Gibbs procedure can be followed more closely by introducing a strain-dependent chemical potential for the membrane-forming lipid in the bilayer resulting in a Gibbs kind of fundamental equation. Completely open bilayer membrane systems can be treated essentially along the conventional Gibbsian lines. Open bilayer membrane systems also offer certain advantages with respect to extracting the thermodynamic information wanted on membrane composition, energy and entropy etc. as will appear from the subsequent treatment.

Description of the bilayer lipid membrane system studied

For the most part we will consider a planar, symmetrical bilayer lipid membrane formed by one water-insoluble lipid component such as a lecithin (Component 2) immersed in a water (Component 1) solution having one solute component such as a salt, a surfactant or a protein (Component 3). Extension to a multicomponent water solution is trivial. Generalization to two or more membrane-forming components is likewise easy for an open bilayer membrane but will need a special study in the case of a partially closed bilayer membrane.

The temperature is always chosen above the gel-liquid crystal (CHAPMAN) transition temperature T_c implying that the lateral (but not the transversal) diffusion within the bilayer is comparatively rapid and in addition that all thermodynamic bilayer properties are isotropic in lateral directions. The pressure is generally supposed to be in the range of atmospheric pressures. Hence we can neglect contributions from pV-terms to the thermodynamic properties of the bilayer membrane system and there is no need to distinguish between the HELMHOLTZ and the GIBBS free energy, F and G, respectively. Indirectly this also means that questions on how mechanical equilibrium is realized at the borders of a bilayer lipid membrane are left out of consideration.

The precise limits of the thermodynamic bilayer membrane system towards the water solution can be conveniently defined by means of a pair of GIBBS dividing surfaces, denoted by $\Sigma_{(1)}$, positioned so as to make the excess of water (Component 1) for each face of the membrane vanish (Fig. 1). Implicitely, we thus make the reasonable assumption that the local water concentration in the central part of the membrane is comparatively low. The thickness of the membrane, $\tau_{(1)}^m$,is defined as the distance between the two $\Sigma_{(1)}$ dividing surfaces.

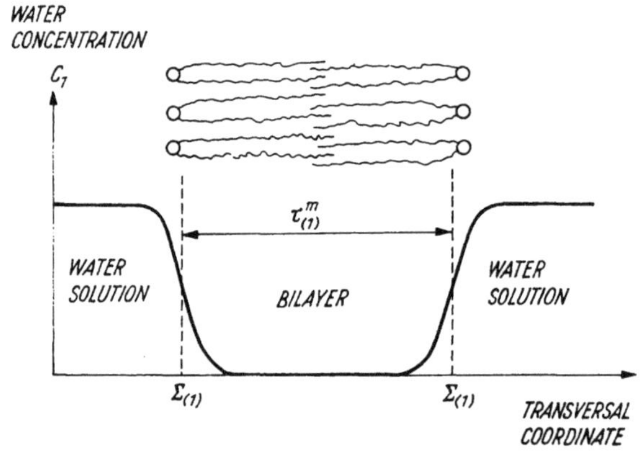

Fig. 1: Definition of the thickness, $\tau_{(1)}^m$, of the bilayer lipid membrane by means of two GIBBS dividing surfaces $\Sigma_{(1)}$. Component 1 is water.

We also imagine a local tension variation across the bilayer membrane in principal accordance with Fig. 2. Thus, when stretched, the bilayer is subject to a resulting tension, γ^m, that we call the membrane tension. For a symmetrical bilayer membrane, the surface of tension is evidently located midway between the $\Sigma_{(1)}$ dividing surfaces.

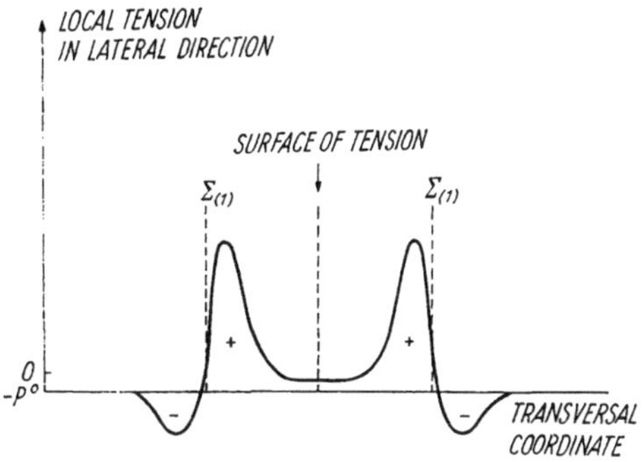

Fig. 2: Hypothetical local tension variation across a stretched bilayer lipid membrane. The area under the curve corresponds to the resultant membrane tension, γ^m. The ambient pressure is p_0.

The energy, $U^m_{(1)}$, entropy, $S^m_{(1)}$, free energy, $F^m_{(1)}$, and mole numbers $n^m_{2(\ )}$, $n^m_{3(1)}$ of the bilayer lipid membrane are evaluated using the ordinary Gibbs convention as indicated by the $_{(1)}$ subscript. Accordingly, the water phase is supposed to extend unchanged right up to the $\Sigma^{(1)}$ dividing surfaces. In view of the very slight water solubility of the membrane-forming lipid component it is evident that $n^m_{2(1)} = n^m_2$, or otherwise expressed, n^m_2 is insensitive for displacements of the two dividing surfaces.

In general the thermodynamic state of a bilayer lipid membrane as described above that is closed for the lipid component 2 is determined by three degrees of freedom. As independent variables of state we may take the temperature and the chemical potentials of the lipid component in the bilayer membrane, μ^m_2, and of the water solute, μ_3. Instead of μ^m_2, either the membrane area A^m, the membrane strain $\varepsilon^m = A^m/A^{mo}$ or the membrane tension γ^m may be taken as an independent variable.

Open and partially closed bilayer lipid membrane systems

As it was mentioned in the introduction, the formal thermodynamic treatment is dependent upon whether the bilayer membrane studied actually is a partially closed or an open system. We can further illustrate this question by referring to the conceptually attractive bilayer membrane preparation procedure devised by Tagaki, Azuma and Kishimoto (Fig. 3), [14].

During the formation of the lipid bilayer from the corresponding monolayer film it is evident that the bilayer is a completely open system. For such a membrane system

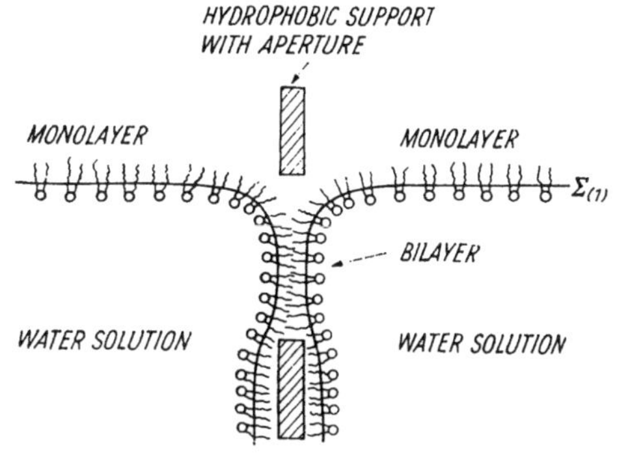

Fig. 3: Preparation of a bilayer lipid membrane from the corresponding monolayer according to the method of Tagaki et al. [14].

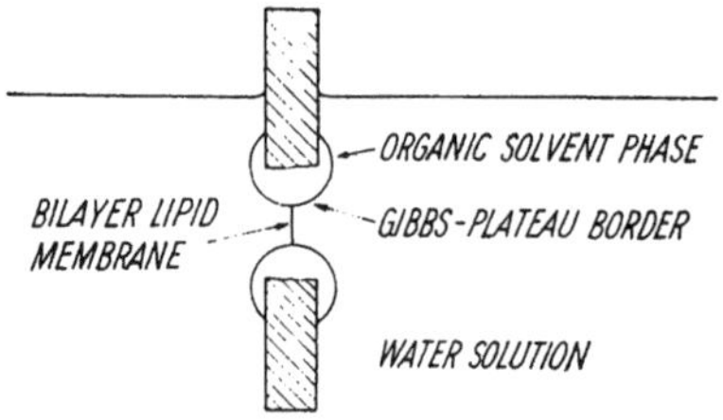

Fig. 4: The solvent method to form bilayer lipid membranes.

the chemical potential of the membrane-forming lipid in the bilayer membrane, μ_2^m, at equilibrium equals the chemical potential in the monolayer film, μ_2^f. Likewise, when the bilayer membrane is formed from a solution of the membrane-forming lipid in an organic solvent (Fig. 4) the bilayer is also an open system and, at equilibrium, μ_2^m, equals the chemical potential of the lipid component in the organic solvent drop. However, once removed from direct mass transfer contact with the monolayer film or the solvent drop, the bilayer lipid membrane is a closed system for the lipid component insofar as the water solubility of this component is low enough. Ordinarily, the intrinsic thermodynamic properties are independent of the membrane area A^m for open bilayer membrane systems whereas the reverse holds true for partially closed bilayer membrane systems.

Definitions

We recognize the fact that a bilayer lipid membrane may exist in a state of mechanical tension (cf. Ref. [1], p. 40). Hence across any plane perpendicular to the membrane there acts a resulting mechanical membrane tension, γ^m, that according to YOUNG's principle, should be attributed to the surface of tension when dealing with mechanical equilibrium problems. In full analogy with the GIBBS treatment of insoluble surface films (cf. Ref. 11) we can then utilize the subsequent relation between integral membrane quantities

$$F_{(1)}^m = \gamma^m A^m + \mu_2^m n_2^m + \mu_3 n_{3(1)}^m \tag{1}$$

to define the chemical potential μ_2^m of the water-insoluble lipid component in the partially closed bilayer lipid membrane. At equilibrium, the chemical potential of component 3 in the membrane is completely determined, of course, by the restriction-free contact with the surrounding water solution, i.e. $\mu_3^m = \mu_3$.

Alternatively, we may introduce a reference bulk state, denoted by superscript[1] for the lipid component. In this context an appropriate reference state is water-solution equilibrated liquid crystalline lipid (cf. Fig. 5) at the same temperature as the bilayer lipid membrane system. The reversible work per unit membrane area, σ^m, of forming the membrane in the water solution from the lipid in the reference state is then given by

$$F_{(1)}^m = \sigma^m A^m + \mu_2^* n_2^m + \mu_3 n_{3(1)}^m \,. \tag{2}$$

It is evident from this definition that σ^m [1] is invariant with respect to relocation of the dividing surfaces which separate the bilayer membrane system from the bulk

[1] In Refs. [10, 11, 12] the notation g_{ex} was used for σ in order to emphasize that this quantity is an excess free energy. In the present context we adhere to the GIBBS notation that is now more widely used (cf. Ref. [13]).

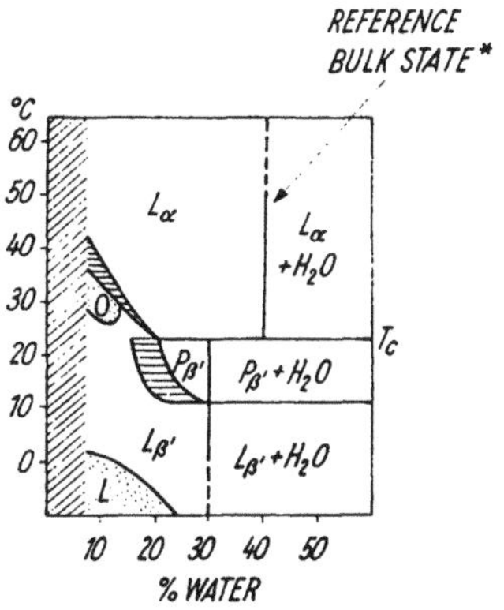

Fig. 5: Gross features of the phase diagram for dimyristoyl lecithin-water. (By courtesy of Dr. JANIAK [20]) where the reference bulk state is marked by* .

water solution. Thus when the dividing surfaces cut through homogeneous bulk water solution at some distance sufficiently remote from the membrane faces, Eq. (2) can be written, equally well, in the more illuminating form

$$\sigma^m A^m = F^m - \mu_1 n_1^m - \mu_2^* n_2^m - \mu_3 n_3^m \,, \tag{3}$$

where no excess properties are involved. From Eq. (3) the work of formation interpretation of σ^m is more readily apparent.

If $\sigma^m < 0$ spontaneous membrane formation would occur at contact between the lipid in the reference bulk state and the water solution or, otherwise expressed, the reference state would be unstable. For lecithin-water systems, however, $\sigma^m > 0$ as a rule, implying that work has to be supplied in order to separate the bilayers stacked in the lamellar liquid crystal at the water swelling limit.

Dividing Eqs. (1) and (2) by A^m results in the expressions

$$\gamma^m = F^m_{(1)}/A^m - \mu_2^m \Gamma_2^m - \mu_3 \Gamma^m_{(3)} \,, \tag{4}$$

$$\sigma^m = F^m_{(1)}/A^m - \mu_2^* \Gamma_2^m - \mu_3 \Gamma^m_{(3)} \,. \tag{5}$$

These relations obviously imply that

$$\sigma^m - \gamma^m = \Gamma_2^m (\mu_2^m - \mu_2^*) \,, \tag{6}$$

i.e. the difference between the area specific work of membrane formation and the membrane tension reflects directly the chemical potential difference $\mu_2^m - \mu_2^*$. In the case of equilibrium formation of the bilayer membrane from the reference bulk state we have $\mu_2^m = \mu_2^*$ and thus, according to Eq. (6),

$$\sigma^m = \gamma^m \,. \tag{7}$$

If a membrane free of tension would form reversibly from the reference bulk state then both γ^m and σ^m would be equal to zero.

It is often natural as well as convenient to compare the bilayer membrane state with the corresponding monolayer state. To this end we also write down the defining relations for μ_2^f, the chemical potential of the lipid component in the monolayer film and σ^f, the work of forming (per unit surface area) the monolayer film. These relations are completely analogous to Eqs. (4) and (5), viz. [11]

$$\gamma^f = F^f_{(1)}/A^f - \mu_2^f \Gamma_2^f - \mu_3 \Gamma^f_{3(1)} \,, \tag{8}$$

$$\sigma^f = F^f_{(1)}/A^f - \mu_2^* \Gamma_2^f - \mu_3 \Gamma^f_{3(1)} \,, \tag{9}$$

where γ^f is the monolayer film tension, i.e. the surface tension of the film-covered water solution surface. The surface pressure of the monolayer film is, as usual, $\gamma^w - \gamma^f$, γ^w indicating the surface tension of the water solution.

As will be further treated below there are two ways of comparing the bilayer and monolayer states at a given temperature which are of particular importance namely i) at the same chemical potential $\mu_2^m = \mu_2^f$ (open bilayer membrane) and ii) at the same head-group density $\Gamma_2^m = 2\Gamma_2^f$ (partially closed bilayer membrane). In the former case we have from Eqs. (4) and (8):

$$\Delta\gamma = \gamma^m - 2\gamma^f = F_{(1)}^m/A^m - 2F_{(1)}^f/A^f - \mu_2^f(\Gamma_2^m - 2\Gamma_2^f) - \mu_3(\Gamma_{3(1)}^f - 2\Gamma_{3(1)}^f) \qquad (\mu_2^m = \mu_2^f)\,. \tag{10}$$

In the latter case we get from Eqs. (5) and (9):

$$\Delta\sigma = \sigma^m - 2\sigma^f = F_{(1)}^m/A^m - 2F_{(1)}^f/A^m - \mu_3(\Gamma_{3(1)}^m - 2\Gamma_{3(1)}^f) \qquad (\Gamma_2^m = 2\Gamma_2^f)\,. \tag{11}$$

The dominant term in Eq. (10) as well as in Eq. (11) should be $F_{(1)}^m/A^m - 2F_{(1)}^f/A^f$ in most cases of interest here.

For a liquid-expanded, rather closely packed bilayer membrane of a pure lecithin we may anticipate that γ^m amounts to a few dyn/cm whereas for a monolayer in physico-chemical equilibrium with such a bilayer, γ^f is likely to be about 20 dyn/cm corresponding to ~ 50 dyn/cm in surface pressure [15]. Thus $\Delta\gamma$ is of the order of -40 dyn/cm as is also $\Delta\sigma$. $\Delta\sigma$ clearly corresponds to the work of reversibly joining two monolayers so as to form a bilayer, i.e. $-\Delta\sigma$ is the work needed per cm² to reversibly cleave a single bilayer to form two monolayer-covered water solution surfaces while keeping the head-group density fixed.

Fundamental equations for a partially closed bilayer lipid membrane

The free energy differential expression for a bilayer lipid membrane that is closed with respect to the lipid component 2 and immersed in a water solution of component 3 is simply

$$dF_{(1)}^m = -S_{(1)}^m\, dT + \gamma^m\, dA^m + \mu_3\, dn_{3(1)}^m\,. \tag{12}$$

Following the Gibbs procedure we combine this expression with the definition given by Eq. (1) to introduce μ_2^m as the central, initially unknown, thermodynamic parameter. The resulting fundamental equation is

$$-d\gamma^m = (S_{(1)}^m/A^m)\, dT + \Gamma_2^m\, d\mu_2^m + \Gamma_{3(1)}^m\, d\mu_3 \tag{13}$$

that has the same form as the familiar Gibbs surface tension equation. Multiplying by the molar membrane area $A_2^m = 1/\Gamma_2^m$ and rearranging yields

$$-d\mu_2^m = s_{2(1)}^m\, dT + A_2^m\, d\gamma^m + y_{3(1)}^m\, d\mu_3\,, \tag{14}$$

where $s_{2(1)}^m$ is the (integral) molar entropy, $S_{(1)}^m/n_2^m$, and $y_{3(1)}^m$ denotes the mole number ratio $n_{3(1)}^m/n_2^m$. It follows from Eq. (14) that stretching of a bilayer membrane at constant T and μ_3 normally involves a decrease of the chemical potential μ_2^m since a stretching process should in general be associated with an increasing γ^m-function.

In a similar manner as for insoluble surface films, $\gamma^m - A_2^m$ isotherms recorded at different T and μ_3 are required for a complete determination of the thermodynamic properties of the closed bilayer lipid membrane. The necessary isotherm integration can be made conveniently from the bilayer state that is in physicochemical equilibrium with the reference state, i.e.

$$\mu_2^m(T, \gamma^m, \mu_3) = \mu_2^*(T, \mu_3) - \int_{\gamma^*}^{\gamma^m} A_2^m \, d\gamma^m , \tag{15}$$

where $\mu_2^*(T, \mu_3)$ is regarded as a known function. Differentiation of $\mu_2^m(T, \gamma^m, \mu_3)$ with respect to T and μ_3 then yields $s_{2(1)}^m$ and $y_{3(1)}^m$ as appears from Eq. (14) whereas Eq. (1) gives $f_{2(1)}^m = F_{(1)}^m/n_2^m$ and the free energy definition the membrane energy $u_{2(1)}^m = U_{(1)}^m/n_2^m$.

Another possibility is to combine the free energy differential expression, Eq. (12), with the defining relation for σ^m, Eq. (2). In this way we obtain the fundamental equation

$$\begin{aligned} d\sigma^m = &- \left[S_{(1)}^m/A^m + \Gamma_2^m \left(\frac{\partial \mu_2^*}{\partial T}\right)_{\mu_3}\right] dT + (\gamma^m - \sigma^m) \, d \ln A^m \\ &- \left[\Gamma_{3(1)}^m + \Gamma_2^m \left(\frac{\partial \mu_2^*}{\partial \mu_3}\right)_T\right] d\mu_3 , \end{aligned} \tag{16}$$

where T, A^m and μ_3 are the independent variables. This is clearly a fundamental equation which includes a SHUTTLEWORTH-HERRING relationship, viz.

$$\left(\frac{\partial \sigma^m}{\partial \ln A^m}\right)_{T, \mu_3} = \gamma^m - \sigma^m \tag{17}$$

that directly accounts for bilayer membrane strain changes.

The following relations result from cross-differentiation of Eq. (16)

$$-\left(\frac{\partial \gamma^m}{\partial T}\right)_{A^m, \mu_3} = S_{(1)}^m/A^m + \left(\frac{\partial (S_{(1)}^m/A^m)}{\partial \ln A^m}\right)_{T, \mu_3} = \Gamma_2^m \left(\frac{\partial s_{2(1)}^m}{\partial \ln A^m}\right)_{T, \mu_3} , \tag{18}$$

$$-\left(\frac{\partial \gamma^m}{\partial \mu_3}\right)_{T, A^m} = \Gamma_{3(1)}^m + \left(\frac{\partial \Gamma_{3(1)}^m}{\partial \ln A^m}\right)_{T, \mu_3} = \Gamma_2^m \left(\frac{\partial y_{3(1)}^m}{\partial \ln A^m}\right)_{T, \mu_3} . \tag{19}$$

From them it is obvious that bilayer membrane tension measurements for a partially closed bilayer membrane kept at fixed membrane area, for instance by means of a solid frame, are insufficient when the object is to evaluate $S_{(1)}^m/A^m$, $\Gamma_{3(1)}^m$, $F_{(1)}^m/A^m$ (or $s_{2(1)}^m$, $y_{3(1)}^m$, $f_{2(1)}^m$ etc.). Knowledge of how stretching affects $S_{(1)}^m/A^m$ and $\Gamma_{3(1)}^m$ (or $s_{2(1)}^m$ and $y_{3(1)}^m$) is also required. The situation in this respect is essentially the same as for solid surfaces (cf. Refs. [10, 12]). All the experimental information needed is contained, of course, in complete $\gamma^m - A_2^m$ isotherm data for the partially closed bilayer lipid membrane system.

The GIBBS elasticity constant of the bilayer membrane, E^m, is defined by

$$E^m = A^m \left(\frac{\partial \gamma^m}{\partial A^m}\right)_{T, \mu_3} \tag{20}$$

Alternative expressions for E^m are obtained by differentiating the Shuttleworth-Herring relation, Eq. (17), with respect to A^m and utilizing Eq. (6):

$$E^m = \frac{\partial}{\partial A^m}\,[A^m(\gamma^m - \sigma^m)]_{T,\mu_3} = -\Gamma_2^m \left(\frac{\partial \mu_2^m}{\partial \ln A^m}\right)_{T,\mu_3}. \tag{21}$$

Hence we may conclude that $E^m \geqq 0$ as is necessary for mechanical stability. E^m is equal to zero when $\gamma^m = \sigma^m$, i.e. when μ_2^m is independent of A.

When $\gamma^m - A_2^m$ isotherm data are available a more convenient form of Eq. (16) is

$$d(\sigma^m A_2^m) = -\left[s_{2(1)}^m + \left(\frac{\partial \mu_2^*}{\partial T}\right)_{\mu_3}\right] dT + \gamma^m\, dA_2^m - \left[y_{3(1)}^m + \left(\frac{\partial \mu_2^*}{\partial \mu_3}\right)_T\right] d\mu_3\,, \tag{22}$$

where $\sigma^m A_2^m$ is the free energy of bilayer membrane formation per mole of the lipid component 2. Eq. (22) can be utilized similarly to Eq. (14) by noting that

$$\sigma^m A_2^m = \gamma^{m*} A_2^{m*} + \int_{A_2^{m*}}^{A_2^m} \gamma^m\, dA_2^m \qquad (dT, d\mu_3 = 0)\,, \tag{23}$$

where γ^{m*} and A_2^{m*} denote the corresponding properties of the particular bilayer membrane that is in physico-chemical equilibrium with the starred reference state.

We remark that for a binary water-lipid system, the derivative $d\mu_2^*/dT$ appearing in the corresponding versions of Eqs. (16) and (22) (without the $d\mu_3$-terms) is given by

$$\frac{d\mu_2^*}{dT} = -(s_2^* - y_1^* s_1^0)\,, \tag{24}$$

where s_2^* is the molar entropy S^*/n_2^*, $y_1^* = n_1^*/n_2^*$ and s_1^0 is the molar entropy of pure water; $s_{2(1)}^m$ for the bilayer lipid membrane and $s_2^* - y_1^* s_1^0$ for the reference bulk phase are evidently analogous properties.

It is obvious that the experimental determination of $\gamma^m - A_2^m$ isotherms for partially closed bilayer lipid membranes presents practical difficulties which in many cases are not easily overcome. For instance, the determination of A_2^m may be difficult to carry out in practice. Film balance techniques, however, are well suited for studies of the monolayer properties of membrane-forming phospholipids [15 — 18] at well-defined head-group densities. A reasonable way to proceed might then be to make use of experimental film balance data supplemented with a theoretical calculation step. Thus knowing the thermodynamic lipid monolayer properties from a complete film balance investigation we wish to deduce the closely related bilayer membrane properties. With this purpose in mind Eq. (11) is a useful starting-point since it involves a comparison between the monolayer and the bilayer at a given head-group density, $\Gamma_2^m = 2\Gamma_2^f$. By differentiating Eq. (11) and combining the resulting expression with Eq. (12) and the analogous expression for $dF_{(1)}^f$ we readily obtain the subsequent fundamental equation

$$\begin{aligned} -d(\Delta\sigma) = {} & (S_{(1)}^m/A^m - 2S_{(1)}^f/A^f)\, dT + (\Delta\gamma - \Delta\sigma)\, d \ln \Gamma_2^f \\ & + (\Gamma_{3(1)}^m - 2\Gamma_3^f)\, d\mu_3 \qquad (\Gamma_2^m = 2\Gamma_2^f)\,. \end{aligned} \tag{25}$$

The point of view here would be to consider $\Delta\sigma(T, \Gamma_2^f, \mu_3)$ as a function known from theoretical calculations for the joining (or disjoining) experiment while the monolayer properties are known from film balance studies and to utilize Eq. (25) to obtain $S_{(1)}^m/A^m$, γ^m, $\Gamma_{3(1)}^m$ and Eq. (11) to get $F_{(1)}^m/A^m$.

In the theoretical disjoining experiment, the distance between the two $\Sigma_{(1)}$ dividing surfaces, $\tau^{(1)}$, is taken as an additional degree of freedom. Hence, in place of Eq. (12), for a bilayer system with a variable distance between the two faces (which remain in water contact) we would have

$$dF_{(1)} = -S_{(1)}\,dT + \gamma\,dA + \mu_3\,dn_{3(1)} - \Pi A\,d\tau_{(1)}\,, \tag{26}$$

where Π is the so-called disjoining pressure introduced by DERJAGUIN. By integrating the free energy differential over $\tau_{(1)}$ from $\tau^m_{(1)}$ to infinite separation at constant T, μ_3 and $A = A^m$ we obtain

$$2F^f_{(1)}/A^f - F^m_{(1)}/A^m = \mu_3(2\Gamma^f_{3(1)} - \Gamma^m_{3(1)}) - \int_{\tau^m_{(1)}}^{\infty} \Pi\,d\tau_{(1)}\,. \tag{27}$$

A comparison with Eq. (11) shows that

$$\Delta\sigma = \int_{\tau^m_{(1)}}^{\infty} \Pi\,d\tau_{(1)}\,, \tag{28}$$

where the disjoining pressure Π is dependent on $\tau_{(1)}$, of course, and on the state variables T, Γ^f_2, μ_3.

Approximate correlations between lipid bilayer and monolayer properties

An oversimplified but, as a first approximation, still reasonable model for a bilayer lipid membrane above T_c involves the assumption that the hydrocarbon chain interior of the membrane behaves similarly to a hydrocarbon liquid. For a partially closed bilayer membrane such a model in effect implies an elastic response from the head-group zones and a fluid behaviour of the hydrocarbon part. Accordingly, contributions to γ^m would chiefly arise due to separating the head-groups whereas the membrane thickness $\tau^m_{(1)}$ would vary with strain so as to keep the membrane volume $\tau^m_{(1)} \cdot A$ approximately constant. It is evident that a simplistic model of this kind is applicable only in the vicinity of the unstrained state, i.e. when γ^m amounts to at most a few dyn/cm. The corresponding monolayer with the same head-group density as the bilayer differs in the first place with respect to the state of the hydrocarbon moiety. For a comparatively densely packed lipid monolayer (45–70 Å²/molecule) kept at or close to the equilibrium spreading pressure a dominant contribution to the resulting surface tension is likely to be due to the reduced density in the hydrocarbon chain and region in a similar way as for the surface zone of a hydrocarbon liquid. For the bilayer membrane and for the monolayer film in a state close to the equilibrium spreading pressure the thermodynamic conditions in the head group zones should be similar and to a large extent determined by the detailed water interaction as proposed by FORSLIND and KJELLANDER [18]. On this admittedly approximate basis we can make the following assumption (Eq. 25)

$$\left(\frac{\partial(\Delta\sigma)}{\partial \ln A^m}\right)_{T,\mu_3} = \Delta\gamma - \Delta\sigma \cong 0\,, \tag{29}$$

i.e. the reversible cleavage work is approximately independent of the state of strain and is moreover related to γ^m and γ^f by the simple relation

$$\gamma^m - 2\gamma^f \cong \Delta\sigma\,. \tag{30}$$

We should stress, however, that Eqs. (29) and (30) presuppose that long-range interactions for the bilayer configuration across the membrane between the head group zones and between the two water solutions are of minor importance. In particular, when $\Gamma_2^m = 2\Gamma_2^f$ is chosen so as to make $\gamma^m = 0$ we have

$$\Delta\sigma + 2\gamma^f = 0 \tag{31}$$

in support of our previous estimate that the order of magnitude of $\Delta\sigma$ is ~ -40 erg/cm^2.

Repeated differentiation of Eq. (30) with respect to ln A^m results in

$$E^m \approx 2E^f \tag{32}$$

implying that the GIBBS elasticity constant for the bilayer membrane is about twice the elasticity constant for the monolayer film at the same head-group density. From the film balance study of L-α-dipalmitoyl lecithin at 45 °C by VILALLONGA [18] we conclude through applying Eq. (32) that E^m typically is equal to ~ 200 dyn/cm in rough agreement with direct measurements on cholesterol-CTAB membranes (cf. Ref. [1]).

Fundamental equations for a bilayer lipid membrane that is open to a monolayer film

In favourable cases at least, open bilayer lipid membranes can be realized above T_c following the membrane preparation procedure of TAKAGI, AZUMA and KISHIMOTO (loc. cit.). Such a bilayer membrane system is open to the (insoluble) lipid monolayer film kept at T, μ_2^f, μ_3 or T, γ^f, μ_3. In this case the free energy differential expression is

$$dF_{(1)}^m = -S_{(1)}^m\,dT + \gamma^m\,dA^m + \mu_2^f\,dn_2^m + \mu_3\,dn_{3(1)}^m\,. \tag{33}$$

Component 2 is supplied, without hindrance, from the monolayer reservoir where its chemical potential is μ_2^f. Eq. (33) can be integrated directly, i.e. stretching the membrane and forming additional membrane area are indistinguishable processes insofar as the monolayer reservoir is large enough. The resulting integrated relation is, of course,

$$F_{(1)}^m = \gamma^m A^m + \mu_2^f n_2^m + \mu_3 n_{3(1)}^m \tag{34}$$

which conforms with Eq. (1), the main difference being that Eq. (34) is not a defining relation. On combining Eqs. (33) and (34) in the usual manner we get

$$-d\gamma^m = (S_{(1)}^m/A^m)\,dT + \Gamma_2^m\,d\mu_2^f + \Gamma_{3(1)}^m\,d\mu_3\,. \tag{35}$$

This fundamental equation is of the GIBBS kind. It differs from Eq. (13) for the partially closed bilayer membrane rather in its utilization than in its formal appearance.

Experimentally, γ^f is preferred as an independent variable in place of μ_2^f. The change of independent variable from μ_2^f to γ^f is readily made by introducing into Eq. (35) the expression

$$-d\mu_2^f = s_{2(1)}^f\,dT + A_2^f\,d\gamma^f + y_{3(1)}^f\,d\mu_3 \tag{36}$$

that holds for the monolayer film. In this way we obtain the fundamental equation

$$A_2^m\,d\gamma^m = -(s_{2(1)}^m - s_{2(1)}^f)\,dT + A_2^f\,d\gamma^f - (y_{3(1)}^m - y_{3(1)}^f)\,d\mu_3\,. \tag{37}$$

This equation obviously includes the following partial derivatives

$$-A_2^m \left(\frac{\partial \gamma^m}{\partial T}\right)_{\gamma^f, \mu_3} = s_{2(1)}^m - s_{2(1)}^f , \tag{38}$$

$$\left(\frac{\partial \gamma^m}{\partial \gamma^f}\right)_{T, \mu_3} = \frac{A_2^m}{A_2^f} = \frac{\Gamma_2^f}{\Gamma_2^m} , \tag{39}$$

$$-A_2^m \left(\frac{\partial \gamma^m}{\partial \mu_3}\right)_{T, \gamma^f} = y_{3(1)}^m - y_{3(1)}^f . \tag{40}$$

The temperature coefficient of the membrane tension as defined by Eq. (38) is expected to have a positive sign since presumably $s_{2(1)}^m < s_{2(1)}^f$ due to the surface entropy associated with the hydrocarbon chain end region in the monlayer case. The rate of change of γ^m with γ^f determines A_2^m. When comparatively densely packed monolayers are involved $(\partial \gamma^m / \partial \gamma^f)$ T, μ_3 should be approximately equal to 2. However, membrane tension measurements for the kind of membrane system considered here enable in principle an exact evaluation of A_2^m. We note that the partial derivative $(\partial \gamma^m / \partial \mu_3)_{T, \gamma^f}$ is expected to be of small magnitude since $y_{(1)3}^m - y_{3(1)}^f$ is likely to be small in most cases.

Equilibrium formation of a bilayer lipid membrane from the reference bulk state of the lipid

A single lipid bilayer can be reversibly formed from the (lamellar) reference bulk state as follows. A crystal of the membrane-forming lipid is placed on the surface of the water solution where swelling takes place and spreading to a lipid monolayer is arranged to occur at the equilibrium spreading pressure. By means of the procedure of TAGAKI et al. (loc. cit.), likewise carried out in a reversible manner, a bilayer membrane is then formed by supplying the work, $\gamma^m A^m$ while keeping γ^f constant. During this course of events $\mu_2^* = \mu_2^f = \mu_2^m$, i.e. the chemical potential of the lipid component 2 is everywhere the same as in the reference state.

The bilayer that is reversibly formed from the reference state at $\mu_2^m = \mu_2^*$ is expected to be slightly stretched with a γ^m-value amounting to a few dyn/cm. Consequently, for an unsupported, planar membrane ($\gamma^m = 0$) we have $\mu_2^m > \mu_2^*$ since stretching is in general associated with a lowering of the chemical potential (Eq. 14). This is consistent with a tendency of bilayer aggregation back to the lamellar reference state.

With respect to the thermodynamic background presented here it seems plausible that the major effects of sonication to prepare vesicle solutions are to separate the lamellae and to stretch the bilayers so that γ^m becomes > 0. The stretching of the bilayer reduces the chemical potential difference $\mu_2^m - \mu_2^*$ that acts as a driving force for bilayer aggregation and may accordingly diminish the vesicle aggregation rate.

A thought-experiment is depicted in Fig. 6 that implies a direct formation of a separate bilayer from a lamellar liquid crystal. This experiment is actually a good illustration of the relation $\sigma^m = \gamma^m$ (Eq. 7) that holds under equilibrium conditions.

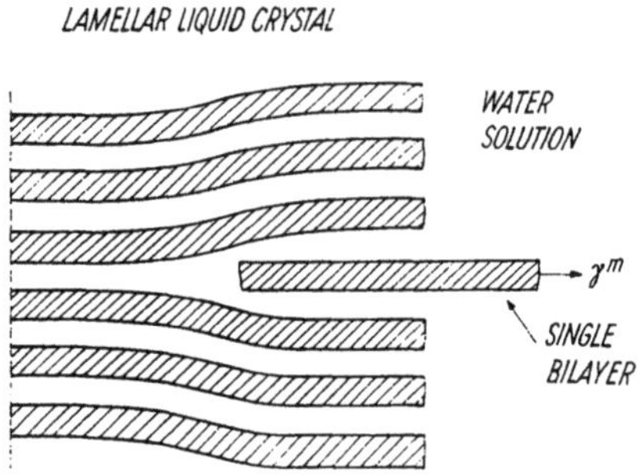

Fig. 6: Drawing illustrating the equilibrium formation of a single bilayer from the lamellar reference bulk state. In order to detach the bilayer the tension γ^m has to be balanced by an external force.

Fundamental equations for a bilayer membrane that is open to a lipid solvent phase

This case corresponds to the most commonly applied bilayer membrane preparation procedure (cf. Ref. [1]). A suitable organic solvent for the lipid component is used (Fig. 4). For the sake of simplicity we assume here that the lipid solvent (Component 4) is a pure solvent. We also presuppose that the solvent phase only contains Components 2 and 4 in appreciable amounts and that Component 3 is the only water solute. According to the phase rule the number of degrees of freedom is 3. Thus T, μ_2, μ_3 is a convenient choice of independent variables. The chemical potential μ_2 is determined in the solvent phase and μ_3, as before, by its value in the water solution. The free energy differential for an open bilayer membrane of this kind is

$$dF^m_{(1)} = -S^m_{(1)}\,dT + \gamma^m\,dA^m + \mu_2\,dn^m_2 + \mu_3\,dn^m_{3(1)} + \mu_4\,dn^m_4\,. \tag{41}$$

By integrating this expression in the usual manner at constant T, μ_2, μ_3 (and hence at constant γ^m, μ_4 as well) and comparing with Eq. (41) we obtain the following fundamental equation

$$-d\gamma^m = (S^m_{(1)}/A^m)\,dT + \Gamma^m_2\,d\mu_2 + \Gamma^m_{3(1)}\,d\mu_3 + \Gamma^m_4\,d\mu_4 \tag{42}$$

which is of the general Gibbs-Duhem kind. Now $d\mu_4$ is clearly not an independent differential since there is a Gibbs-Duhem relation for the solvent phase (superscript L) which can be written

$$0 = s^L_2\,dT + d\mu_2 + y^L_4\,d\mu_4\,, \tag{43}$$

where $s^L_2 = S^L/n^L_2$ and $y^L_4 = n^L_4/n^L_2$. By introducing Eq. (43) into Eq. (42) we get

$$-d\gamma^m = (S^m_{(1)}/A^m - \Gamma^m_4 s^L_2/y^L_4)\,dT + (\Gamma^m_2 - \Gamma^m_4/y^L_4)\,d\mu_2 + \Gamma^m_{3(1)}\,d\mu_3\,. \tag{44}$$

In its formal appearance this equation is closely related to the Gibbs surface tension equation. We should note, in particular, that the derivative $(\partial\gamma^m/\partial\mu_2)_{T,\mu_3}$ does not yield Γ^m_2 unless it can be ascertained that the bilayer content of organic solvent, Γ^m_4, is very small when y^L_4 is of the order of 1. However, it appears that the surface density $\Gamma^m_{3(1)}$ can be determined from γ^m measurements at variable water solute concentrations since

$$\left(\frac{\partial\gamma^m}{\partial\mu_3}\right)_{T,\mu_2} = -\Gamma^m_{3(1)}\,. \tag{45}$$

Examples of the application of Eq. (45) are available in the literature (cf. Ref. [1]).
An alternative form of Eq. (44) is

$$-d\gamma^m = (S^m_{(1)}/A - \Gamma^m_2 \tilde{s}^L_2 - \Gamma^m_{3(1)} \tilde{s}^W_3 - \Gamma^m_4 \tilde{s}^L_4)\, dT + \left[\Gamma^m_2 \left(\frac{\partial \mu_2}{\partial x^L_2}\right)_T + \Gamma^m_4 \left(\frac{\partial \mu_4}{\partial x^L_2}\right)_T\right] dx^L_2 + \Gamma^m_{3(1)} \left(\frac{\partial \mu_3}{\partial x^W_3}\right)_T dx^W_3 \,, \tag{46}$$

where T, x^L_2, x^W_3 are used as independent variables. x^L_2 and x^w_2 stand for mole fractions in the solvent and water phases. The tilde notation is used to denote a partial molar property. A positive temperature coefficient for γ^m has been observed for dodecylacid-phosphate-cholesterol-dodecane in 0,1 M NaCl [3]. According to Eq. (46) this observation would mean that

$$S^m_{(1)}/A < \Gamma^m_2 \tilde{s}^L_2 + \Gamma^m_{3(1)} \tilde{s}^W_3 + \Gamma^m_4 \tilde{s}^L_4 \tag{47}$$

or, which is the same,

$$s^m_{2(1)} < \tilde{s}^L_2 + y^m_{3(1)} \tilde{s}^W_3 + y^m_4 \tilde{s}^L_4 \tag{48}$$

as seems rather likely on qualitatively comparing the states of the lipid component in the organic solvent and in the bilayer.

Asymmetrical bilayer lipid membranes

The symmetrical bilayer lipid membranes treated in the previous sections can be turned into asymmetrical membranes by assuming that the water solutions in contact with the two membrane faces have different solute concentrations and that the membranes are impermeable to the water solute component 3. In the case of a bilayer membrane that is open to a lipid solvent phase the thermodynamic state of an asymmetrical membrane is determined by the variables T, μ_2, μ_3', μ_3'' where $'$ and $''$ refer to the two membrane sides. It follows that the fundamental equation (44) has to be modified as follows

$$-d\gamma^m = (S^m_{(1)}/A^m - \Gamma^m_4 s^L_2/y^L_4)\, dT + (\Gamma^m_2 - \Gamma^m_4/y^L_4)\, d\mu_2 + \Gamma^{m'}_{3(1)}\, d\mu_3' + \Gamma^{m''}_{3(1)}\, d\mu_3'' \,, \tag{49}$$

where $\Gamma^m_2 = \Gamma^{m'}_2 + \Gamma^{m''}_2$ and $\Gamma^m_4 = \Gamma^{m'}_4 + \Gamma^{m''}_4$. This equation obviously implies that the superficial densities of component 3 on the $'$ and $''$ faces of the membrane are given by the partial derivatives

$$-\left(\frac{\partial \gamma^m}{\partial \mu_3''}\right)_{T,\mu_2,\mu_3''} = \Gamma^{m'}_{3(1)} \,; \qquad -\left(\frac{\partial \gamma^m}{\partial \mu_3''}\right)_{T,\mu_2,\mu_3'} = \Gamma^{m''}_{3(1)} \,. \tag{50}$$

The cross-differentation relationship

$$\left(\frac{\partial \Gamma^{m'}_{3(1)}}{\partial \mu_3''}\right)_{T,\mu_2,\mu_3'} = \left(\frac{\partial \Gamma^{m''}_{3(1)}}{\partial \mu_3'}\right)_{T,\mu_2,\mu_3''} \tag{51}$$

accounts for the interdependence across the membrane.

In the case of a partially closed bilayer membrane that is asymmetric and in contact with water solutions at μ_3' and μ_3'' on the $'$ and $''$ faces, respectively, it is most convenient to extend the definition of σ^m as follows

$$\sigma^m A^m = F^m_{(1)} - \mu_2^{*'} n_2^{m'} - \mu_2^{*''} n_2^{m''} - \mu_3' n^{m'}_{3(1)} - \mu_3'' n^{m''}_{3(1)} . \quad (52)$$

In this definition two reference bulk states are involved with $\mu_2^{*'}(T, \mu_3')$ and $\mu_2^{*''}(T, \mu_3'')$. By proceeding similarly as before we deduce the following fundamental equation

$$d\sigma^m = -\left[S^m_{(1)}/A^m + \Gamma_2^{m'}\left(\frac{\partial \mu_2^{*'}}{\partial T}\right)_{\mu_3'} + \Gamma_2^{m''}\left(\frac{\partial \mu_2^{*''}}{\partial T}\right)_{\mu_3''}\right] dT$$
$$+ (\gamma^m - \sigma^m)\, d \ln A^m - \left[\Gamma^{m'}_{3(1)} + \Gamma_2^{m'}\left(\frac{\partial \mu_2^{*'}}{\partial \mu_3'}\right)_T\right] d\mu_3'$$
$$- \left[\Gamma^{m''}_{3(1)} + \Gamma_2^{m''}\left(\frac{\partial \mu_2^{*''}}{\partial \mu_3''}\right)_T\right] d\mu_3'' \quad (53)$$

which includes an essentially unchanged SHUTTLEWORTH-HERRING equation (Eq. 17) and the cross-differentiation relationships:

$$-\left(\frac{\partial \gamma^m}{\partial T}\right)_{A^m, \mu_3', \mu_3''} = S^m_{(1)}/A^m + \left(\frac{\partial (S^m_{(1)}/A^m)}{\partial \ln A^m}\right)_{T, \mu_3', \mu_3''} = \Gamma_2^m \left(\frac{\partial s^m_{2(1)}}{\partial \ln A^m}\right)_{T, \mu_3', \mu_3''}, \quad (54)$$

$$-\left(\frac{\partial \gamma^m}{\partial \mu_3'}\right)_{T, A^m, \mu_3''} = \Gamma^{m'}_{3(1)} + \left(\frac{\partial \Gamma^{m'}_{3(1)}}{\partial \ln A^m}\right)_{T, \mu_3', \mu_3''} = \Gamma_2^{m'} \left(\frac{\partial y^{m'}_{3(1)}}{\partial \ln A^m}\right)_{T, \mu_3', \mu_3''}, \quad (55)$$

$$-\left(\frac{\partial \gamma^m}{\partial \mu_3''}\right)_{T, A^m, \mu_3} = \Gamma^{m''}_{3(1)} + \left(\frac{\partial \Gamma^{m''}_{3(1)}}{\partial \ln A^m}\right)_{T, \mu_3', \mu_3''} = \Gamma_2^{m''} \left(\frac{\partial y^{m''}_{3(1)}}{\partial \ln A^m}\right)_{T, \mu_3', \mu_3''}, \quad (56)$$

$$\left(\frac{\partial \Gamma^{m'}_{3(1)}}{\partial \mu_3''}\right)_{T, A^m, \mu_3'} = \left(\frac{\partial \Gamma^{m''}_{3(1)}}{\partial \mu_3''}\right)_{T, A^m, \mu_3'} . \quad (57)$$

Likewise a fundamental equation can be deduced for an asymmetrical bilayer membrane that is open to two monolayer films kept at different surface pressures. It is readily shown that the asymmetric version of Eq. (35) is:

$$d\gamma^m = -(S^m_{(1)}/A^m - \Gamma_2^{m'} s^{f'}_{2(1)} - \Gamma_2^{m''} s^{f''}_{2(1)})\, dT$$
$$+ \Gamma_2^{m'} A_2^{f'}\, d\gamma^{f'} + \Gamma_2^{m''} A_2^{f''}\, d\gamma^{f''} - \Gamma_2^{m'} (y^{m'}_{3(1)} - y^{f'}_{3(1)})\, d\mu_3'$$
$$- \Gamma_2^{m''} (y^{f''}_{3(1)} - y^{f''}_{3(1)})\, d\mu_3'' . \quad (58)$$

In particular, from this equation it appears that $\Gamma_2^{m'}$ and $\Gamma_2^{m''}$ can be evaluated separately from $\gamma^m(\gamma^{f'}, \gamma^{f''})$ measurements by means of the derivatives

$$\left(\frac{\partial \gamma^m}{\partial \gamma^{f'}}\right)_{T, \gamma^{f''}, \mu_3', \mu_3''} = \Gamma_2^{m'} A_2^{f'} , \quad (59)$$

$$\left(\frac{\partial \gamma^m}{\partial \gamma^{f''}}\right)_{T, \gamma^{f'}, \mu_3', \mu_3''} = \Gamma_2^{m''} A_2^{f''} . \quad (60)$$

Vesicle bilayer lipid membranes

As the final topic we consider lipid bilayers in vesicles which are large enough so that curvature effects can be neglected. To gain in simplicity we restrict the treatment to the symmetric case. A vesicle bilayer is presumably closed for the lipid component.

Thus the fundamental equation (16) should be applicable in principle. In the vesicle case, however, the membrane area is no longer an independent variable but a function of T and μ_3. When this circumstance is taken into account explicitly we get the proper form of the fundamental equation for the present application, viz.,

$$-d\sigma^m = \left[S^m_{(1)}/A^m + \Gamma^m_2\left(\frac{\partial \mu_2}{\partial T}\right)_{\mu_3} + (\sigma^m - \gamma^m)\left(\frac{\partial \ln A^m}{\partial T}\right)_{\mu_3}\right] dT + \left[\Gamma^m_{3(1)} + \Gamma^m_2\left(\frac{\partial \mu_2}{\partial \mu_3}\right)_T + (\sigma^m - \gamma^m)\left(\frac{\partial \ln A^m}{\partial \mu_3}\right)_T\right] d\mu_3\,. \quad (61)$$

As yet nothing very definite can be stated about the relative magnitudes of the $(\sigma^m - \gamma^m)$ terms in Eq. (61). However, the expected order of $(\partial \ln A^m/\partial T)_{\mu_3}$ is 10^{-4}/K as evaluated for pure water. Furthermore it follows from our previous reasoning in the subsection on equilibrium formation of a bilayer lipid membrane that $|\sigma^m - \gamma^m|$ should be small, say at most 1 erg/cm². Thus we estimate

$$|\sigma^m - \gamma^m|\left(\frac{\partial \ln A}{\partial T}\right)_{\mu_3} \lessapprox 10^{-4}\ \mathrm{erg/(cm^2\,K)}\,. \quad (62)$$

The two leading terms in the dT coefficient of Eq. (61) implay essentially an entropy difference for a single bilayer as compared with a bilayer in the lamellar reference bulk state. A rough guess would be that this entropy difference is of the order of 10^{-1} to 10^{-2} erg/(cm² K). Accordingly the dT term of Eq. (62) might well be of minor importance.

From an experimental point of view the derivative relations for γ^m corresponding to Eqs. (18) and (19) are probably of some interest. These are

$$-\left(\frac{\partial \gamma^m}{\partial T}\right)_{\mu_3} = S^m_{(1)}/A^m + \left(\frac{\partial (S^m_{(1)}/A^m)}{\partial \ln A^m}\right)_{T,\mu_3} - E^m\left(\frac{\partial \ln A^m}{\partial T}\right)_{\mu_3}, \quad (63)$$

$$-\left(\frac{\partial \gamma^m}{\partial \mu_3}\right)_T = \Gamma^m_{3(1)} + \left(\frac{\partial \Gamma^m_{3(1)}}{\partial \ln A^m}\right)_{T,\mu_3} - E^m\left(\frac{\partial \ln A^m}{\partial \mu_3}\right)_T. \quad (64)$$

Even here it is likely that the E^m containing terms have the nature of small correction terms. Whether this actually is true or not has to be ascertained by further experimental work.

Concluding remarks

At the present stage of development the most important purpose of applying classical thermodynamics to bilayer lipid membrane systems is probably to survey the experimental possibilities and to obtain guidance for carrying through measurements which can result in new information as to the thermodynamic properties of bilayer membranes. From the present formal investigation it appears to be of particular significance to devote further work to developing techniques for membrane tension measurements in conjunction with the bilayer lipid membrane preparation procedure of Takagi et al. Eqs. (34), (38)--(40) show that when the monolayer properties are regarded as known, a complete knowledge of the thermodynamic bilayer membrane properties can be gained from $\gamma^m(T, \gamma^f, \mu_3)$ measurements on a bilayer lipid membrane that is open towards the corresponding lipid monolayer film.

The second best possibility would imply $(\gamma^m - A_2^m)$ measurements at variable T and μ_3 for a partially closed bilayer lipid membrane analogous to film balance measurements on insoluble surface films. In this case either of the fundamental equations (13) and (22) are applicable. We should note, however, that isotherm integration would be required in order to evaluate all the thermodynamic bilayer properties.

The bilayer lipid membrane system that is open to an organic solvent phase is probably the least useful arrangement when the purpose is to derive relevant thermodynamic information mainly because of the presumable presence of organic solvent in the bilayer.

Also when it comes to systematic investigations of asymmetrical bilayer lipid membranes it would seem that the open bilayer-monolayer system, if then at all a realistic possibility in practice, offers definite advantages.

The author wishes to thank Drs. B. TOSHEV and R. KJELLANDER for valuable discussions.

References

[1] H. T. TIEN, "*Biblayer Lipid Membranes*", Marcel Dekker, New York, 1974
[2] H. WENNERSTRÖM and G. LINDBLOM, Quart. Reviews in Biophysics, **10**, 1 (1977), 67
[3] H. T. TIEN, J. Phys. Chem., **72** (1968), 2723
[4] R. J. GOOD, J. Colloid Interface Sci., **31** (1969), 540
[5] E. EVANS and S. SIMON, J. Colloid Interface Sci., **51**, (1975), 266
[6] E. EVANS and R. WAUGH, J. Colloid Interface Sci., **60** (1977), 286
[7] B. V. TOSHEV and I. B. IVANOV, Colloid & Polymer Sci., **253** (1975), 558
[8] I. B. IVANOV and B. V. TOSHEV, Colloid & Polymer Sci., **253** (1975), 593
[9] J. W. GIBBS, "*The Collected Works of J. W. Gibbs*", Vol. 1, Longmans Green and Co., London 1928, p. 219
[10] J. C. ERIKSSON, Surface Sci., **14** (1969), 221
[11] J. C. ERIKSSON, J. Colloid Interface Sci., **37** (1971), 659
[12] J. C. ERIKSSON, Acta Chem. Scand., **31** (1977), 235
[13] A. I. RUSANOV, J. Colloid Interface Sci., **63** (1978), 330
[14] TAGAKI, AZUMA and KISHIMOTO, Ann. Rept. Biol. Osaka University **13** (1965), 107
[15] M. C. PHILLIPS and H. J. HAUSER, J. Colloid Interface Sci., **49** (1974), 31
[16] L. W. HORN and N. L. GERSHFELD, Biophys. Journ., **18** (1977), 301
[17] N. L. GERSHFELD and K. TAJIMA, J. Colloid Interface Sci., **59** (1977), 597
[18] F. VILALLONGA, Biochem. Biophys. Acta, **163** (1968), 290
[19] E. FORSLIND and R. KJELLANDER, J. Theor. Biol., **51** (1975), 97
[20] M. J. JANIAK, Thesis, Boston Univ., 1977

An experimental determination of the curvature dependence of surface tension from fluid flow studies

J. A. Wingrave

(Continental Oil Company, Ponca City, Oklahema /USA)

and

R. S. Schechter and W. H. Wade

(The University of Texas at Austin, Austin, Texas/USA)

Abstract

The time required to saturate initially evacuated mesoporous Vycor spheres by hydro dynamic liquid imbibition was measured gravimetrically for water and five n-alkanes Using these imbibition times in conjunction with the equations of continuity and capillarity, it was possible to calculate the curvature dependent surface tensions for the imbibing liquids. The resulting tensions were greater than flat interface surface tensions by 5 to 50 percent, in good agreement with theoretical calculations. The characteristic interfacial thickness parameter, δ, calculated from the curvature dependence, ranged from 1 to 10 Å for the various liquids studied.

Introduction

The curvature of an interface between two contiguous phases affects the physical properties of both the bulk phases and the interface. The Young-Laplace equation, developed in the early nineteenth century [1, 2], established the relationship between the interfacial curvature, J, and the bulk phase pressure. In 1878 Gibbs [3] employed thermodynamics to corroborate the Young-Laplace equation and, in addition, considered that even interfacial properties were dependent on J. This dependence of interfacial properties on J (henceforth referred to as the curvature effect) has, however, been the subject of very little experimental research.

The theory of the curvature effect has been extended and elucidated by Buff's [4—6] hydrostatic treatment of interfaces and Defay, Prigogine, et al. [7, 8] nonequilibrium thermodynamic interfacial analyses. Several semiquantitative estimates of the curvature effect [9—15] have corroborated these theories and provided an assessment of its magnitude. However, to date there have been only eight attempts [16—23] utilizing only four different experimental techniques to measure the curvature effect. From these eight studies, the multicomponent, *l–l* nucleation technique of Nielsen et al. [21, 22] emerges as the only one precise enough to detect curvature effects. While this work is pioneering, the *l–l* nucleation requires two phases and a minimum of three components and is, therefore, not applicable to measuring curvature effects in simple one component, *l–v* systems. Therefore, the following study was undertaken to develop and exploit a method for measuring the curvature effect for single pure liquids (i.e., two phase, one component, *l–v* systems).

Experimental

Chemicals: The lower n-alkanes (n-pentane and n-heptane) used in this study were Matheson, Coleman, and Bell Spectrograde chemicals. The higher n-alkanes (n-decane, n-tridecane, and n-hexadecane) were Phillips Petroleum Company 99+ mole percent grade purity. All n-alkanes were dried a minimum of one month over activated (400 °C in air) Linde 4A Molecular Sieve (1/8 inch pellet form) and vacuum distilled before use. The water was obtained initially as deionized tap water and vacuum distilled before use. The porous Vycor glass (No. 7930) was supplied courtesy of Corning Glass in the form of $\sim$ 1 m lengths of 0.72 cm diameter rods. For the liquid imbibition experiments, these rods were cut and ground into spheres with radii, ca. 0.35 cm and radial tolerances for each sphere of $< \pm 5.7 \times 10^{-4}$ cm as described previously [24, 25]. For the liquid permeability experiments, ca. 3.0 mm thick discs (0.72 cm diameter) were cut and the circular faces polished on Pyrex glass plates. The range of deviations for four thickness measurements around the perimeter of each disc was never more than 1.3 percent of the disc thickness [24, 25].

Liquid Imbibition Apparatus: To measure the imbibition of liquid into the porous Vycor spheres, a Cahn RG microbalance was modified as shown in Figure 1 and described previously [24, 25]. Initially a sphere, which had been cleaned at 400 °C in air until all color from carbonaceous decomposition products was absent, was placed on one arm of the balance and evacuated to $< 5 \times 10^{-6}$ Torr at 400 °C. The liquid sample was degassed and distilled over activated molecular sieve into the second sample bulb which was in turn vacuum distilled into the reservoir bulb. The porous Vycor sphere was cooled and thermostated at 20.00 $\pm$ 0.02 °C. Then in rapid sequence the reservoir bulb was rotated to its upright position (as shown by dotted outline), the vacuum valve to the Vycor chamber closed, and the liquid allowed to flow in and completely immerse the porous Vycor sphere.

From the admission of the first vapors until total immersion generally took about 5 seconds and in no case more than 10 seconds. The time required for saturation by liquid imbibition of the porous Vycor sphere was from 250 seconds (n-pentane) to 5000 seconds (n-hexdecane). The recorder trace provided the mass of liquid imbibed

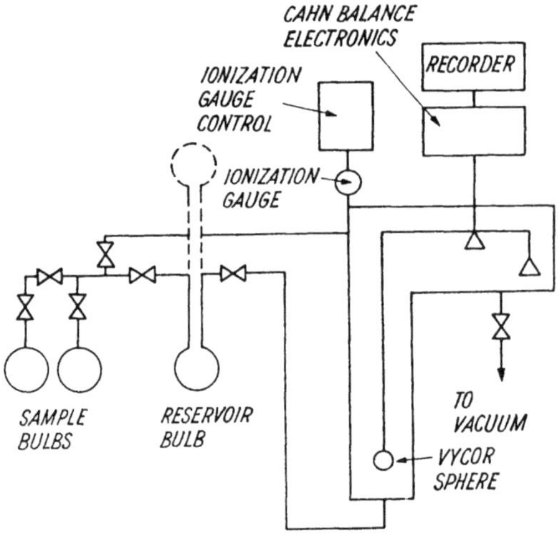

Fig. 1: Liquid imbibition apparatus

versus time data. Due to the initial hydrodynamic transients, the mass imbibed base line was indeterminant. Therefore, the saturated liquid pore volume of each sphere (and hence porosity) was determined as that mass of vapor which condensed on a sphere at saturation vapour pressure. This vapour adsorbed mass was then subtracted from the final liquid imbibition mass to establish the tare mass (or base line for the recorder trace of imbibition versus time) of the immersed porous Vycor sphere prior to all liquid and vapor imbibition. The 95 percent confidence interval for saturation imbibition times (corrected for differences in sphere radii) for different spheres was ± 2.6 percent.

Liquid Permeability Apparatus: The DARCY's Law permeability constants, k, were determined using the apparatus shown in Figure 2 and described elsewhere [24, 25]. Initially each porous Vycor disc was heated at 400 °C in air until all color disappeared. Then each disc was coated with epoxy resin around its rim and inserted in a porous Vycor sleeve. When the epoxy had cured, the Vycor sleeve was assembled in the permeability chamber. The system was evacuated to $< 5 \times 10^{-6}$ Torr overnight. Liquid was cryogenically vacuum distilled into the sample bulb (distillation apparatus not shown) and with the liquid at 77 K, the sample bulb containing frozen liquid was attached to the permeability apparatus in the down position, Figure 2. After a brief outgassing the liquid temperature was raised to room temperature, the sample bulb inverted, the vacuum pumps isolated, and the liquid allowed to fill and thermally equilibrate in the thermostated system. The isolation valve was shut, the vacuum cap removed, and when the liquid level in the microliter pipette indicated thermal equilibrium at 24.40 ± 0.1 °C, the permeability chamber was pressured with N_2. For each liquid studied, the flow rate measured in the microliter pipette was deter-

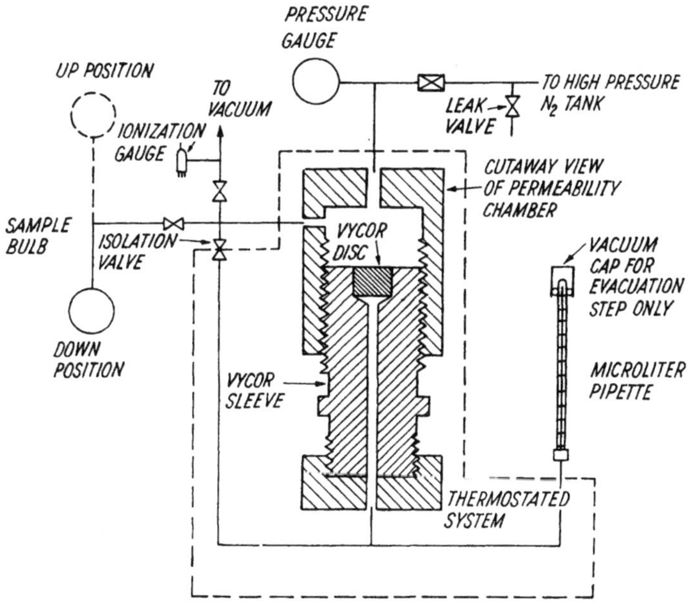

Fig. 2: Liquid permeability apparatus

mined as the average of five least square fit flow rate determinations; two at 700 psig and three at 500 psig. The 95 percent confidence interval for flow rate measurements with different Vycor sleeves was ± 6.8 percent of the mean permeation rate.

Theory

When liquid contacts an initially evacuated porous medium, the capillary forces induce a pressure in the liquid which, in the case of wetting ($\theta < 90°$), causes the liquid to imbibe into the pores. The magnitude of this capillary pressure as given by the Young-Laplace equation [1, 2] is just the product of the l–v interfacial curvature, J, and the liquid surface tension, γ. When imbibition occurs in porous media where the principal radii of curvature of the l–v interface approach molecular dimensions, the curvature effect manifests itself as a change in γ [3–6]. A change in γ will subsequently alter the time required for imbibition saturation of the porous medium as a result of changes in the capillary driving pressure. Hence, from a knowledge of the mass imbibition rate equation and the curvature of the imbibing l–v interface, a determination of γ and the resulting curvature effects can be made from liquid imbibition experiments in porous media having sufficiently small pores.

For the curvature effect/liquid imbibition experiments, porous Vycor (Corning No. 7930) was chosen because previous studies [24, 25] have shown that the pores should be small enough to induce curvature effects but large enough that imbibing liquids behave as continuum fluids. In the same studies, the imbibing l–v interfaces were found to be geometric surfaces at local equilibrium which changed in curvature, J, during imbibition. In other words, the change in curvature of each l–v interface occurred along a Laplace stable path [26] during imbibition. With this knowledge of the dynamic mechanism of liquid imbibition, the hydrodynamic liquid imbibition equation was derived [25] from the continuity equation [27] and Darcy's law [28]. For a rigid porous sphere of radius R, the imbibition equation relates the radial position of the l–v front, η, after imbibition has proceeded for a time, t, under a driving pressure, ΔP, in the form,

$$2\eta^3 - 3R\eta^2 + R^3 - (c\,\Delta P)(2\eta^3 - 4R\eta^2 + 2R^3) = \frac{3k\,\Delta P R}{\varphi}[2 - (c - b)\,\Delta P]\,t\,, \qquad (1)$$

$\Delta P = P_R - P_\eta \doteq -P_\eta$
P_R = pressure at $r = R$ (vapor pressure + hydrostatic head)
P_η = capillary pressure at $r = \eta$
c = isothermal liquid compressibility
b = viscosity analog of c [27]
k = Darcy's Law permeability coefficient
φ = porosity of porous medium
μ_0 = liquid viscosity at $r = R$

In equation (1), the quantities b and c correct the flow equation for pressure induced changes in liquid viscosity and density, respectively, and are defined in the equations,

$$\varrho = \varrho_0 \exp(-cP_\eta)\,, \qquad (2)$$

$$\mu = \mu_0 \exp(-bP_\eta) \qquad (3)$$

ϱ_0 = liquid density at $r = R$
ϱ = liquid density at $r = \eta$
μ_0 = liquid viscosity at $r = R$
μ = liquid viscosity at $r = \eta$

For a complete liquid imbibition experiment, the radial position, η, of the l–v front vanishes in a finite time, t_I. If these final conditions, i.e.,

$$\eta = 0 \quad \text{when } t = t_1$$

are imposed on the imbibition equation, an expression relating the imbibition saturation time, t_1, to the capillary driving pressure, P_η results

$$R^2(1 + 2cP_\eta) = \frac{3kP_\eta}{\varphi}[(b - c)\,P_\eta - 2]\,t_1\,. \tag{4}$$

Dividing both sides of equation (4) by the bracketed term on the r.h.s., dropping terms in P_η^2 (since second and higher order terms in P_η were not used in the derivation of equation 1, references [24, 25]), and solving for P_η gives,

$$P_\eta = -\frac{2\varphi R^2}{12kt_1 + R^2\varphi(b + 3c)}\,. \tag{5}$$

Elimination of the capillary driving pressure, P_η, by combining equation (5) with the YOUNG-LAPLACE equation (1, 2),

$$P_\eta = \gamma J\,, \tag{6}$$

provides a relationship between γ and the parameters of the imbibition experiment,

$$\gamma = -\frac{1}{J}\left[\frac{2\varphi R^2}{12kt_1 + \varphi R^2(b + 3c)}\right]. \tag{7}$$

When all eight unknowns on the r.h.s. of equation (7) have been evaluated, the curvature dependent value of γ can be calculated. However, the interfacial thickness parameter is a more fundamental measure of the curvature effect since it is in principle independent of J [15, 30, 31].

This parameter is defined [10] as,

$$\delta \equiv \lambda(\Gamma = 0) - \lambda(C_s = 0)\,, \tag{8}$$

$\lambda(\Gamma = 0)$ = principal radius of curvature of the equimolar dividing surface located at $\Gamma = 0$
$\lambda(C_s = 0)$ = principal radius of curvature of the surface of tension located at $C_s = 0$

and is related to γ by the TOLMAN equation [10]:

$$\gamma = \gamma_\infty(1 - \delta J)\,. \tag{9}$$

Eliminating γ between equations (7) and (9) and solving for δ gives,

$$\delta = \frac{1}{J} + \frac{1}{\gamma_\infty J^2}\left[\frac{2\varphi R^2}{12kt_1 + \varphi R^2(b + 3c)}\right]. \tag{10}$$

Six of the eight variables on the r.h.s. of equation (10) (γ_∞, φ, R, t_1, b, and c) can either be measured or readily obtained from other literature sources. However, the remaining two (k and J) require independent measurement which must be examined further.

The DARCY's Law permeability coefficient, k, was determined from the liquid permeability rate data obtained using the technique already described. For steady state axial flow of a liquid through the circular face of the Vycor disc of area, A, and thickness, L, under a driving pressure, P_L, the volume flow of liquid, Q, from DARCY's Law is,

$$Q = kA \frac{P_L}{L}. \tag{11}$$

Solving for k gives,

$$k = \frac{QL}{AP_L}, \tag{12}$$

from which k can be calculated directly using liquid permeability rate data.

The multitude of individual l–v interfaces, that are constantly changing in both size and number during the liquid imbibition process, are at all instances of time surfaces of constant curvature, J [24, 25]. These individual l–v interfaces proceed through all pore voids and hence the capillary driving force for liquid imbibition is dependent upon the mean cross-sectional pore dimension. The two conventional methods for measuring such dimensions, sorption isotherms and mercury porosimetry, are not applicable in mesopores since they are not independent of the curvature effect. Steady state KNUDSEN gas flow (SSKF) [32] is independent of this effect but is primarily dependent upon the mean pore constrictions or necks. Thus, another procedure was developed, based on Knudsen gas flow experiments which measures the mean, of the entire cross-sectional pore dimension.

For purposes of demonstration, consider the pore system shown in Figure 3. If the gas pressures $P = P_0$ and $P = 0$ are suddenly imposed at the capillary ends, a transient state gas flow will result.[1]

During transient state KNUDSEN flow (TSKF) in each pore, gas molecules which enter through each pore neck into the adjoining pore body will not be influenced by neighbouring molecules as specified by KNUDSEN diffusion theory [33]. Nevertheless, the net direction of molecular flight will necessarily have a radial component caused by gas molecules filling the radial extremities of the pore bodies. This is illustrated in Figure 3. As SSKF is approached, the net radial flow of gas molecules will vanish. The usefulness of this behaviour is that under TSKF conditions, the radial gas flow process depends on the pore body volumes. Under SSKF, only pore necks affect gas flow [32]. A method of determining mean l–v interfacial curvatures in porous media based on TSKF and SSKF experimental data can now be developed.

Since SSKF is dependent on pore neck dimensions and TSKF is dependent on total pore dimensions, i.e., pore body and pore neck dimensions, the equations of flow require different definitions of porosity (void volume/total volume of porous medium). The concept of SS porosity, φ_s, and TS porosity, φ_t, for KNUDSEN flow was first introduced by GOODKNIGHT et al. [34, 35]. Using the Barrer time lag equation [36]

[1]) Three requirements on the gas flow system for this procedure will be imposed: (a) the time the gas molecules spend in the adsorbed state must be short compared to the time spent in the gaseous state, (b) gas phase transport must be by KNUDSON flow, and (c) quantum effects must be negligible.

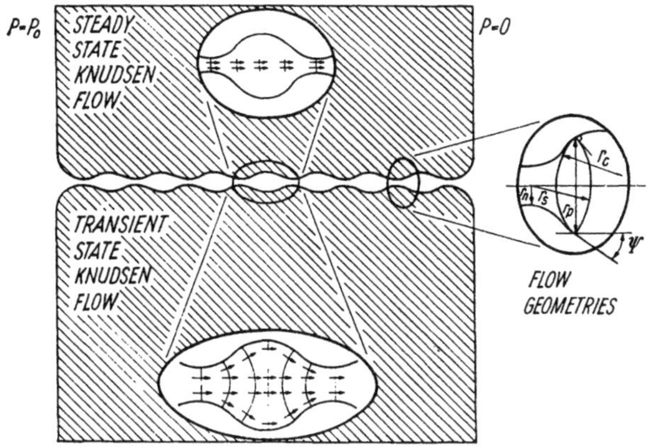

Fig. 3: Capillary gas flow diagram

for TSKF, one writes,

$$T_{LB} = \frac{\tau^2 L^2 \varphi_t}{6D\varphi_s}. \tag{13}$$

T_{LB} = abscissa intercept of SSKF data on volume of gas transmitted vs time plot, or BARRER time lag.
τ = tortuosity of porous media
L = thickness of porous media (one-dimensional flow)
D = diffusion coefficient

From the expressions developed for φ_s and φ_t in rigid isotropic porous media in Appendix I one obtains

$$\frac{\varphi_t}{\varphi_s} = \frac{\bar{r}_c^2}{\bar{r}_n^2} \tag{14}$$

r_c = average mean radius of curvature for all pores
r_n = average hydraulic radius of pore necks or constrictions

Equation (13) becomes,

$$T_{LB} = \frac{\tau^2 L^2 \bar{r}_c^2}{6D\bar{r}_n^2}. \tag{15}$$

If the flow is KNUDSEN, then the diffusion coefficient from gas kinetics is,

$$D = \frac{2\bar{r}_n}{3}\left(\frac{8R_g T}{\pi M_\omega}\right)^{1/2}, \tag{16}$$

which can be combined with equation (15) to give,

$$T_{LB} = \frac{\tau^2 L^2 \bar{r}_c^2}{4\bar{r}_n^3}\left(\frac{\pi M_\omega}{8R_g T}\right)^{1/2}. \tag{17}$$

Solving for $\bar{r}_c$ yields,

$$\bar{r}_c = \bar{r}_n \sqrt{\frac{4T_{LB}\bar{r}_n}{\tau^2 L^2}\left(\frac{8R_gT}{\pi M_\omega}\right)^{1/2}} . \tag{18}$$

The tortuosity as defined in the above equations will be equal to $\sqrt{2}$ in isotropic porous media. An examination of this statement will be made in the Discussion section. Equation 18 becomes,

$$\bar{r}_c = \bar{r}_n \sqrt{\frac{2T_{LB}\bar{r}_n}{L^2}\left(\frac{8R_gT}{\pi M_\omega}\right)^{1/2}} . \tag{19}$$

The l–v interfacial curvature, J, can then be determined as,

$$J = -\frac{2}{\bar{r}_c - t_0}\cos\theta , \tag{20}$$

t_0 = thickness of immobile adsorbate film immediately ahead of the l–v interface [24, 25].

The variables in equation (19) can be determined from SSKF and TSKF experiments as explained elsewhere [24, 25, 34, 35, 36]. The determination of the values of t_0 and $\bar{r}_c$ will be described in the Discussion section along with a discussion of the resulting value for J in porous Vycor.

Discussion

General

The TOLMAN equation [10], shown in truncated form in equation (9), has long provided a means of calculating the curvature dependence of surface tension if only the value of the characteristic interfacial thickness parameter, δ, was known. However, no method of measuring δ in one component l–v interfacial systems has ever been developed. In this study a method for measuring δ for single component, l–v interfaces is presented. The method is based on experiments of liquid imbibition in rigid mesoporous media where the imbibition rate is related directly to the capillary driving pressure and hence surface tension. The interfacial thickness parameter can be calculated from equation (10) when the eight experimentally obtained variables on the r.h.s. have been determined for the liquid/porous medium of interest. However, evaluating some of these variables in mesopores is not a simple task and deserves closer examination. Therefore, our discussion will first examine the evaluation of the variables in equation (10) prior to a discussion of the curvature effect.

Porosity measurements

The porosity was calculated from equation 21,

$$\varphi = \frac{M_I}{\frac{4}{3}\pi R^3 \varrho_0} , \tag{21}$$

M_I = mass of liquid condensed in porous Vycor sphere at saturated vapor pressure
R = radius of porous Vycor sphere (measured with micrometer calipers)

The values of porosity were found to be dependent on liquid type in agreement with the results of liquid immersion experiments by WADE [23]. From the results shown in Table 1, it can be seen that water gives the greatest porosity values. Since water is the smallest molecule studied, it is possible that a molecular sieve effect is operative where larger molecules are excluded from micropore filling. However, this cannot be the case for the n-alkanes since n-pentane is the smallest molecule for this group but has the smallest porosity. Thus, while molecular sieving may or may not be the reason for the large porosity for water, the n-alkanes indicate that other phenomenological effects must be operative.

Table 1: Porosity of Porous Vycor

Liquids	R (cm)	M_I (mg)	ϱ^{20} (g/cm^3)[a])	φ
n-pentane	0.3220	24.09	0.6262	0.275
n-heptane	0.3220	27.31	0.6838	0.286
n-decane	0.3410	34.54	0.7300	0.285
n-tridecane	0.3220	30.13	0.7564	0.285
n-hexadecane	0.3220	30.79	0.7733	0.285
water	0.3390	53.48	0.9982[b])	0.328

[a]) Reference [52], 20 °C
[b]) Reference [53], 20 °C

One potential explanation might be volume changes in the porous Vycor induced by sorbed water which were found to occur by AMBERG and MCINTOSH [37]. However, the porosity difference between water and the n-alkanes is ca. 15 percent, and the work by AMBERG and MCINTOSH only predicts ca. 6 percent total volume change between pristine and water saturated porous Vycor. This volume change should be even less for n-alkanes. Thus, while some of the porosity difference might be explained by such a pore volume increase mechanism it is apparently not of sufficient magnitude to account solely for the porosity differences for the different liquids shown in Table 1.

The interaction between the liquid and solid in mesoporous liquids could conceivably induce liquid density changes which could be the result of surface order caused by short-range interactions or longer-range forces which might induce bulk liquid structuring. However, it seems unlikely that specific short-range chemical interactions between liquid and Vycor could be causing this behaviour, since the porosities for the n-alkanes are not all the same while dispersion-type n-alkane/Vycor interactions would be nearly identical. The possibility of long-range forces causing molecular ordering in the bulk liquid can neither be confirmed nor rejected on the basis of the available data.

Regardless of the cause of the porosity dependence on liquid type, the correct value of porosity to be used in equation (10) is the one given in Table 1. However, the possibility that this variable porosity is indicative of hydrodynamic liquid anomalies, which could alter the interpretation of the liquid imbibition experiments, cannot be tacitly discounted and will be discussed in the following section.

The liquid imbibition mechanism

Implicit in the derivation of equation (10) are five assumptions about the mechanism of liquid imbibition into porous media: (a) net transport by gas or surface diffusion ahead of the l–v interfaces is negligible; (b) l–v interfaces are constant curvature surfaces, in spite of constant curvature changes during imbibition; (c) bulk liquid transport occurs by creeping flow; (d) bulk liquid behaves as a continuum fluid in mesopores; and (e) vapour bubble nucleation is absent in the imbibing liquid which is stressed by the capillary driving pressure. For liquid n-alkane and water, imbibition in mesoporous Vycor at 20 °C, it has been shown that these criteria are individually satisfied [24, 25]. In addition, the experimental imbibition rate data [24, 25] agreed with the calculated rate theory to within the experimental error of ± 2 percent, further corroborating the preceding assumptions. The imbibition times, t_I, are given in Table 2 along with error estimated as differences between experiments.

Table 2: Liquid imbibition data

Liquid	No. of Experiments	t_1	Range of t_1
n-pentane	2	250	4.1
n-heptane	7	472	3.2
n-decane	2	1124	1.7
n-tridecane	2	2483	1.1
n-hexadecane	2	5053	1.4
water	2	474	0.2

(Porous Vycor spheres with different radii were used; therefore, no average value can be computed.)

Permeability experiments

The permeability of porous Vycor to n-alkanes and water was measured as Darcy's Law permeability coefficients, k, which are related to the permeability constant, K, by,

$$K = k\mu_0 . \qquad (22)$$

The experiments were purposely run at low enough pressures so that pressure induced effects on liquid density and viscosity were negligible (0.3 to 3 percent for the liquids studied). When the permeabilities were calculated by equation (12,) the values of k shown in Table 3 were reasonably quite different for different liquids since k is a function of the properties of both the liquid and the porous medium. However, the values of K, which should only be dependent upon the porous medium, are influenced by the liquid also. Similar results were found by Elias and Schlumpf

[38], who noted that permeability constants for a homologous series of liquids in porous Vycor fit an equation of the form,

$$K = \frac{B}{M_w} + K_0 , \tag{23}$$

M_w = molecular weight
B and K_0 = constants

Table 3: Permeability of Porous Vycor

Liquid	Experimental data				Least Squares fit Experimental data	
	k (cm²/poise) $\times 10^{14}$		$\mu^{20°}$ (cpoise)	K (cm²) $\times 10^{16}$	K (cm²) $\times 10^{16}$	k (cm²/poise) $\times 10^{14}$
n-pentane	18.85	R2.5	0.235a)	4.43	4.47	19.02
n-heptane	9.16	R3.1	0.418a)	3.83	3.80	9.09
n-decane	3.75	R2.6	0.928a)	3.48	3.29	3.55
n-tridecane	1.56	R2.4	1.886a)	2.95	3.01	1.60
n-hexadecane	0.80	R3.2	3.484a)	2.80	2.84	0.82
water	3.89	R2.1	1.002b)	3.90	11.67	11.65

a) Data from Reference [52], 20 °C
b) Data from Reference [53], 20 °C
R Percent relative range for 1 experiment (2:700 psig: 3:500 psig runs)

For the data in Table 3 a least squares fit gives,

$$B = 1.728 \times 10^{-14}\ \mathrm{g\ cm^2\ mol^{-1}} , \tag{24}$$

$$K_0 = 2.08 \times 10^{-16}\ \mathrm{cm^2} , \tag{25}$$

and the resulting least squares fit data shown in the table are in excellent agreement with a relationship of this form for the n-alkanes. However, water would obviously not obey equations (22—25.)

The explanation for this behaviour is probably related to the nonslip of liquids near solid surfaces. In mesopores the liquid molecules along the wall constitute a significant fraction of the pore cross-sectional area available for flow. Therefore, in the case of the n-alkanes, the larger molecules might be predicted to restrict flow more and give a lower K which agrees with the experimental data. Furthermore, the stationary or restricted molecular layer near the pore wall may be several molecular layers deep if longer-range bonding forces are present. If this reasoning is applied to water which hydrogen bonds readily to Vycor and other water molecules, one would expect the equations (23—25) (derived for nonhydrogen bonding n-alkanes) to predict too large a value of K. Data in Table 3 corroborate this theory. Extending these concepts further, one can make an heuristic calculation of the free-flowing cross section of pores in porous Vycor filled with water relative to n-alkanes. Assuming K is proportional to pore cross-sectional area [39], the ratio of the radii would be,

$$\frac{\bar{r}_p\text{water})}{\bar{r}_p(\text{n-alkanes})} = \left[\frac{K(\text{water})}{K(\text{n-alkanes})}\right]^{1/2} = \left(\frac{4.47 \times 10^{-16}}{11.67 \times 10^{-16}}\right)^{1/2} = 0.62 \tag{26}$$

or if K is proportional to volume (molecular weight) as suggested by equation (23), then

$$\frac{\bar{r}_p(\text{water})}{\bar{r}_p(\text{n-alkanes})} = \left[\frac{K(\text{water})}{K(\text{n-alkanes})}\right] = \left(\frac{4.47 \times 10^{-16}}{11.67 \times 10^{-16}}\right)^{1/3} = 0.73\,. \tag{27}$$

From these results one would conclude that the effective radius of the fluid core in porous Vycor available to water flow is reduced to around 70 percent by the interaction forces (mainly hydrogen bonding) present in water but absent in n-alkanes.

Thus, the anomalous behaviour of K has a reasonable explanation based on the physical properties of the permeant liquid. Nevertheless, the measured value of K is the correct value to use in equation (10) to calculate S, independent of what causes K to be dependent on liquid properties.

The average *l-v*-interfacial curvature

The determination of the average l-v interfacial curvature, J, was determined by a gas flow technique and equations (19) and (20). The method is based on the concepts put forth in the Theory section, that in isotropic porous media, steady state Knudsen gas flow will depend on pore necks, while in transient state KNUDSEN flow, the complete pore structure will affect gas flow. This difference in KNUDSEN gas flow leads to different porosities for the two types of flow: φ_s for steady state and φ_t for transient state. When the concepts of these porosities are combined with the BARRER time lag theories in the manner of GOODKNIGHT et al. [34, 35], the equation

$$T_{LB} = \frac{\tau^2 L^2 \varphi_t}{6D\varphi_s}, \tag{13}$$

results instead of the classical time lag equation developed by BARRER [36]

$$T_{LB} = \frac{\tau_0^2 L^2}{6D}. \tag{28}$$

Since the time lags are identical in equations (13) and (28), it is apparent that,

$$\tau_0^2 = \tau^2 \frac{\varphi_t}{\varphi_s}. \tag{29}$$

Equation 29 is a very significant result. It shows that tortuosity, τ_0, as measured by the time lag method, is actually two effects combined. The quantity τ compensates for crooked pore channels and has historically been accepted as the true tortuosity; interpretated as the ratio of mean axial pore length to macroscopic flow length, L. However, time-lag-measured tortuosity has a second component, expressed in equation (29) as a porosity ratio, φ_t/φ [34, 35]. This ratio allows for the increased radial flow of gas molecules in transient compared to steady state flow. It arises from the fact that gas molecules, flowing through pores whose radial dimensions vary along their axes, must divert radially from the axis of each pore to a greater degree during transient than during steady state flow [34, 35]. For isotropic porous media, the very definition of isotropy required, $\tau = \sin 45° = \sqrt{2}$, since the axis of the mean pore would necessarily lie at a 45° angle to the macroscopic direction of flow along L [40]. In fact such behaviour was observed as early as 1937 by CARMAN [41]. However, historically the distinction between τ and τ_0 has not been recognized and since, $(\varphi_t/\varphi_s) > 1$, the time lag measurements of tortuosity have always been $> \sqrt{2}$. Based on the preceeding discussion the reason for such a discrepancy is now clear. For example, isotropic porous media, with undulating ink bottle shaped pores with alternate small necks and large bodies would have a larger τ_0 (determined from time lag measurements) than an isotropic porous media with pores of nearly constant crossection. However, since both were isotropic, τ for both would be $\sqrt{2}$.

By incorporating these concepts on tortuosity into the BARRER time lag equation an expression for calculation of the average interfacial curvature, J, of the imbibing l-v interfaces was derived in the Theory section, equation (20). Calculation of J from (19) and (20) was made for porous Vycor with the $\bar{r}_n$ value taken from BASMADJIAN and CHU [32], the time lag data of BARRER and BARRIE [42], and the adsorbate film thickness, t_0, taken from sorption isotherm studies [43—46] of n-alkanes on silica. The results are shown in Table 4. The average mean hydraulic l–v interfacial radius, $\bar{r}_{pk}$, corrected for the adsorbate film is found to be 43 Å for a curvature of -4.65×10^6 cm^{-1}. A check on the accuracy of the calculated values for J will be made in the next section.

Table 4: Liquid-vapor interfacial curvature determination

Symbol	Value	Source
$\bar{r}_n$	28 Å	32
T_{LB}	714 sec (Neon)	42
L^2	7.24 cm^2	42
T	292 K	42
M_ω	20.183 g/ mol	—
R_g	8.134×10^7 erg/K mol	—
t_0	6 Å	43—46
r_p	49 Å	calculated
$\bar{r}_{pk}$	43 Å	calculated
J	-4.65×10^6 cm^{-1}	calculated

The ideal liquid equations

In the previous sections the variables necessary for the solution of equation 10 have been discussed except for the ideal liquid parameters b and c. The values for these quantities were determined at 20 °C and one atmosphere pressure by an interpolation method discussed elsewhere [24]. The results are shown in Table 5. Since these values are to be used to describe imbibing liquid under stress not at a pressure of one atmosphere as calculated, these values b and c for imbibition in mesopores may be in error by several percent. However, the form of equation (10) indicates that such errors will have a minimal effect on the calculated value of δ. The values b and c appear only in the second term of a two-term denominator. The value of these two denominator terms is shown in Table 5. Since in the most extreme case the second term containing b and c is only 8.4 percent of the first term, it is clear that even large errors in estimating b and c will have only negligible effects on the total denominator and hence δ. Also from the form of this denominator, it is apparent that the first and second denominator terms are the zeroth and first-order correction terms for capillary pressure-induced changes in liquid imbibition. It is of interest to note that in spite of their coefficients (3 for c and 1 for b) $b > 3c$; pressure induced changes in viscosity, not density, have a greater effect on liquid imbibition dynamics. In summary, however, the pore size in mesoporous Vycor is sufficiently diminutive to cause negligible small pressure-induced effects in liquid density and viscosity during hydrodynamic liquid imbibition.

Table 5: Ideal liquid parameters and pressure corrections for liquid imbibition

	b (20°) (bar^{-1}) $\times 10^5$	c (20°) (bar^{-1}) $\times 10^5$	$12kt_{135}$ (sec cm^2/poise) $\times 10^{11}$	$\varphi R^2(b + 3c)$ (sec cm^2/poise) $\times 10^{11}$
n-pentane	72.1	20.0	57.21	4.46
n-heptane	76.0	13.6	51.49	4.10
n-decane	76.0	10.8	45.32	3.79
n-tridecane	76.0	9.3	47.38	3.63
h-hexadecane	76.0	8.3	49.12	3.52
water	−1.4	4.58	22.13	0.50

Interfacial curvature effect on surface tension

The data for the curvature dependence of surface tension are shown in Table 6 for the six liquids listed in Column 1. Columns 2—7 list experimental data used in equations (7) and 10 to calculate the curvature dependent surface tension, γ (Column 8), and the characteristic interfacial thickness parameter, δ (Column 10), respectively. The values of δ are positive and on the order of Ångströms in agreement with theoretical estimates of this quantity [9—15]. From all these estimates, the only liquid in common with those in Table 6 is for δ calculated by Tolman [10] for water of 0.99 Å. This is in good agreement with the experimental value of 1.1 Å obtained in this study.

In another study Nielsen et al. [21, 22] developed a three-component l-l-l nucleation technique to measure curvature effects for l-l interfaces. Their results on droplets with a radius of 32 Å which are mostly water showed a 17 percent decreases in interfacial tension (for drops, $J > 0$ and by the Tolman equation $\gamma < \gamma_\infty$). This would give a value for δ from equation (9) of 2.7 Å. This is in reasonably good agreement with the present study where concave l-v interfaces with mean radius of curvature gave an 8 percent increase in surface tension and $\delta = 1.1$ Å.

For the n-alkanes a pronounced inverse relationship between δ and molecular weight is noted. Such a result is particular interesting since δ and interfacial transition zone thicknesses have historically, in the absence of any experimental data, been assumed to be on the order of one or two molecular diameters [47]; a direct relationship between δ and molecular weight. However, more recent statistical mechanical theories [12—15]

Table 6: Interfacial curvature dependence of surface tension

Liquids	t_{135} (sec)	J (cm²/poise) $\times 10^{14}$	φ	b (20°) (bar⁻¹) $\times 10^5$	c (20°) (bar⁻¹) $\times 10^5$	γ_∞[a] (dyn/cm)	γ (dyn/cm)	γ/γ_∞	δ[b] (Å)	T/T_c	T_F^c (°C)	Γ_{ST} (mol/cm) $\times 10^{10}$
n-pentane	250.0	19.02	0.275	72.1	20.0	16.00	23.5	1.47	10.1	0.624	−129.7	8.8
n-heptane	472.0	9.09	0.286	76.0	13.6	20.30	27.1	1.33	7.2	0.543	−91.0	4.6
n-decane	1124.0	3.55	0.285	76.0	10.8	23.92	30.6	1.28	6.0	0.474	−29.7	3.2
n-tridecane	2483.0	1.60	0.285	76.0	9.3	26.1	30.0	1.15	3.2	0.433	−5.5	1.4
n-hexa-decane	5053.0	0.82	0.285	76.0	8.3	27.6	28.5	1.03	0.7	0.404	18.0	0.25
water	474.0	3.90	0.328	−1.4	4.58	72.25	76.5[c]	1.05	1.1	0.453	0.0	6.1

[a]) Reference [52].
[b]) $J = -4.65 \times 10^6$ cm⁻¹.
[c]) Reference [53].

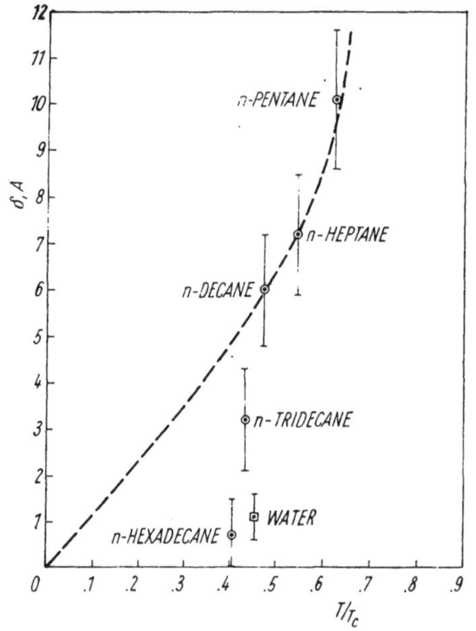

Fig. 4: Relationship between the characteristic interfacial thickness parameters and the relative critical temperature

are in good agreement with the experimental results of this work wherein δ and molec-cular weight for homologous series vary inversely.

Another interesting feature of the characteristic interfacial thickness parameter is its relationship to the reduced critical temperature. When the relative critical temperature approaches 1, statistical thermodynamic arguments indicate that δ asymptotically approaches infinity [12]. The data from Table 6 are plotted on a δ vs. T/T_c axis in Figure 4.

The error bars represent $\pm$ the cumulative relative ranges of all the variables in Columns 2–7, Table 6. The three lowest molecular weight n-alkanes suggest a smooth curve relationship (dashed line) intersecting the origin with an asymptote at $T/T_c = 1$ in qualitative agreement with statistical mechanical theories [12, 13]. However, data points for the other three liquids fall significantly below this line. This behaviour cannot be an artifact of the liquid imbibition method for measuring δ since all variables in equation (10) were experimentally measured in mesopores. An alternate explanation would be that in mesopores the larger n-alkanes and the hydrogen bonding water may be undergoing solid-like ordering effects [48] not present with the low molecular weight n-alkanes. Such a hypothesis is corroborated by the fact that the δ values for all the liquids correlate qualitatively with freezing point temperatures, T_F, shown in Table 6. However, on the basis of the limited data presented in Figure 4, the hypothesized statistical mechanical relationship between δ and T/T_c can be neither confirmed nor rejected.

For the curvature dependent surface tensions, γ, of the n-alkanes, an unexpected relationship between γ and molecular weight is observed. In the case of flat liquid interfaces, γ_∞ and molecular weight will vary directly. However, γ shows a maximum value for n-decane. The explanation for this is apparent from the Tolman equation. As the molecular weight increases, the quantity, $1 - \delta J$, decreases (note, $J < 0$), whereas, γ_∞ increases. Because of these two competing effects n-decane exhibits a maximum value of γ.

In the last column of Table 6 are pure liquid, molar surface excess quantities defined relative to the GIBBS' surface tension. They are calculated from,

$$\Gamma_{ST} = \left(\frac{\varrho_l - \varrho_v}{M_\omega}\right)\delta . \tag{30}$$

For the n-alkane homologous series, the trends in molecular weight and interfacial thickness parameter combine to augment the increase in Γ_{ST} with decreasing molecular weight and vice versa. As a result of these relationships, Γ_{SR} ranges over an order of magnitude for the liquids studied with the value of water falling within the range of values for the n-alkanes.

The liquid imbibition method of measuring curvature effects is generally applicable and can be used with different porous media of larger or smaller pores than the mesoporous Vycor used in this study. However, numerous problems can arise.

If the pores are too large, calculation difficulties arise with equation (10) where the two terms on the r. h. s. become large and nearly equal leaving the δ calculation as a small difference between two large terms. If porous media with very small pores are used, then any errors in the pressure correction terms become large as these terms increase. Also equation (10) is a truncated form of the general TOLMAN equation. The next term in the series would be $+3/4(\delta J)^2$. For liquid imbibiton (concave l-v interfaces), the signs in the TOLMAN equation alternate which tends to suppress truncation errors. However, for n-pentane in porous Vycor with its large δ value, this second-order term is ~ 0.15, which far exceeds experimental error. For the other liquids in this study, this correction was less than experimental error due to smaller values of δ.

The value of J plays an important role in determining δ by the liquid imbibition method. For the liquids in this study with porous Vycor, the contact angle was 0° in every case [49, 50]. Also, the precision of J as determined by the KNUDSEN gas flow technique is important. However, assessing its precision is difficult. On the basis of agreement between δ values in this study and others, the value of J seems reasonable. Also, if δ values are assumed to always be > 0, then the J value of -4.65×10^6 cm^{-1} giving a very small but positive δ for n-hexadecane suggests J was accurately measured. With regard to the accuracy of J, it is important to note that J should be practically he same for all n-alkanes and water and, as such, is just a parameter in equation (10). Therefore, the fact that the values of δ in Table 6 for the n-alkanes are different is truly a phenomenological result independent of the J value chosen. The value of J only affects the absolute accuracy of the δ values.

As an extension of the liquid imbibition curvature effect technique, a series of experiments were done with porous Vycor modified with preadsorbed monolayer of n-heptanol. The purpose of the monolayer was to reduce the pore diameter and evaluate the liquid imbibition mechanism and the curvature effect in a medium with smaller pores than the pristine porous Vycor. If successful, it was hoped to learn whether δ was curvature dependent. The relationship between δ and J has been theoretically discussed [10, 15], but the question remains unresolved.

For both the porous Vycor sphere in the imbibition apparatus (Figure 1) and the Vycor sleeve (Figure 2), the n-heptanol was allowed to saturate the porous Vycor before being evacuated at 20 °C for 25 hours which was determined independently to give precisely monolayer average. The liquid imbibiton and permeability experiments were then carried out as before with the pristine porous Vycor. The four liquids listed

Table 7: Interfacial curvature depend tension–determination in modified Porous Vycor[a])

Liquids	t_{135} (sec)	k (cm²/poise) × 10^{14}	K (cm²) × 10^{16}	φ	b (20 °C) (bar^{-1}) × 10^5	c (20 °C) bar^{-1}) × 10^5	b γ_∞ (dyn/cm)	γ (dyn/cm)
n-pentane	562.0	7.30	1.65	0.275	72.1	20.0	16.00	24.6
n-heptane	950.0	4.33	1.73	0.286	76.0	13.6	20.30	25.6
n-decane	1878.0	2.16	1.88	0.285	76.0	10.8	23.92	26.1
n-tridecane	3419.0	1.20	2.08	0.285	76.0	9.3	26.1	26.0

[a]) $J = -5.1 \times 10^6$ cm^{-1}.
[b]) Reference [52].

in Table 7 were studied with the indicated results. The significant feature of these data is the fact that $\gamma \sim 26$ dyn/cm for all liquids studied. The equivalence of γ for these four liquids is not a reasonable result. It seems likely that the n-heptanol does not remain as a monolayer on the pore walls. Instead, it would seem that the equivalence of γ for these four different liquids could only be caused by enough n-heptanol adsorbing into the l-v interface to completely mask the effect of the imbibing liquids. This hypothesis is further corroborated by the fact that 26 dyn/cm would be a reasonable value of γ for n-heptanol ($\gamma_\infty = 26.09$ dyn/cm, Reference [51]). Therefore, no values of δ can be calculated.

Appendix

Transient and steady state KNUDSEN gas flow porosities [34, 35]

For the flow geometry given in Figure 3, the area available to flow in SKKF, A_n, is through the pore necks of radius, r_n,

$$A_n = \pi r_n^2 \,. \tag{I-1}$$

In TSKF the gas flow will be proportional to the mean equipotential surface area A_s. At any position in the capillary with radius, r_p, the surface A_s has a radius of curvature, r_s, and a related solid angle ω_s, so that,

$$A_s = \omega_s r_s^2 = \pi r_p^2 (1 + \sin \psi) \,. \tag{I-2}$$

For any pore (wetting liquids only, $\theta = 0$), the area, A_c, with radius of curvature, r_c, and solid angle, ω_c, gives a similar equation (see Figure 3),

$$A_c = \omega_c r_c^2 = \pi r_p^2 (1 + \cos \psi) \,. \tag{I-3}$$

Now in isotropic porous media (note, the capillary in Figure 3 is not isotropic) the angle between the pore wall and the macroscopic flow direction, ψ, should range from 0° to 90°. From the definition and concept of isotropy, one would conclude the ψ averaged over all pores would give a mean value, ψ, of 45°. Based on this assumption which will be examined further in the Discussion section, equations (I-2) and (I-3) simplify to,

$$A_s = \pi r_s^2 = 1.7 \pi r_p^2 \,, \tag{I-4}$$

$$A_c = \pi r_c^2 = 1.7 \pi r_p^2 \,, \tag{I-5}$$

where,

$$\left.\begin{aligned} \omega_s &= \pi \\ \omega_c &= \pi \end{aligned}\right\} \text{ if } \quad \psi = \frac{\pi}{2} \,.$$

From equations (I-1), (I-4), and (I-5), the hydraulic radius, r_i, can be defined generallyas,

$$r_i = \sqrt{A_i/\pi} \,, \tag{I-6}$$

where the mean hydraulic radii for all three radii in equations (I-1), (I-4), and (I-5) can then be defined generally as the average over all flow paths, x,

$$\bar{r}_i = \frac{\int r_i \, dx}{\int dx} \,. \tag{I-7}$$

Finally, the mean pore areas for SSKF, TSKF, and liquid imbibition in rigid isotropic porous media can be written, respectively,

$$\bar{A}_n = \pi \bar{r}_n^2 , \tag{I-8}$$

$$\bar{A}_s = \pi \bar{r}_s^2 , \tag{I-9}$$

$$\bar{A}_c = \pi \bar{r}_c^2 , \tag{I-10}$$

where $\bar{A}_s = \bar{A}_c$.

If a surface within the porous medium perpendicular to the macroscopic direction of flow with an area, A_T, and N_T pores exposed, then the total cross-sectional void area is $N_T \pi \bar{r}_p^2$ and the total surface area of all the SSKF equipotential surfaces would be $N_T \pi \bar{r}_n^2$ and the total TSKF equipotential surface area would be $N_T \pi \bar{r}_s^2 = N_T \pi \bar{r}_c^2$. The SS and TS porosities can then be defined as,

$$\varphi_s = \frac{N_T \bar{A}_n}{A_T} = \frac{\pi N_T \bar{r}_n^2}{A_T} , \tag{I-11}$$

$$\varphi_t = \frac{N_T \bar{A}_c}{A_T} = \frac{\pi N_T \bar{r}_c^2}{A_T} . \tag{I-12}$$

So that,

$$\frac{\varphi_t}{\varphi_s} = \frac{\bar{r}_c^2}{\bar{r}_n^2} .$$

List of symbols

A = area of porous Vycor disc for permeability experiment
A_c = area of constant mean curvature surface eminating from pore wall at any point, πr_c^2
$\bar{A}_c$ = average of A_c over all pores
A_i = πr_i^2
A_n = cross section area of pore neck or constrictions, πr_n^2
$\bar{A}_n$ = average of A_n over all pores
A_s = area of equipotential flow surface in a pore in transient state KNUDSEN gas flow, πr_s^2
$\bar{A}_s$ = average of A_s over all pores
A_T = macroscopic area of porous medium perpendicular to fluid flow
b = pressure coefficient of viscosity
B = permeability constant parameter
c = isothermal liquid compressibility
C_s = coefficient of dJ in GIBB's thermodynamic equation, Reference [3]
D = diffusion coefficient
J = curvature of l–v interface $= \frac{1}{\lambda_1} + \frac{1}{\lambda_2}$
k = DARCY's Law permeability coefficient
K = DARCY's Law permeability constant
K_0 = permeability constant parameter
L = thickness of porous medium
M_I = mass of liquid condensed in porous Vycor sphere at saturation vapour pressure

M_ω = liquid molecular weight
N_T = number of pores in area A_T
P = pressure
P_η = liquid pressure at imbibing liquid from $r = \eta$
P_R = liquid pressure at surface of porous Vycor sphere $r = R$
P_0 = gas pressure for KNUDSEN gas flow experiments
ΔP = $P_R - P_\eta \doteq -P_\eta$
P_L = driving pressure for permeability experiment
Q = DARCY's Law volume flow rate
r = radial coordinate for porous Vycor sphere
r_c = mean radius of curvature of constant mean curvature surface eminating from pore wall
$\underline{r}_c$ = average of r_c over all pores
$\bar{r}_f$ = mean pore radius of effective fluid core in liquid filled pore
r_i = hydraulic radius for any type of pore radius
$\underline{r}_i$ = average of r_i over all pores
r_n = hydraulic radius of pore necks or constrictions
$\bar{r}_n$ = average of r_n over all pores
r_p = pore radius at any point along the pore axis
$\bar{r}_p$ = average of r_p over all pores
$\bar{r}_{pk}$ = $\bar{r}_p - t_0$
r_s = radius of curvature for equipotential flow surface in pore in transient state KNUDSEN gas flow
$\bar{r}_1$ = average of r_s over all pores
R = radius of porous Vycor sphere
R_g = universal gas constant
t = time
t_1 = time required for liquid imbibition to saturate porous Vycor sphere
t_{135} = time required for liquid imbibition to saturate porous Vycor sphere of $R = 0.35$ cm
t_0 = thickness of immobile adsorbate film immediately ahead of the imbibing l–v interfaces
T = absolute temperature
T_c = absolute critical temperature
T_{LB} = Barrer time lag, Reference [36]
T_F = freezing point temperature, centigrade
X = distance measured along pore axis
γ = liquid surface tension of curved l–v interface
γ_∞ = liquid surface tension of flat l–v interface
Γ = general GIBB's surface excess adsorption
Γ_{ST} = GIBB's surface excess adsorption defined at surface of tension
δ = characteristic interfacial thickness parameter
η = position of macroscopic l–v front measured radially outward in porous Vycor sphere
θ = equilibrium advancing contact angle
$\lambda, \lambda_1, \lambda_2$ = principle radii of curvature of l–v interface
$\lambda(\Gamma = 0)$ = radius of curvature of equimolar dividing surface
$\lambda(C_s = 0)$ = radius of curvature of GIBBS' surface of tension
μ = liquid viscosity, pressure dependent
μ_0 = liquid viscosity at one atmosphere pressure
π = 3.14159

ϱ = liquid density, pressure dependent
ϱ_0, ϱ_l = liquid density at atmospheric pressure
ϱ_v = vapor density
τ = tortuosity, ratio of mean axial pore length to L
τ_0 = tortuosity as measured by Barrer time lag method
φ = porosity of porous Vycor sphere
φ_s = steady state porosity
φ_t = transient state porosity
ψ = angle between pore wall and pore axis
ω_c = solid angle subtended by A_c
ω_s = solid angle subtended by A_s

References

[1] T. Young, "*Miscellaneous Works*", G. Peacock, ed., J. Murray, London Vol. I, p. 418 (1855)
[2] P. S. de Laplace, "*Mechanique Celeste*", Supplement to Book 10 (1806)
[3] J. W. Gibbs, "*The Collected Works of J. W. Gibbs*", Longmans Green, New York Vol. I, p. 228—29 (1931)
[4] F. P. Buff, "*Handbuch der Physik*", Berlin: Springer-Verlag, p. 281—304 (1960)
[5] F. P. Buff, J. Chem. Phys., **23**, 419 (1955)
[6] F. P. Buff, J. Chem. Phys., **25**, 146 (1956)
[7] R. Defay, I. Prigogine, A. Bellemans, and D. H. Everett, "*Surface Tension and Adsorption*", Wiley, New York (1966)
[8] I. Prigogine and R. Defay, "*Chemical Thermodynamics*", D. H. Everett, trans., Longmans Green, London (1954)
[9] J. G. Kirkwood and F. P. Buff, J. Chem. Phys., **17**, 338 (1949)
[10] R. C. Tolman, J. Chem. Phys., **17**(**2**), 118 (1949)
[11] T. L. Hill, J. Chem. Phys., **20(1)**, 141 (1952)
[12] S. Toxvaerd, Mol. Phys., **26**(**1**), 91 (1973)
[13] I. W. Plesner and O. Platz, J. Chem. Phys., **48**, 5361 (1968)
[14] C. A. Croxton and R. P. Ferrier, J. Phys., C**4**, 2433 (1971)
[15] A. I. Rusanov and E. N. Brodskaya, J. Colloid Interface Sci., **62**(3), 542 (1977)
[16] D. J. Woodland and E. Mach, J. Am. Chem. Soc., **55**, 3149 (1933)
[17] L. H. Cohan and G. E. Meyer, J. Am. Chem. Soc., **62**(4), 2715 (1940)
[18] N. N. Fedyakin, Trudy Tekhnol. Inst. Pishchevoi Prom., **8**, 37 (1957)
[19] V. K. LaMer and G. M. Pound, J. Chem. Phys., **17**, 1337 (1949)
[20] G. M. Pound and V. K. LaMer, J. Chem. Phys., **19**, 506 (1951)
[21] A. E. Nielsen and P. S. Bindra, Croat. Chem. Acta (Ark. Kem.), **45**, 31 (1973)
[22] A. E. Nielsen and S. Sarig, J. Cryst. Growth, **8**, 1 (1971)
[23] W. H. Wade, S.P.E.J., April, 139 (1974)
[24] J. A. Wingrave, Ph. D. Dissertation, The University of Texas at Austin, Austin, Texas, p. 32 (1975)
[25] J. A. Wingrave, W. H. Wade, and R. S. Schechter, "*Wetting, Spreading, and Adhesion*", J. Padday, ed., Academic Press, Chapter 12 (1978)
[26] D. H. Everett and J. M. Haynes, J. Colloid Interface Sci., **38(1)**, 125 (1972)
[27] R. B. Bird, W. E. Stewart, and E. N. Lighfoot, "*Transport Phenomena*", John Wiley, New York, 74—83 (1968)
[28] H. Darcy, "*Les Fontaines Publiques de la Ville de Dijon*", Victor Dalmont, Paris (1856)

[29] P. W. Bridgman, "*The Physics of High Pressure*", London, G. Bell and Sons, Ltd. (1949)
[30] R. C. Tolman, J. Chem. Phys., **17(3)**, 333 (1949)
[31] J. C. Melrose, I.E.C., **60(3)**, 53 (1968)
[32] D. Basmadjian and K. P. Chu, Can. J. Chem., **42**, 946 (1964)
[33] R. D. Present, "*Kinetic Theory of Gases*", New York, McGraw-Hill Co., Inc., p. 55 (1958)
[34] R. C. Goodknight, W. A. Klikoff, Jr., and I. Fatt, J. Chem. Phys., **64**, 1162 (1960)
[35] R. C. Goodknight and I. Fatt, J. Phys. Chem., **65**, 1709 (1961)
[36] R. M. Barrer, "*Diffusion In and Through Solids*", London, Cambridge University Press, p. 18 (1951)
[37] C. H. Amberg and R. McIntosh, Can. J. Chem., **80**, 1012 (1952)
[38] H. Elias and H. Schlumpf, Helv. Chim. Acta, **47**, 1503 (1964)
[39] M. K. Hubbert, Pet. Trans. AIME, **207**, 222 (1956)
[40] R. M. Barrer, J. Phys. Chem., **57**, 35 (1953)
[41] P. C. Carman, Trans. Inst. Chem. Eng., **15**, 150 (1937)
[42] R. M. Barrer and J. A. Barrie, Proc. Roy. Soc. London, **213**, 250 (1952)
[43] R. W. Hopper and D. R. Uhlmann, J. Chem. Phys., **56(8)**, 4043 (1972)
[44] K. S. Lyalikov, Acta Physicochim, U.S.S.R., **12**, 42 (1940)
[45] D. H. Everett and W. I. Whitton, Trans. Faraday Soc., **48**, 749 (1952)
[46] J. C. Melrose, J. Colloid Interface Sci. **38**, 312 (1972)
[47] N. K. Adam, "*Physics and Chemistry of Surfaces*", Oxford Presss, London, p. 5 (1941)
[48] D. H. Everett and G. H. Finderegg, J. Chem. Thermod., **1**, 573 (1969)
[49] L. S. Hersh and M. P. Teter, J. Phys. Chem., **76**, 3633 (1972)
[50] M. K. Bernett and W. A. Zisman, J. Colloid Interface Sci. **29**, 413 (1969)
[51] C. O. Timmons and W. A. Z_3sman, J. Colloid Interface Sci. **28(1)**, 106 (1968)
[52] "*Handbook of Chemistry and Physics*", 49th ed. CRC Publishing Co., Cleveland, Ohio (1968)
[53] F. D. Rossini, "*Selected Values of Physical and Thermodynamic Properties of Hydrocarbons and Related Compounds*", Pittsburg, Pennsylvania, Carnegie Press (1953)

Acknowledgements

The authors would like to thank the Robert A. Welch Foundation for their support and interest in this project. The assistance of Continental Oil Company in the preparation of this manuscript is also gratefully acknowledged.

Bei Fragen zur Produktsicherheit wenden Sie sich bitte an:
If you have any questions regarding product safety, please contact:

Walter de Gruyter GmbH
Genthiner Straße 13
10785 Berlin
productsafety@degruyterbrill.com